Studies in Systems, Decision and Control

Volume 104

Series editor

Janusz Kacprzyk, Polish Academy of Sciences, Warsaw, Poland
e-mail: kacprzyk@ibspan.waw.pl

About this Series

The series "Studies in Systems, Decision and Control" (SSDC) covers both new developments and advances, as well as the state of the art, in the various areas of broadly perceived systems, decision making and control- quickly, up to date and with a high quality. The intent is to cover the theory, applications, and perspectives on the state of the art and future developments relevant to systems, decision making, control, complex processes and related areas, as embedded in the fields of engineering, computer science, physics, economics, social and life sciences, as well as the paradigms and methodologies behind them. The series contains monographs, textbooks, lecture notes and edited volumes in systems, decision making and control spanning the areas of Cyber-Physical Systems, Autonomous Systems, Sensor Networks, Control Systems, Energy Systems, Automotive Systems, Biological Systems, Vehicular Networking and Connected Vehicles, Aerospace Systems, Automation, Manufacturing, Smart Grids, Nonlinear Systems, Power Systems, Robotics, Social Systems, Economic Systems and other. Of particular value to both the contributors and the readership are the short publication timeframe and the world-wide distribution and exposure which enable both a wide and rapid dissemination of research output.

More information about this series at http://www.springer.com/series/13304

Šárka Hošková-Mayerová
Fabrizio Maturo · Janusz Kacprzyk
Editors

Mathematical-Statistical Models and Qualitative Theories for Economic and Social Sciences

 Springer

Editors
Šárka Hošková-Mayerová
Department of Mathematics and Physics,
    Faculty of Military Technology
University of Defence
Brno
Czech Republic

Janusz Kacprzyk
Systems Research Institute
Polish Academy of Sciences
Warsaw
Poland

Fabrizio Maturo
Department of Business Administration
University "G. d'Annunzio" of
    Chieti-Pescara
Pescara
Italy

All chapters were reviewed.

ISSN 2198-4182   ISSN 2198-4190   (electronic)
Studies in Systems, Decision and Control
ISBN 978-3-319-85492-2   ISBN 978-3-319-54819-7   (eBook)
DOI 10.1007/978-3-319-54819-7

Preface

The book "Mathematical-Statistical Models and Qualitative Theories for Economic and Social Sciences" is part of the important series "Studies in Systems, Decision and Control" published by Springer. This is the result of a scientific collaboration, in the field of economic and social systems, among experts from "University of Defence" in Brno (Czech Republic), "G. d'Annunzio" University of Chieti-Pescara (Italy), "Pablo de Olavide" University of Sevilla (Spain), and "Ovidius University" in Constanța, (Romania).

The variety of the contributions developed in this volume reflects the heterogeneity and complexity of economic and social phenomena; thus, in this book, there is a convergence of many research fields, such as statistics, decision making, mathematics, complexity, psychology, sociology, and economics. The different studies included in this book, selected using a peer-review process, present also empirical interesting researches conducted in various countries. Each chapter was peer-reviewed by two independent referees (e.g., J. Beránek, J. Čermák, D. Řezáč, and V. Voženílek, (CZ); M. Grega, S. Filip, and J. Klučka (SK); A. Porrovecchio (FR); N. Bortoletto, M. Squillante, and A. Ventre, (IT); E. Barrera Algarín, O. Vazquez-Aguado, and C.M. Vicente (ES)). The volume is divided into two parts: The first one is "Recent Trends in Mathematical and Statistical Models for Economic and Social Sciences," whereas the second one is "Recent Trends in Qualitative Theories for Economic and Social Sciences."

Part I collects research of scholars and experts on quantitative matters, who propose mathematical and statistical models for social sciences, economics, finance, and business administration.

The book opens with the contribution of Veronika Mitašová, Ján Havko, and Tomáš Pavlenko. They propose a study titled "Correlational Research of Expenditure Spend on Slovak Armed Forces Participation in Peace Support Operations Led by NATO" in which a multifactor single-equation econometric model is created and tested.

Salvador Cruz Rambaud, Fabrizio Maturo, and Ana María Sánchez Pérez aim to develop three approaches for obtaining the value of an n-payment annuity, with payments of one unit each, when the interest rate is random. To calculate the value of these annuities, the authors assume that only some non-central moments of the capitalization factor are known. The first technique consists in using a tetra parametric function which depends on the arctangent function. The second expression is derived from the so-called quadratic discounting, whereas the third approach is based on the approximation of the mathematical expectation of the ratio of two random variables by Mood et al. (1974). A comparison of these methodologies through an application, using the R statistical software, shows that all of them lead to different results.

Josef Navrátil and Veronika Sadovská concentrate on the health risk assessment of selected pollutants derived from residential fire simulated in fire container. The key interest of their study is dedicated to polycyclic aromatic hydrocarbons (PAHs) and their toxicity and harmful effects on the health of firefighters, whose protection by breathing apparatus is insufficient.

Qualitative and quantitative comparison of the results of the entrance draft tests and the entrance tests of mathematics is illustrated by Radovan Potůček. His contribution focuses on the applicants for the bachelor and master study at the Faculty of Military Technology of the University of Defence in Brno. He refers to tests organized by the Department of Mathematics and Physics and presents the results of the applicants (from the military secondary school and civilian secondary schools), comparing and evaluating them from the qualitative and quantitative point of view.

Recent trends in digital ethnography are presented by Vanessa Russo. She shows theories, models, and case studies with the aim of defining the boundaries of digital ethnography. Finally, with the help of the comparison between empirical cases, she tries to understand critical points, limits, and research prospects for digital ethnographers.

Lenka Hrbková, Jozef Zagrapan, and Roman Chytilek analyze the demand side of negativity and privatization in news with an experimental study of news consumer habits. They remark that negativity in media and emphasis on personal side of politics are often cited as a common journalist practice, which is harmful to democratic processes. Journalists and media houses are often held accountable for these phenomena because they prioritize profit over the quality of content. Then, they offer an analysis focused on demand side of both negativity and privatization of political news. Using the Dynamic Process Tracing Environment (DPTE), they test the assumption that both of these features of political media coverage may be driven by audience demand for negative and personal news.

Martin Hubacek and Vladimir Vrab propose a constructive simulation for cost assessment of training. They highlight that constructive simulation, which is used as a tool for training of commanders and staffs of military units, has important benefits for a higher quality of training. Furthermore, constructive simulation gradually penetrates into other spheres such as the training of emergency staff. However, relevant studies about the economical benefits of the use of constructive simulation for training are relatively rare. The presented cost comparison of the exercises is based on the authors' experience gained during the implementation of various types of exercise, at the Center of Simulation and Training Technologies of Brno, with the use of constructive simulation OneSAF.

Social problems and decision making for teaching approaches and relationship management in an elementary school are the topics of interest of Luciana Delli Rocili and Antonio Maturo. Their chapter illustrates teaching experiences in the primary school and applications of theories regarding the choice of teaching methods in this particular context. In order to decide the most appropriate intervention strategies, both in terms of teaching and for an efficient management of relationships within the class and with the students' families, they consider that the first step is to discover the students' social and environmental background. The experience described herein is at the basis of the final proposal on how to concretely implement some decisional procedures at school, as for instance those linked to the limited rationality and the analytic hierarchy process theorized by Saaty.

Ferdinando Casolaro and Alessandra Rotunno propose a chapter titled "From the pictorial art to the linear transformations." In particular, the authors suggest a path for the teaching of geometry in Italy that reflects the development, which took place over the past two centuries. Moreover, they highlight the social aspects of a teaching based on the graphical visualization as required by the projective geometry.

Bekesiene, Hošková-Mayerová, and Diliunas focus on the identification of effective leadership indicators in army forces of Lithuania. They remark that leadership is of overriding importance in the military sphere because the foundation for leading a unit consists in influence, motivation, and soldiers' inspiration by the leader's personal example. The Lithuanian Army seeks to develop a military leadership identity as a way to promote mission success. This study is sought to identify the effective leadership style, which is appreciated by soldiers in the Lithuanian Armed Forces. The authors adopt the Leader Behavior Description Questionnaire (LBDQ) for measuring the behavior of leaders. The data collected from military personnel, holding different ranks and doing their professional military service of all the units of the Lithuanian Armed Forces, were analyzed using structural equation models (SEM).

"Why We Need Mathematics in Cartography and Geoinformatics" is the title of the chapter of Václav Talhofer. He highlights that mathematics is necessary for understanding of many procedures that are connected to modeling of the Earth as a celestial body, to ways of its projection into a plane, to methods and procedures of modeling of landscape and phenomena in society, and to visualization of these models in the form of electronic as well as classic paper maps. Not only general mathematics, but also its extension of differential geometry of curves and surfaces, ways of approximation of lines and surfaces of functional surfaces, mathematical statistics, and multicriteria analyses seem to be suitable and needful. Moreover, he suggests that the underestimation of the significance of mathematical education in cartography and geoinformatics is inappropriate and lowers competences of cartographers and geoinformaticians to solve problems.

Hana Svatoňová and Radovan Šikl investigate the cognitive aspects of interpretation of image data. Interpretation of image data is a complex of complicated intellectual operations, which is based on visual. The theoretical part of their study summarizes the scientific knowledge of processes of visual perception applied in the process of visual interpretation of satellite, aircraft, and map image data. Author presents partial phases of image data interpreting process: from the initial recording of the image to detection, identification, and objects classification. The complexity of the cognitive process with regard to biological and psychological characteristics of the individual is highlighted. The research section presents the results of image data interpretation research according to gender of individuals/research respondents.

Engin Baysen, Šárka Hošková-Mayerová, Nermin Çakmak, and Fatma Baysen study the misconception regarding providing citations. Their research aims at finding out citation understandings of Czech and Turkish secondary and high school students. Except for few students, secondary and high school students have misconceptions concerning providing citations. Students are unintentionally vulnerable to plagiarize while reporting. The study shows that only secondary and high school education is not enough for implementing honesty regarding citation. Therefore, the authors remark the importance of educating and informing students about honesty in research and plagiarism.

Subjective preconditions and objective evaluation of interpretation of image data are analyzed in the chapter of Hana Svatoňová and Šárka Mayerová-Hošková. In learning and teaching, there is an ongoing teaching relationship with specific and bidirectional relations between the teacher and student. Teacher can have either a strongly positive or, on the contrary, a strongly negative impact via his communication and interactions with students. From the wide idea of attitudes and values that constitute the relationship between the teacher and the student, their article is focused on a part of subjective assumptions about the success of students in a specific task: in this case, interpretation of aerial and satellite images and maps. The respondents of the researcher were elementary school teachers and students aged between 11 and 15. The subjective assumptions of teachers were compared with assumptions of students, and subsequently, all subjective assumptions were compared with objective data.

Fabrizio Maturo, Stefania Migliori, and Francesco Paolone analyze the influence of institutional and foreign shareholders on national board diversity of companies. Investigating the external antecedents of board diversity, they suggest the use of functional data analysis for diversity assessment in corporate governance studies. Focusing on a sample of 1230 Italian medium–large firms, their results show that institutional shareholders do not influence national board diversity, while foreign shareholders strongly affect it, especially when they hold more than 50% of shares. Thus, the authors address the research gap on the determinants of national board diversity and enrich comparative European research on this topic.

Francesco Paolone and Matteo Pozzoli investigate the effect of financial crisis of Earnings Manipulation by adopting the Beneish model. Specifically, empirical evidence from the "Top World Enterprises," ranked by "Sales Revenues" in the fiscal year 2013, is presented. Their results show that there has been a greater propensity for manipulating earnings in the first year of the global crisis:

Companies have had a tendency to increase creation of social wealth, in terms of generating higher profits. This would mean that the crisis has had a positive effect on handling of income by the largest companies in the world because the crisis itself has restricted the earnings manipulation policies.

Reasoning and decision making in practicing counseling are considered by Antonio Maturo and Antonella Sciarra. The counseling procedure is considered as a dynamical decision-making problem, where the awareness of alternatives and objectives and their evaluations are maieutically induced by the counselor. After presenting some relevant practices of counseling and related decision-making procedures, this study shows the use of the mathematical theory of decisions for a formalization of the counseling methods, in order to model, clarify, and make rigorous procedures of decision. Finally, it is shown that fuzzy reasoning can give a useful formal help to the task of the counselor because of its flexibility.

Ana Vallejo Andrada, Šárka Hošková-Mayerová, José Luis Sarasola Sanchez-Serrano, and Josef Krahulec deal with how society views the current wave of migration, specifically in Andalusia (Spain) and Czech Republic. The problem is described in a pre-case study, which covers results concerning citizens' approach to an urgent social topic, i.e., migration and immigration and risks related to these questions. First, the research presents a summary about the history of migrations in both regions; then, the current situation in those regions is characterized; after that, the questionnaire was prepared with the idea of how people feel this phenomenon, and survey was made. Finally, based on the results, possible risks are presented and some strategies on how to deal with inconvenient situations, which might arise, are suggested.

Domenico Di Spalatro, Fabrizio Maturo, and Lorella Sicuro face the issue of inequalities in the provinces of Abruzzo making a comparative study through the indices of deprivation and principal component analysis. The indices of deprivation are a valuable tool to measure the socioeconomic disadvantage in certain geographical areas of interest. This study aims to compare inequalities between the provinces of Abruzzo over the last two decades suggesting some indices of deprivation to capture the key aspects of the great wealth of information relating to population census. Specifically, they propose three indices of deprivation to measure the material and social disadvantage. Moreover, a principal component analysis is performed using the most known indicators of deprivation. Using these methods, their results show an increase in the proportion of disadvantaged areas in the Abruzzo region from 1991 to 2011 in its four provinces.

Part II "Recent Trends in Qualitative Theories for Economic and Social Sciences" collects research of scholars and experts on qualitative matters, who propose and discuss on social, economic, and teaching issues.

Part II opens with the chapter of Grazia Angeloni which highlights the reasons why educational institutions should be considered complex systemic organizations. Specifically, she suggests a multidisciplinary approach tending to make use, on the one hand, of different lenses, in order to appreciate the organizational phenomenon taken into consideration, and on the other, it is an effort to join different epistemes for practical purposes.

Jose Luis Sarasola Sánchez-Serrano, Ana Vallejo Andrada, and Alberto Sarasola Fernandez in their study "Sociability and Dependence" illustrate the results of a survey conducted for studying elders sociability in the urban area and its relation with the dependency degree.

Stefania Fantinelli presents her research titled "Knowledge Creation Processes Between Open Source Intelligence and Knowledge Management." She adopts some interviews on a sample of Italian analysts and experts to reveal what is the common use of "Open Source Intelligence" methods and how they are linked to the knowledge creation and knowledge management processes. Furthermore, she explores this method in a social psychology perspective and in relation to knowledge management in organizations.

Advanced technologies for social communication are proposed by Roberto Salvatori, who highlights the speed of change and spread of ICT in Education, which put the entire educational system in the position of continuously redesigning new methods and teaching models adapted to a globalized and interconnected society, where knowledge is distributed, easily accessible, and constantly updated.

Valentina Savini deals with the concept of social distance as an interpretation of a territory. She shows that the physical and spatial aspects of social distance identified by Simmel disappeared from the theoretical setting when the issue was analyzed in America. Moreover, the approaches that combine the physical and relational dimensions, based on Italian and English sociologists of the last few years, should be preferred.

Gabriele Di Francesco presents a socio-vital areas analysis with a qualitative approach to sociological analysis of urban spaces and social life. He presents the theoretical synthesis and the technical and methodological setting of a research, aimed to analyzing and evaluating the urban places and spaces of social life, where the human interactions take place and the city come to life.

Cultural and natural heritage challenges are illustrated by Zdena Rosická. The author remarks that every cultural heritage object and its content has its unique character and calls for an individual approach considering safety, protection, security, risk-preparedness, and further viable use. The public usually know about high-value losses of cultural property caused by burglary, fire of flood when the mass media report them; however, physical care, including environmental and conservation control, property transfer and transport, personal access, thefts from exhibits during the day, and incidents of smash bring about higher-cost internal losses, which are sometimes not reported at all. In case any disaster strikes, harm to cultural treasure is sometimes serious and losses irreplaceable unless relevant measures are taken in time.

Fiorella Paone aims to contribute to reflection about new form of emerging literacy starting from communicational changes, which characterize new social and cultural paradigm due to the use of electronic media. The narrative paradigm thought as linguistic practice of construction of liquid identities is considered as a possible operative strategy able to promote personal development in a holistic perspective, enforcing both personal and social self.

Vincenzo Corsi takes into account the sociological methods and construction of local welfare in Italy. He describes the importance of the study of the social care needs of the population for the construction of the local welfare system. Then, he shows some aspects of the local welfare system in Italy.

Daniel Flaut and Enache Tuşa analyze some aspects of social life in Romanian villages in the interwar period. At the same time, they describe the standard of living in rural areas, which varied depending on the ethnicity of the residents, using examples taken from the Dobrudja region.

A large spectrum of problems related to statistics, mathematics, teaching, social science, and economics has been presented in this volume. Therefore, a broad range of tools and techniques that may be used to solve problems on these topics has been presented in detail in this book, which is an ideal reference work for all those researchers interested in recent quantitative and qualitative tools. Due to the wide range of topics of the research results collected in this book, it is addressed, in equal measure, to mathematicians, statisticians, sociologists, philosophers, and specialists in the fields of communication, social, and political sciences.

Brno, Czech Republic
Pescara, Italy
Warsaw, Poland
July, 2017

Šárka Hošková-Mayerová
Fabrizio Maturo
Janusz Kacprzyk

Contents

Summary

The book "Mathematical-Statistical Models and Qualitative Theories for Economic and Social Sciences" is part of the important series "Studies in Systems, Decision and Control" published by Springer. This is the result of a scientific collaboration, in the field of economic and social systems, among experts from "University of Defence" in Brno (Czech Republic), "G. d'Annunzio" University of Chieti-Pescara (Italy), "Pablo de Olavide" University of Sevilla (Spain), and "Ovidius University" in Constanța, (Romania).

The first part of the book deals with "Recent Trends in Mathematical and Statistical Models for Economic and Social Sciences," whereas the second one concerns "Recent Trends in Qualitative Theories for Economic and Social Sciences." The variety of the contributions developed in this book reflects the heterogeneity and complexity of economic and social phenomena; thus, a large spectrum of problems related to statistics, mathematics, teaching, social science, and economics has been presented in this volume. Therefore, a broad range of tools and techniques that may be used to solve problems on these topics has been presented in detail.

Due to the wide range of topics of the presented research results, this peer-review book is addressed, in equal measure, to mathematicians, statisticians, sociologists, philosophers, and specialists in the fields of communication, social, and political sciences.

Part I
Recent Trends in Mathematical and Statistical Models for Economic and Social Sciences

Chapter 1
Correlational Research of Expenditure Spend on Slovak Armed Forces Participation in Peace Support Operations Led by NATO

Veronika Mitašová, Ján Havko and Tomáš Pavlenko

Abstract The expenditure which Ministry of Defence of the Slovak Republic spends on Armed Forces members' participation in peace support operations led by the North Atlantic Treaty Organisation is dependent on many factors. The main aim of this chapter is to find out the dependence level of this parameter from two selected factors, namely the level of defence expenditure and the GDP of the Slovak Republic. For this purpose the multifactor single-equation econometric model is created and tested. Based on the tests results, conclusions and proposals for further examination are formulated.

Keywords Peace support operation · Expenditure · Correlation · Econometric model

1.1 Introduction

Nowadays, peace support operations are an effective tool for conflict prevention and conflict resolution in problematic areas of the world. Mentioned type of operation is under the auspices of the international crisis management organizations. International organization of collective security, the North Atlantic Treaty Organisation ("NATO"), has undoubtedly a significant place among them.

V. Mitašová (✉) · J. Havko · T. Pavlenko
Faculty of Security Engineering, University of Žilina, Univerzitná 1, Žilina Slovak Republic
e-mail: veronika.mitasova@fbi.uniza.sk

J. Havko
e-mail: jan.havko@fbi.uniza.sk

T. Pavlenko
e-mail: tomas.pavlenko@fbi.uniza.sk

© Springer International Publishing AG 2017   3
Š. Hošková-Mayerová et al., *Mathematical-Statistical Models and Qualitative Theories for Economic and Social Sciences*, Studies in Systems, Decision and Control 104,
DOI 10.1007/978-3-319-54819-7_1

This contribution is focused on the assessment of factors, which affects participation of the Armed Forces of the Slovak Republic in NATO peace support operations. It is necessary to identify and understand what influences the amount of finance spent on these operations from state budget of the Slovak Republic ("the SR"). Based on the input data the multifactor single-equation econometric model is created and tested. According to the results of each test, the hypothesis, that the amount of the expenditure which Ministry of defence ("MoD") of the SR spend on Armed Forces members' participation in peace support operations led by NATO is dependent on the defence expenditures and GDP of the SR, is confirmed or refuted.

1.2 Qualitative Analysis and Input Data

Before econometric model creation, it is necessary to do qualitative analysis, which means assessing relations between variables. In the contribution it will be verified, if the estimated correlation among the amount of finance spend on the Slovak Armed Forces members participation in NATO peace support operations, the defence expenditures and the level of GDP exists or not.

It can be assumed, that the amount of finance, which states spend on participation of their armed forces in peace operations, is affected by a lot of factors. Table 1.1 shows some of them.

For the purpose of creating and testing econometric model we should work with factors mentioned above, but because of many reasons, we will work only with data listed in the Table 1.2. First factor, variable x_1, is defence expenditure of the SR. This kind of data is publicly available on the official website of MoD of the SR, but only since 2001. Second factor, variable x_2—GDP of the SR, is available on the Eurostat portal. Third one, variable x_3, could be number of completed and ongoing peace support operations led by NATO. Although the information about number of

Table 1.1 Factors affecting the amount of expenditure spend by MoD of the SR on Armed Forces members' participation in peace support operations led by NATO

Variable	Factors
Amount of expenditure spend by MoD of the SR on Armed Forces members' participation in peace support operations led by NATO	Defence expenditure of the SR
	GDP of the SR
	Number of completed and ongoing peace support operations led by NATO
	Global security environment
	Number of military personnel in the SR
	Number of Slovak troops deployed in peace operations led by international crisis management organizations
	Costs of military personnel training
	Others

Table 1.2 Input data

Year	Amount of expenditure spend by MoD of the SR on Armed Forces members' participation in peace support operations led by NATO in million Euro (y)	Defence expenditure of the SR at current prices in million Euro (x_1)	GDP of the SR at current prices in million Euro (x_2)
2001	4.35943	632	23 572.9
2002	3.414446	662	25 971.7
2003	12.809034	762	29 489.2
2004	15.170682	762	33 994.6
2005	14.720513	848	38 489.1
2006	23.293318	898	44 501.7
2007	16.069106	929	54 810.8
2008	17.62134	994	64 413.5
2009	32.80862	967	62 794.4
2010	37.572209	853	65 897.0
2011	34.334504	763	68 974.2
2012	38.209909	790	71 096.0

Source Eurostat (2015), Ivančík (2013), SIPRI (2015)

completed and ongoing operations is available, it is difficult and also slightly mistakenly uses these data as a total number of peace support operations led by NATO in the individual year. Some operations are ongoing throughout the year, others only in certain months. More appropriate would be, in this case, to examine the correlation between the amount of expenditure spend by MoD of the SR on Armed Forces members' participation in peace support operations led by NATO and the number of peace support operations during particular months. However, because of other data distribution, it is not possible. The same problem is in the case of another factor, variable x_6—the number of Slovak troops deployed in peace operations led by international crisis management organizations. The number of troops in peace support operations may change through the operation, and this number is often slightly different from the mandate. Even though variable x_4, which is global security environment, affects the amount of expenditure spend by MoD of the SR on Armed Forces members' participation in peace support operations led by NATO, it is not a numerical quantity, so it would not be possible to use it in our econometric multifactor model. Other two variables, x_5—the number of military personnel in the SR and x_7—costs of military personnel training, would be appropriate to examine, but they are not available (Hošková-Mayerová 2017).

The small range of data is influenced by the fact, that the SR is relatively young country, but also a lot of needed information are confidential or they are not publicly available.

Based on the data in the Table 1.2, it is possible to notice a growing trend of expenditure for peace support operations led by NATO. The expenditure had increased in 12 years more than nine times. A similar trend is in the development of

GDP, which increased three times during the 12-year period. On the contrary, defence expenditures of the SR maintain approximately the same level.

1.3   The Construction of Econometric Model and Its Testing

It is possible to create and to test econometric model in different ways, from laborious and difficult calculation through different software. The results in this contribution were gained through MS Office Excel tools and STATISTICA software. Data sources were SIPRI database (Stockholm International Peace Research Institute) and Eurostat.

The particular regression equation, which expresses the relation between dependent variable y (expenditure of MoD on participation of the Slovak Armed Forces in peace support operations led by NATO in million Euro) and explanatory variables x_1 and x_2 (defence expenditure at current prices in million Euro and GDP of the SR at current prices in million Euro), can be written in the form:

$$y^\wedge = 4.1035 - 0.0199x_1 + 0.0007x_2 + u \qquad (1.1)$$

The value of the parameter $b_1 = -0.0199$ indicates, that the increase of defence expenditure by 1 million Euro will reduce the amount of expenditure on the Slovak Armed Forces participation in NATO peace support operations by 19,900 Euro, while GDP remains unchanged. The value of the parameter $b_2 = 0.0007$ indicates, that increase of GDP by 1 million Euro will increase the expenditure on the Slovak Armed Forces participation in NATO peace support operations by 700 Euro, providing defence expenditure remained unchanged.

Multiple $R = 0.8949$ means, that there occurs a very strong positive correlation in the econometric model. The value of reliability $R^2 = 0.8008$ can be interpreted, that about 80% of the explaining variable variability is explained by the variability of explanatory variables. Remaining 20% of the explaining variable variability, which is amount of expenditure spend by MoD of the SR on Armed Forces members' participation in peace support operations led by NATO, is explained by other variables (Table 1.3).

Graphical representation of variables y, x_1, x_2 distribution is in the Figs. 1.1, 1.2 and 1.3. On the right side of these figures are values characterizing variables, namely minimum, maximum, standard deviation or median.

Table 1.3 Basic statistical values of the econometric model	Summary statistics	
	Multiple R	0.8949
	Multiple R^2	0.8008
	Adjusted R^2	0.7566

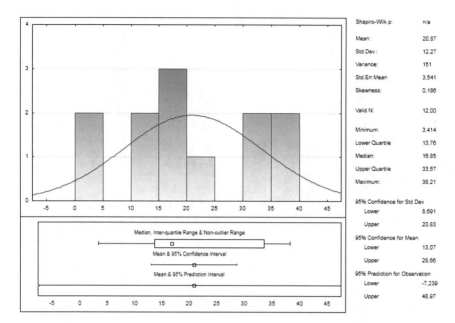

Fig. 1.1 Graphical summary for variable y

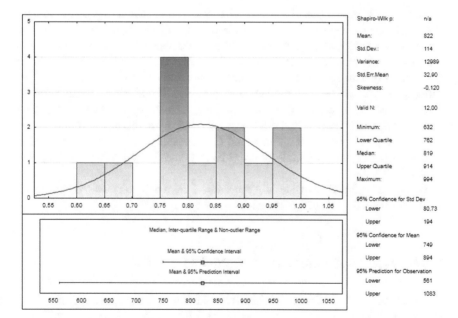

Fig. 1.2 Graphical summary for variable x_1

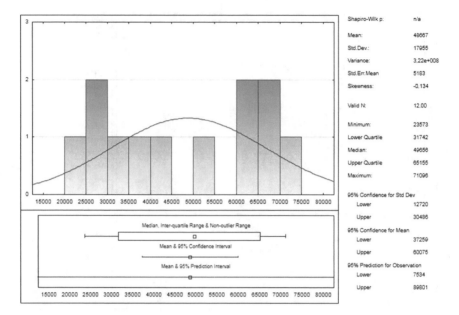

Fig. 1.3 Graphical summary for variable x_2

Econometric model must be tested after its creation, thus it is possible to conclude whether the model is applicable for practical needs. These tests are testing model as a whole through the coefficient of determination R^2 tested by Fisher's statistics (F-statistic), testing parameters of variables through Student's probability distribution (t-statistics), testing of residuals autocorrelation (we have <15 observations, so we will use Von-Neumann ratio and D-statistics). Verification of heteroscedasticity, respectively homoscedasticity, is made by Goldfield and Quandt test. The last test is used to find out the presence of multicollinearity in the model. For this purpose, we will use Farrar and Glauber method, where the correlation matrix is tested through Chi-square (Mikolaj and Vančo 2004; Ristvej and Kampová 2009).

Because of comprehensive testing procedures, there will be stated only particular resultant values of tests and conclusions related to them.

1.3.1 Testing of Econometric Model as a Whole

The value of determination coefficient R^2 is 0.8008. It means, that approximately 80% of the amount of expenditure spend by MoD of the SR on Armed Forces members' participation in peace support operations led by NATO variability is explained by the variability of defence expenditure and GDP in the SR. Remaining nearly 20% of the explaining variable variability is affected by other factors, which are not included in the model. F-statistic test was performed at a significance level 0.05.

Therefore, if the model would be significant as a whole, the inequality (1.2) has to be valid.

$$F_r > F_{\alpha; k; [n-(k+1)]} \tag{1.2}$$

The value of $F_r = 18.0938$ and $F_{0.05, 2, 9} = 4.2565$, consequently, the determination coefficient of on the significance level 0.05 is considered as a significant and the model as a whole is considered as a statistically significant too.

1.3.2   Testing of Variables Parameters Through t-Statistics

Both of parameters, $b_1 = 0.0199$ and $b_2 = 0, 0.0007$, are tested at the significance level 0.05. To consider parameter as a statistically significant, the inequality (1.3) has to be valid.

$$|t_i| > t_{\alpha; [n-(k+1)]} \tag{1.3}$$

In the case of parameter b_1, after substituting into inequality $0.9811 < 2.2622$, it does not apply. It means that the parameter b_1 at the significance level 0.05 is not considered as statistically significant. After substituting into inequality in the case of parameter b_2, inequality form is $5.2772 > 2.2622$. Based on it, we can formulate conclusion that inequality applies, therefore, the parameter b_2 at significance level 0.05 is considered as statistically significant.

1.3.3   Testing of Residuals Autocorrelation

Testing of residuals autocorrelation for <15 observations is carried out by Von-Neumann ratio. To autocorrelation verification the calculated value is compared to the tabulated one, while the inequality (1.4) has to apply.

$$D^+ < D < D^- \tag{1.4}$$

Value of $D^+ = 1.2301$, $D^- = 3.1335$ and $D = 2.0705$, therefore, the inequality applies. We conclude autocorrelation of residuals absents in our econometric model.

1.3.4 Testing of Heteroscedasticity

It is necessary to test heteroscedasticity for each variable x (x_1 and x_2). For the hypothesis "scatter of residuals is constant, there is homoscedasticity of residuals in the model" acceptation, inequality (1.5) has to apply, whereby the value v means degree of freedom.

$$F_{2,1} \leq F_{\alpha(v,v)} \tag{1.5}$$

It is calculated according to Eq. (1.6). The value $v = 2$ and tabulated value $F_{0.05;(2,2)} = 19$.

$$v = [\frac{(n-M)}{2}] - (k+1) \tag{1.6}$$

After substituting into inequality in the case of variable x_1, it is $1.6852 < 19$, in the case of variable x_2, it is $1.4757 < 19$. Therefore, the inequality applies in both cases, we conclude, that heteroscedasticity of residuals absents in our model. For better expression of values, in the Fig. 1.4 are shown predicted and observed values of variable y.

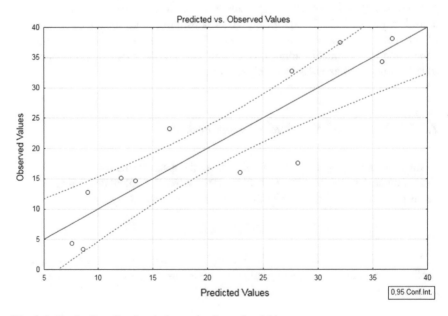

Fig. 1.4 Graph of predicted and observed values of variable y

Table 1.4 Matrix of correlational coefficients

	Defence expenditure of the SR at current prices in million Euro (x_1)	GDP of the SR at current prices in million Euro (x_2)
Defence expenditure of the SR at current prices in million Euro (x_1)	1.0000	−0.6167
GDP of the SR at current prices in million Euro (x_2)	−0.6167	1.0000

1.3.5 Testing of Multicollinearity

For testing a presence of multicollinearity in the model is used Farrar and Glauber method. This method allows to assess the overall multicollinearity in the set of explanatory variables and through determinant of correlation matrix testing by χ^2 (chi-square) test, it is possible to find out, which variables cause the multi-collinearity (Table 1.4).

To prove that there is not multicollinearity in the model, the following inequality must apply:

$$\chi_R^2 \leq \chi_{\alpha(v)}^2 \tag{1.7}$$

After substituting tabulated value and value of determinant $|R|$ converted to empirical value, the inequality is 4.4550 > 3.8415. However, the inequality does not apply. Therefore, the multicollinearity is in the model. We adopt the conclusion about interdependence of explanatory variables x_1 and x_2 at significance level 0.05.

1.4 Results and Discussion

The contribution is focused on multifactor single-equation econometric model creation. Through that we were able to fulfil stated objective and to find relations and dependencies among chosen variables. The aforementioned variables were the amounts of expenditure spend by MoD of the SR on Armed Forces members' participation in peace support operations led by NATO as explaining variable, defence expenditure of the SR and GDP of the SR as explanatory variables.

The prerequisite was that between mentioned explaining and the two explana-tory variables, there will be a strong dependence. After creating and testing model as a whole, we concluded, that about 80% of the explaining variable variability is explained by the variability of explanatory variables. However, what is interesting and somewhat surprising, is fact, that parameter b_1 at the significance level 0.05 is not statistically significant. We expected that the amount of expenditure spend by MoD of the SR on Armed Forces members' participation in peace support operations led by NATO will certainly depend on defence expenditure of the SR. However, based on tests results, we ascertained our assumption was not correct.

Due to other tests we can comprehensively assess created model and its suitability for application. The first was residuals autocorrelation test. We found out, that there is not residuals autocorrelation, which is a positive finding. Another test confirmed that heteroscedasticity of residuals absents in the econometric model. But the last test, which is implemented in the case of multifactor econometric model testing, multicollinearity test, demonstrated an adverse finding. The explanatory variables are interdependent; therefore, multicollinearity is present in our model.

Based on the results of the individual tests, the created econometric model can be described as inappropriate for using in the present form. The main reason for this statement is fact, that one of the explanatory variables, the amount of defence expenditures in the SR, does not explain the variable y sufficiently. Another reason is the presence of multicollinearity, interdependence of explanatory variables.

Interestingly, we created one-equation econometric model from the input data again, but this time, it was only one factor econometric model. We wanted to ascertain, whether after a modification of the original model, in which the variable x_1 (defence expenditure in the SR) will not be, the new one will prove as reliable and useful. New form of regression equation is:

$$y^\wedge = -8.4932 + 0.0006x + u \qquad (1.8)$$

The value of multiple R = 0.8829 indicates is a very strong positive correlation. The value of reliability $R^2 = 0.7795$ means, that about 78% of the explaining variable variability is explained by the variability of explanatory variable. This single-factor econometric model was tested; all results are in the Table 1.5.

The single-factor model successfully passed all tests. Their results confirmed that the model as a whole, and also the parameter b are at the significance level 0.05 statistically significant. There is not residuals autocorrelation; also the hypothesis about homoscedasticity was confirmed. Thereby was proven, that after removal statistically insignificant variable—the amount of defence expenditure in the SR, we created new model which is usable in practice.

Certainly, this claim would be justified in the context of further research in this field and original econometric model. It would be definitely appropriate to adjust or

Table 1.5 The results of single-factor econometric model tests

	Critical/tabulated value ($\alpha = 0.05$)	Comparative criterion calculated value (inequality)	
Testing of econometric model as a whole unit	$F_\alpha = 4.9646$	$F_r > F_\alpha$	$F_r = 35.3581$
Testing of variable parameter	$t_\alpha = 2.2281$	$|t_i| > t_\alpha$	$t_i = 5.9463$
Testing of residuals autocorrelation	$D^+ = 1.2301$ $D^- = 3.1335$	$D^+ < D < D^-$	$D = 1.7760$
Testing of heteroscedasticity	$F_\alpha = 9.2766$	$F_i \leq F_\alpha$	$F_i = 6.0031$

modify this model. Further research of dependence level with other variables could increase the percentage of the explaining variable variability explained by the variability of explanatory variable. In other words, it would be suitable to focus on this model and clarify the dependence. Despite of the model reliability, there are a lot of factors, which are logically more accurate to explain variable y, the amount of expenditure spend by MoD of the SR on Armed Forces members' participation in peace support operations led by NATO. This variable certainly depends on costs related to preparation of the Armed Forces members. Unfortunately, such data are not available.

In conclusion, it is necessary to mention, that our model was created from a relatively small range of data. It also could misrepresent and influence the results. Its reliability would increase if we could use wider range of variables data (Maturo and Hošková-Mayerová 2016; Ristvej and Kampová 2009). It is necessary to mention that because of the test results, the need of research a lot of other variables was highlighted. However, that is directly related to the processing dissertation thesis dealing with these issues.

1.5   Conclusions

Activities of international crisis management organizations are often the focus of media and public attention. That is why it is interesting to research factors which influence the amounts of financial sources spend on these activities.

The aim of this chapter was to reveal dependence among chosen variables. We expected strong correlation among them, so we created and tested econometric model. Based on test results, we have to definitely acknowledge that chosen explanatory variables are not able to adequately explain the dependent variable y— amount of expenditure spend by MoD of the SR on Armed Forces members' participation in peace support operations led by NATO.

Nevertheless, the scope for further research was created, thus it will be a part of processing dissertation thesis.

References

Eurostat. *Gross domestic product at market prices.* [on line]. Available at: http://ec.europa.eu/eurostat/tgm/refreshTableAction.do?tab=table&plugin=1&pcode=tec00001&language=en (2015).

Hošková-Mayerová, Š. (2017) Education and Training in Crisis Management. In: *The European Proceedings of Social & Behavioural Sciences EpSBS, Volume XVI.* Future Academy, 2017, p. 849–856.

Ivančík, R.: *Financovanie operácií medzinárodného krízového manažmentu (Financing of Operations led by International Crisis Management Organisations)*—lecture materials. General Staff of the Armed Forces of the Slovak Republic (2013).

Maturo, F., Hošková-Mayerová, Š. (2016) Fuzzy Regression Models and Alternative Operations for Economic and Social Sciences Recent Trends in Social Systems: Quantitative Theories and Quantitative Models, Decision and Control, Vol. 66, Maturo (Eds.), 235–248.

Mikolaj, J., Vančo, B.: *Ekonometria pre manažérov (Econometrics for Managers)*. Žilina: Fakulta špeciálneho inžinierstva ŽU (2004).

Ristvej, J., Kampová, K.: *Ekonometria pre manažérov—návody na cvičenia (Econometrics for Managers—exercises)*. Žilina: EDIS- vydavateľstvo ŽU (2009). 140 pp. ISBN 978–80-554-0107-2.

SIPRI Military Expenditure Database. [on line]. Available at: http://www.sipri.org/research/armaments/milex/milex_database (2015).

Chapter 2
Health Risk Assessment of Combustion Products from Simulated Residential Fire

Josef Navrátil, Veronika Sadovská and Irena Švarcová

Abstract The chapter concentrates on the health risk assessment of selected pollutants derived from residential fire simulated in fire container. The key interest is dedicated to Polycyclic Aromatic Hydrocarbons (PAHs) and their toxicity and harmful effects on the health of firefighters, whose protection by breathing apparatus is insufficient. The final evaluation suggests measurements regarding the issue of the protection of people dealing with fire.

Keywords Fire · Combustion products · Polycyclic aromatic hydrocarbons · Toxic effects · Health risks

2.1 Introduction

Currently, fires belong among the most significant risk events because of high number of injuries and deaths caused by fires. In 2014, there were 114 dead and 1179 injured people at all fires in the Czech Republic. The most of fires with tragic consequences begins at family houses and residential houses, where 63 persons died and 694 persons were injured. The residential fires add up to 20% from the total number of firefighting events (Vonásek and Lukeš 2015).

The health consequences from fires are mostly caused by burns, injuries derived from mechanical damages, high heat or lower oxygen content, then by burning products intoxication and finally by psychosocial impacts on residents and members of rescue units. The combustion products intoxication has absolute major contribution to the total number of deaths caused by fires. It was estimated that up to 30% of burned persons were also intoxicated by products of combustion (Brown 1990). Toxic combustion products represent higher risk than other causes of injuries and

J. Navrátil (✉) · V. Sadovská · I. Švarcová
University of Defence, Kounicova 65, 662 10 Brno, Czech Republic
e-mail: josef.navratil@unob.cz

V. Sadovská
e-mail: sadovskaveronika@seznam.cz

© Springer International Publishing AG 2017   15
Š. Hošková-Mayerová et al., *Mathematical-Statistical Models and Qualitative Theories for Economic and Social Sciences*, Studies in Systems, Decision and Control 104,
DOI 10.1007/978-3-319-54819-7_2

deaths of people occurred in vicinity of fire. The monitoring of combustion products during simulated fires at the Flashover fire container or other special spaces enables to acquire important information about their chemical make-up and their toxicity, which helps to predict health risks of fire participants.

2.2 Analysis of the Current State

The amount, chemical properties and toxicity of combustion products arisen from fires depend on the type of burning material and other factors resulting from dynamical nature of the given fire. The toxicity of combustion products from concrete materials is significantly influenced by the amount of oxygen present in the fire environment. It is known that the combustion products become more toxic with the decreasing concentration of oxygen. These conditions occur mainly in closed areas, in which more significant toxic substances emerge more probably than in open areas. Nowadays, the interiors of residents and flats are equipped with synthetic materials, which may highly contribute to the emerging of toxic substances during fires.

Therefore, the experiments with simulation of residential fire are carried out in order to gather comprehensive information about qualitative and quantitative combustion products composition. The attention is given to low molecular chemical compounds such as CO_2, CO, HBr, HCl, HCN, NOx and volatile organic compounds (VOC) such as benzen and styrene and mainly to polycyclic aromatic hydrocarbons (PAHs), polychlorinated and polybrominated dibenzo-p-dioxins/furans (PCDD/F, PBDD/F) and fragments of brominated flame retardants (BFRs) e.g. from tetrabromobisphenol A (TBBPA) or dekabromodiphenyl ether (deka-BDE) and, last but not least, particulate matters (PM). Despite the fact that CO_2 and HCN mostly contribute to intoxication of persons, the attention is especially given to obtain data about organic substances.

Particulate matters are known for their mutagenic, carcinogenic and genotoxic effects. The inhalation and effect of PM on human organism are directly influenced by their size. Only such PM smaller than 10 micrometers in diameter reach deeply respiratory tract. Increased PM deposition in lungs may lead to acute cardiovascular events (Donaldson and Borm 2007). Increased morbidity and mortality due to cardiovascular and respiratory system illnesses belong among health effects of the short-term exposure to PM. The long-term exposition may result in augmented incidence of respiratory tract cancer and in shortened life expectancy as a consequence of cardiovascular, pulmonary and oncological diseases (Lippman 2010).

The majority of PAHs belongs to indirectly acting genotoxic substances. Diarrhea, nausea, lack of appetite and vomiting may be considered as short-term effects of PAHs. The mostly emphasized member of PAHs is benzo(a)pyrene (BaP). The typical rout of entry for BaP and other PAHs is by inhalation or skin absorption. The exposure to BaP potentially leads to fetal development disorders, irritation and burning of skin and disease cancer risk. Repeated exposure causes reduced and chapped skin. International Agency for Research on Cancer (IARC) classifies BaP as carcinogenic to human into Group 1, then naphthalene, chrysene and benzo(b) fluoranthene as possibly carcinogenic to humans into Group 2B (International Agency on Research of Cancer 2015).

PCB are those substances which may cause cancer diseases and also may induce other harmful effects on human health, e.g. effects on immune, reproductive, nerve and endocrine system of individual organism.

Harmful health effects of PCDD/F are connected with chloracne development, which may occurs during several days up to months after the exposure (International Programme on Chemical Safety 1989). Further impacts on human health after the acute exposure to dioxins are mostly hepatotoxic and neurotoxic effects and hypertension. Research studies involving animals indicate dioxins as harmful substances for reproductive system and as teratogenic agent. IARC classifies 2, 3, 7, 8-TCDD as carcinogenic to human (Group 1). TBDD/F are considered as toxic to reproduction of animals. Bromide compounds are potentially toxic to human skin, liver and gastrointestinal tract.

2.3   Applied Methods and Devices

Combustion products derived from fire container were picked up by passive samplers filled with porous polymer Tenax TA (with specific surface area of $35 \text{ m}^2 \text{ g}^{-1}$) and by samplers with coal sorbent Carbopack ($12 \text{ m}^2 \text{ g}^{-1}$) and by the portable concentrator (Ministry of the Environment of the Czech Republic 2005). Passive samplers were clipped on the garment of firefighters training in fire container. The average exposure duration of passive samplers to combustion product lasts about 20 min, combustion products have been captured by the portable concentrator for 3 min.

After the exposure, the adsorbent from passive samplers was extracted by mixture of hexan-dichloromethane in closed vial and the extract was examined by gas chromatography and mass spectrometry (GC-MS) with the help of devices Agilent GC7890-A and MS 5775C Agilent. Analysis were especially oriented on the identification of PAHs. The sample of combustion products captured by the portable concentrator was also examined by the GC-MS method.

From the aspect of possible assessment of health risks, the data from residential fire simulations presented at specialized literature (Večeřa et al. 2010) were applied. The issue was primarily discussed with the author of research study investigating residential fire simulation named Per Blomqvist from Swedish National Testing and Research Institute. After the personal consultation (Blomqvist et al. 2004), the study data was carefully compiled and the exposure scenario for the health risk assessment was created.

At the health risk assessment, the methodical directive prepared by the Ministry of the Environment of the Czech Republic (MoE) called "The Risk Analysis of the Contaminated Area" was applied. Following equations were used for the estimation of health risks of carcinogenic substances (Ministry of the Environment of the Czech Republic 2005):

$$ILCR = LADD \cdot ICPF. \tag{2.1}$$

where *ILCR (Individual Lifetime Cancer Risk)* indicates increasing probability of cancer diseases number over the general average, *LADD (Lifetime Average Daily Dose)* means average lifelong daily exposure [mg kg^{-1} den^{-1}] and *ICPF (Inhalation Cancer Potency)* represents inhalation slope factor [mg kg^{-1} den^{-1}].

In the case of inhalation, the following equation for calculation of *LADD* is used:

$$LADD = CA \cdot IR \cdot ET \cdot EF \cdot BW^{-1} \cdot AT^{-1}. \tag{2.2}$$

where *CA (Average Concentration)* is concentration of the pollutant in the air [mg m^{-3}], *IR (Intake Rate)* represents the volume of inhaled air [m^{-3} h^{-1}], *ET (Exposure Time)* [h day^{-1}], *EF (Exposure Frequency)* [den year^{-1}], *ED (Exposure Dose)* [year], *BW (Body Weight)* [kg] a *AT (Average Time)* [day].

The value 10^{-6} is regarded as acceptable rate of risk. This value represents increase of the individual lifelong carcinogenic risk about one case per million exposed persons.

2.4 Results and Discussion

2.4.1 Fire Container

Selected chromatogram of determined combustion products originated from wood materials burned in the fire container, which were captured in passive samplers filled with various sorbents and picked up by portable concentrator is introduced at the Fig. 2.1. The comparison of capture efficiency with respect of determined compounds is also demonstrated. Chemical compounds assigned to particular peaks are introduced at the Table 2.1. With the respect of NIST database, other peaks have probability lower than 10%. Therefore, the identification is not possible. The origin of the intense peak of mono(2-ethylhexyl)phthalate is obviously caused by decomposition of plastic remnants of pallet packaging. By passive sampler usage, peaks of organic compounds emerge at chromatograms with retention time corresponding to identified compounds given earlier. Concentrations were extremely low thus their identification was complicated and unreliable.

Naphthalene was identified in examined samples taken by the portable concentrator. This compound is categorized by United States Environmental Protection Agency (US EPA) into category C and by IARC into Group 2B for its possible carcinogenic effects. Naphthalene acts as neurotoxic compound, may cause serious chronical diseases of the respiratory tract and lungs, asthmatic bronchitis, serious chronical skin and eyes diseases. Naphthalene was also identified at samples taken

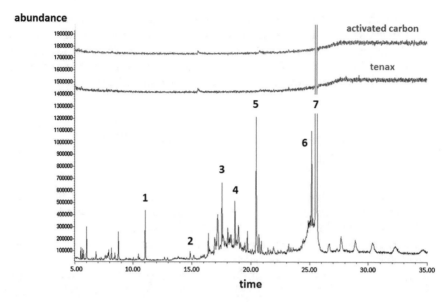

Fig. 2.1 Chromatogram of sample taken by passive samplers and portable concentrator

Peak	Assigned compound	Probability
Peak 1	Naphthalene	Identical RT standard
Peak 2	Acenaphthylene	Identical RT standard
Peak 3	Heptadecane	18.8%
Peak 4	Phenanthrene	Identical RT standard
Peak 5	Di-n-butyl phthalate	33.1%
Peak 6	Diisooctyl phthalate	32.1%
Peak 7	Mono(2-ethylhexyl) phthalate	37.1%

Table 2.1 Chemical compounds assigned to particular peaks

from cotton face piece, which firefighters wear during the training. Therefore it is obvious, that given substances are captured in these firefighter gadgets and garment and such substances are in direct contact with their skin on face area. Instructors of the training program are repeatedly exposed to such substances and they may suffer from listed chronical effects of naphthalene and other PAHs.

US EPA classifies acenaphthylene and phenanthrene into Group D as Not Classifiable as to Human Carcinogenicity. Heptadecane is indicated as substance hazardous for human. Dibutyl phthalate is substance with relatively low acute and chronical toxicity and in according to US EPA, it is categorized as substance which cannot be classified. Diisooctyl phthalate is applied as softening agent in plastic used predominantly at households. It is extensively used in processing polyvinyl chloride and ethylcellulose resins to produce cable wearer, imitation leather and electric wire. There is no sufficient data for classifying diisooctyl as carcinogenic substance and there is no evidence of its acute or chronic effects on humans.

2.4.2 Simulated Room Fire Scenario and Health Risk Assessment

Results arisen from the experimental study on the simulation of room fire carried out at Swedish National Testing and Research Institute are used as the ground for health risks calculation. The room fire simulation was conducted by the study author (Blomqvist et al. 2004) at rooms with floor area of 4 × 4 m^2 and the height to the ceiling was 2.4 m. Rooms were equipped with 303 kg of typical domestic furniture including sofa, armchair, books, carpet, shelving etc. Rooms were equipped with identical furniture except various types of televisions (TV set). The experiment EX1 employed a TV set with non-fire retarder enclosure material produced at European Union (EU) and the experiment EX2 employed a TV set containing brominated flame retardants produced at USA. The fire was initiated by a lighted candle. The combustion gases were determined with the help of infrared spectroscopy Fourier transformation method (FTIR). The Table 2.2 contains comprehensive results of combustion gases mass.

The particular combustion products and their mass are listed at the research project final report (Navrátil et al. 2013). On basis of results from conducted experiments EX1 and EX2, benzene contributes to the total mass of detected VOCs with 21%, resp. 38% and naphthalene with 16%, resp. 18%. Benzene acts as precursor of PAHs and soot particle. According to IARC, benzene is classified as

Table 2.2 Mass of the detected combustion products from the simulated room fires (Blomqvist et al. 2004)

Detected products	Experiment 1—EX1	Experiment 2—EX2
	$\frac{\text{Total}}{\text{g}}$	
Inorganic gases	4.490E+05	4.740E+05
VOC	4.100E+02	1.420E+03
PAH	3.329E+02	8.670E+02
PCDD/PCDF	4.200E-06/3.280E-06	13.972E-06/6.350E-06
PBDF	8.100E-06	2.400E-06

Legend *VOC*—volatile organic compounds, *PAH*—polycyclic aromatic hydrocarbons, *PCDD*—polychlorinated dibenzo-p-dioxins, *PCDF*—polychlorinated dibenzofurans, *PBDD*—polybrominated dibeznodioxins, *PBDF*—polybrominated dibenzofurans carcinogenic to human (Group 1) causing especially leukemia and lung cancer. Toluene, phenol and styrene rank among the next volatile compounds determined during the experiment. Their characteristics contribute to their toxicity and irritant effects of produced smoke during a room fire. Naphthalene had the most contribution to the PAHs detected at the experiment (EX1–43.9%; EX2–43.5%). Phenanthrene, acenaphthylene, fluoranthene and pyrene were also detected in considerable quantity.

The calculation of health risks was accomplished with help of results derived from chemical analysis connected with mentioned study on simulated room fire. At the suggested scenario, it was assumed that a room fire conditions are identical as the simulated fire conditions. It was also assumed that the fire is conducted at modular home (prefab house), where approximately 10% of combustion gases volume leaks out to staircase (20 × 5 × 10 m^3). There is an expectation, that smoke enter the staircase area during the fire, where persons without breathing apparatus spend no more than 5 min (Kukleta 2014, personal communication). The exposure scenario was focused on the health risk assessment of benzene and benzo(a)pyrene. It was also expected that rescuers will arrive at the scene of fire with 30 min delay caused by late fire notification, which means that the majority of furnishing have burned down.

Benzene and benzo(a)pyrene were selected on basis of examined combustion products from simulated room fire due to their carcinogenic effects on humans proven by IARC.

The health risk assessment of benzene and benzo(a)pyrene was proceeded according to the methodical directive created by MoE. The LADD was calculated with the following values: $IR = 0.83$ m^{-3} h^{-1}, $ET = 0.083$ h day^{-1}, $EF = 0.130$ den year^{-1}, $ED = 30$ years, $BW = 85$ kg and $AT = 3900$ days.

The risk rate ILCR was obtained with help of calculated LADD and ICPF values listed in particular documents published by the Office of Environmental Health Hazard Assessment (OEHHA) in 2009: $ICPF_B = 1,00E$–0.1 and $ICPF_{BaP} = 3.90E$ $+0.1$ (California Environmental Protection Agency, Office of Environmental Health Hazard Assessment 2001). The Table 2.3 contains the assessment of ILCR for benzene and benzo(a)pyrene.

According to the Table 2.3, the value of ILCR for both substances are greater at the experiment EX2 (TV USA) than at the experiment EX1 (TV EU). The carcinogenic risk of benzene at experiment EX1 is balancing the acceptable rate of risk. The acceptable rate of risk was exceeded 4.4 times in case of benzene

Table 2.3 The calculation of LADD and ILCR for benzene and benzo(a)pyrene (Navrátil et al. 2013)

Pollutant	Experiment	$\frac{C}{\text{mg m}^{-3}}$	$\frac{LADD}{\text{mg kg}^{-1}\text{ day}^{-1}}$	ILCR
Benzene	EX 1 (TV EU)	8.610E+00	0.698E-05	0.698E-06
	EX 2 (TV USA)	5.396E+01	4.373E-05	4.373E-06
Benzo(a)pyrene	EX 1 (TV EU)	5.162E-01	4.183E-07	1.631E-06
	EX 2 (TV USA)	1.301E+00	1.055E-06	4.112E-06

determined at the experiment EX2. The ILCR values for benzo(a)pyrene exceeded the acceptable rate of risk at both experiments EX1 and EX2: 1.6 times, resp. 4.1 times. The probability of cancer development resulted from inhalation exposure to bezno(a)pyrene is highly significant. Therefore those rescuers operating in vicinity of fire without breathing apparatus are exposed to combustion products with carcinogenic effects. With respect to the ILCR values, it is necessary to accept certain measures for those rescuers who do not wear breathing apparatus and operate in vicinity of fire or in area which is secondary filled with smoke.

2.4.3   The Analysis of Health Risk Uncertainties

During the health risk assessment, range variety of uncertainties appears and it is required to respect them at the final result interpretation. Passive samplers chosen inappropriately and incorrect sample collecting may rank among main causes of those uncertainties. The choice of chromatographic column may be connected with the uncertainty of the concentration determination during the analysis conducted on the standard operation procedures (Navrátil 2017; Hošková-Mayerová 2017).

Scenarios related to the health risk assessment have suggested the full burning down of the room and no fire progress to other housing units. The combustion gases concentration were not constant for whole duration of fire. During the clearance operations, combustion gases concentration was lowered by dilution with surrounding air.

Synergetic effects of benzene, benzo(a)pyrene and other substances were not included at the cancer risk calculation. The inhalation exposure was the only one reflected route of entry. The results may be overestimated due to inaccurate calculation of LADD values. Other combustion products which may emerge during fire were not included in the assessment.

2.5   Conclusion

Beside the heat radiation, the combustion product toxicity ranks among obvious fire danger, which may threaten people afflicted with fire and rescuers without breathing apparatus operating indirectly and repeatedly at fire environment.

High toxicity of combustion products is often influenced by application of new synthetic materials used in house building and for furnishing production. Mutual combination of material decomposition products from fire may result in formation of products with considerable toxicity in high concentration. It is required to accept the preliminary precaution principal and to implement useful measures in order to eliminate health risks consequent upon exposure to combustion products from fire primarily at operational management (during fire) and secondary at organization management (after transfer of firefighter gear to fire station). Rescuers are exposed to long-term effects of harmful combustion products and exhalations. Therefore, it is desirable to observe those rescuers repeatedly with respect to their state of health and irrespective to their duty description. The monitoring of PAHs as combustion products is also determining factor for selection of effective decontamination process of personal protective equipment.

The experiment conducted in order to take samples at fire container pointed out the necessity of further study on combustion products from residential fires. Further period of study is going to focus on colleting samples from rescuers gear and the inner part of the fire container or room after a fire.

References

Blomqvist, P., Rosell, L., Simonson M. (2004). Emissions from Fires Part II: Simulated Room Fires. *Fire Technology*, 40, 59–73 (2004). doi:10.1023/B:FIRE.0000003316.63475.16.

Brown, N. J. (1990). Health Hazard Manual for Firefighters, Manual and User Guides, Cornell University ILR School. http://digitalcommons.ilr.cornell.edu/. Accessed 25 June 2015.

California Environmental Protection Agency, Office of Environmental Health Hazard Assessment (2001). A guide to Health Risk Assessment. In: Office of Environmental Health Hazard Assessment. http://sfrecpark.org/wp-content/uploads/HRSguide2001.pdf. Accessed 14 April 2015.

Donaldson, K., Borm, P. (2007) *Particle Toxicology*. Tailor and Francis Group, CRC Press. ISBN 0-8493-5092-1.

Hošková-Mayerová, Š. (2017) Education and Training in Crisis Management. In: *The European Proceedings of Social & Behavioural Sciences EpSBS*, *Volume XVI.*: Future Academy, 2017, p. 849–856.

International Agency on Research of Cancer (2015). IARC Monographs, Volumes 1–116. http://monographs.iarc.fr/ENG/Classification/ClassificationsAlphaOrder.pdf. Accessed 16 July 2015.

International Programme on Chemical Safety (1989). Polychlorinated dibenzo-p-dioxins and dibenzofurans. Environmental Health Criteria 88. WHO: Geneva. http://www.inchem.org/documents/ehc/ehc/ehc88.htm. Accessed 1 April 2015.

Kukleta, P. (2014) Personal communication. 24 June 2014.

Lippman, M. (2010). Targeting the Components Most Responsible for Airborne Particulate Matter Health Risks. *Journal of Exposure Science and Environmental Epidemiology*, 20(2), 117–118.

Ministry of the Environment of the Czech Republic (2005). Metodický pokyn pro analýzu rizik kontaminovaného území č. 12. Věstník MŽP, ročník XV, článek 9. ISSN – 0862-9013.

Navrátil J., Sadovská V., Přibylová M. (2013). The Identification of Harmful Substances Originated from Fire and Measures for People Protection. *Final Report from the Specific Research Project*. University of Defence, Brno.

Švarcová, I. Hošková-Mayerová, Š. Navrátil, J. (2017) Crisis Management and Education in Health. In: *The European Proceedings of Social & Behavioural Sciences EpSBS, Volume XVI.*: Future Academy, 2017, 255–261.

Večeřa, Z., Mikuška, Pavel, Kellner, J., Navrátil, J. (2010). Portable Continual Aerosol Concentrator (type G). University of Defence, Brno.

Vonásek V., Lukeš P. et al. (2015). Statistická ročenka 2014. MV-GŘ HZS ČR. http://www.hzscr.cz/clanek/statisticke-rocenky-hasicskeho-zachranneho-sboru-cr.aspx. Accessed 10 October 2015.

Chapter 3
Qualitative and Quantitative Comparison of the Entrance Draft Tests and the Entrance Tests Results in Mathematics

Radovan Potůček

Abstract This contribution deals with the qualitative and quantitative comparison of the entrance draft tests and the entrance tests in mathematics for the bachelor and master study applicants at the Faculty of Military Technology at the University of Defence in Brno. These tests have been organized by the Department of Mathematics and Physics since 2010. The chapter focuses on the results and the rates of 353 individuals applying for the study as well as the future students' results obtained at the entrance examinations and proceeding voluntary draft tests within the period of the last six years 2010–2015. The results of the Military secondary school and secondary civilian schools applicants are compared and evaluated from the qualitative and quantitative points of view.

Keywords Entrance examination · Entrance draft test · Entrance proceeding success rate

3.1 Introduction

The task of social studies consists in reflecting and describing the ongoing trends within the, studying them and drawing conclusions and consequences that might be expected. The students' approach to studies reflects changes apparent in today's society. Relevant information about study at universities can be found not only through media but especially on the Internet. This information includes questions about the range of fields and specializations offered by particular universities to

R. Potůček (✉)
Faculty of Military Technology, Department of Mathematics and Physics,
University of Defence, Kounicova 65, 662 10 Brno, Czech Republic
e-mail: radovan.potucek@unob.cz

© Springer International Publishing AG 2017
Š. Hošková-Mayerová et al., *Mathematical-Statistical Models and Qualitative Theories for Economic and Social Sciences*, Studies in Systems, Decision and Control 104,
DOI 10.1007/978-3-319-54819-7_3

their students. In addition, universities also provide the most detailed information about the admission procedure, its content and range using various printed materials and especially through their websites.

The Faculty of Military Technology at the University of Defence in Brno (hereinafter referred to as FMT UD) follows the trend of growing awareness and provides the information about requirements, expectations and demands for its future students, as well as about study specializations, and, last but not least, about the admission procedure. And these are the Open days (see http://www.unob.cz/sluzby_zarizeni/stranky/den_otevrenych_dveri_fvl_fvt.aspx), which are an ideal opportunity for contacting future applicants and providing them with information about the studies, quality and demands as well as information about the admission procedure. The Department of Mathematics and Physics at the FMT UD also takes regularly part in these activities, which are usually organized three times a year. Representatives of the department inform not only about the entrance tests in mathematics but the applicants are also offered the chance to take part in the draft entrance tests in mathematics for free.

3.1.1   History and Goals of Draft Entrance Tests in Mathematics

The history of draft entrance tests in mathematics at the FMT UD began in 2010. The idea of draft tests is not entirely new; the similar opportunity has already been offered by, e.g., the Faculty of Civil Engineering at the Czech Technical University (CTU) in Prague (see https://mat.fsv.cvut.cz/kurzy/nanecisto) or the Faculty of Finance and Accounting at the University of Economics (UE) in Prague (see http://kbp.vse.cz/testy/). The main goals of draft tests in mathematics are to make sure the future students are well-prepared for the entrance exams, know the level and content of the tests and are familiar with the real entrance exam tests environment.

The idea of offering the draft entrance test those applying for the study at the FMT UD was initiated by the teachers of mathematics from the Military Secondary School and College in Moravská Třebová (hereinafter referred to as MSS). Therefore, the first draft tests in mathematics organized in 2010 were designed only for MSS students. In the following years, the draft tests in mathematics were designed for everybody interested in the study, i.e., not only for MSS applicants but also for applicants from secondary civilian schools (hereinafter referred to as SCS). The draft entrance tests in mathematics have been organized regularly since 2010 every year: usually in February or at the beginning of March: thus, almost two months before organizing the entrance examinations at the end of March: they consist of the entrance test in mathematics, the English language and also testing the physical fitness.

Having passed a draft entrance exam, each participant is given the former draft entrance test with answers, the overview of important formulas covering the secondary school mathematics, the self-study material and final individual revision for entrance exams in mathematics with 20 sample tasks solved in detail. In order to cover various versions of assignments, some tasks are sometimes more difficult than the entrance exams ones. Solved draft entrance tests are corrected and sent applicants via email.

3.1.2 *History and Content of Entrance Tests in Mathematics*

The entrance examinations in mathematics have been organized since the very beginning of the University of Defence foundation in September 2004; that time three existing institutions merged: the Military Academy Brno (established in 1951), the Military University of the Ground Forces Vyškov (established in 1947) and the Military Medical Academy Hradec Králové (re-established in 1988). The author of this chapter has been dealing with education of mathematics as well as the admission process for a long time, since 1988; that time he became an assistant at the Department of Mathematics at the Military Academy in Brno, later on, at the Department of Mathematics and Physics at the Faculty of Military Technology. Therefore, he was able to observe in detail trends of knowledge and mathematical abilities of future students for almost thirty years. Another teacher of the Department of the Mathematics and Physics, Šárka Hošková-Mayerová, has also been dealing with the issues of examination, entrance exams and students' personalities. Some of her educational results and reports while co-operating with other authors, are stated in the papers (Hošková-Mayerová 2010, 2011, 2014; Hošková-Mayerová and Rosická 2012; Cvachovec 2011; Račková and Hošková-Mayerová 2013a, b, 2014; Hošková and Račková 2010) and (Baysen et al. 2016): the author also deals with the draft entrance tests (see Hošková-Mayerová and Potůček 2017).

The entrance examinations in mathematics at the FMT UD are intended not only for bachelor curricula applicants, but also for five-year master's degree curricula ones (in the last two years). The entrance examinations consist of a test made up of 20 tasks covering the secondary school mathematics (see Potůček and Račková 2008); there are 5 multiple choice answers and only one of them is correct. The level of difficulty is not very high because the time for taking the test is limited by 50 min and the applicants solve the tasks by circling the answer. The applicants are not allowed to use any help, neither a calculator. The applicant gets 5 points for each correctly marked answer, 0 point for an incorrect answer; therefore, having taken the test in mathematics, it is possible to get 100 points at maximum. Having finished the admission process, results of all formerly used versions of entrance test are available both for applicants and the general public on the UD website (http://www.unob.cz/fvt/studium/Stranky/zkusebni_otazky.aspx).

The types of tasks of the entrance tests in mathematics are the same as the draft entrance tests ones. There have been made just small changes in the past few years; now, these tests are designed using the following 20 types of tasks covering the basic secondary school mathematics:

1. Modification of a formula with powers and radicals,
2. Modification of an exponential expression,
3. Solution of an algebraic formula with the help of formulas and reduction,
4. Expression of the unknown from the formula,
5. Solving a quadratic equation, relative position of parabola and axis,
6. Discriminant of a quadratic equation depending on the parameter,
7. Solving linear inequalities with absolute value,
8. Domain of a composite function,
9. Number term logarithm, solving of a logarithmic equation,
10. Solving exponential equations, logarithmic rules,
11. Relations among goniometric functions,
12. Determination of the smallest period of goniometric functions,
13. Solution of a goniometric equation,
14. Operation with complex numbers,
15. Task from plane geometry,
16. Task from solid geometry,
17. Modification of combinatorial expression formula or arithmetic or geometric sequence,
18. The relative positions of two lines in the plane,
19. Classification of conics given by a general formula,
20. Word problems (verbal tasks) about co-operative work or leading to the system of two linear equations.

As an example of such tasks, there can be given three following tasks 3, 7, and 9 (tasks are originally expressed in the Czech language):

3. After simplification of the expression $\frac{a^4-a}{a^2-1}$ we get
   (a) $\frac{a^3-a^2+a}{a-1}$ (b) $\frac{a^2+2a+1}{a-1}$ (c) $\frac{a^3+a^2+a}{a+1}$ (d) $\frac{a^2-2a+1}{a+1}$ (e) a^2+a

7. Solution of the inequality $|x-1|>2$ are all the real numbers for which it holds
   (a) $x\in\mathbb{R}\setminus\langle-1,3\rangle$ (b) $x\in(-1,3)$ (c) $x\in\langle-1,3\rangle$ (d) $x\in\langle0,4\rangle$
   (e) $x\in(-3,1)$

9. Solution of the equation $\frac{3+\log x}{3-\log x}=5$ is
   (a) the equation has no solution (b) $x=\sqrt{10}$ (c) $x=10$ (d) $x=10^2$
   (e) $x=\pm\sqrt{10}$

3.2 Overview of Success Rate of Applicants from MSS and SCS at Draft Entrance Tests in Mathematics

The information about the success rate achieved by the applicants at solving individual tasks at draft entrance tests in mathematics from 2010 to 2015 are presented in the following Table 3.1: the abbreviation M stands for MSS students, C stands for SCS, SE stands for successfully solved examples, T stands for a task, R stands for weighted mean of success rate, NA stands for the number of applicants. The tasks with success rate less than 50% are highlighted in dark blue, and the tasks with success rate higher or equal 50% are in light blue.

This table, which gives numbers and percentage success rate of applicants when solving individual types of tasks in 2010–2015 with distinction of SCS and MSS students, shows that the average success rate of solving tasks is 49%; therefore, it is possible to state that the level of difficulty of draft entrance tests in mathematics is chosen appropriately. MSS students were more successful than SCS students at solving 16 tasks. The biggest differences in favour of MSS students proved to be with tasks to express unknown from the formula (task no. 04, the difference of 24%) and tasks to classify conic section (task no. 19, the difference of 18%). SCS students reached a higher success rate than MSS students only with 4 tasks, i.e., a task to solve a logarithmic equation (task no. 09, the difference of 4%), a task to determine the smallest period of a goniometric function (task no. 12, the difference of 1%), a task of plane geometry (task no. 15, the difference of 8%), and a task to modify a combinatorial expression formula or a task of arithmetic or geometric sequence (task no. 17, the difference of 4%).

Table 3.1 Success rate of applicants from MSS and SCS, success rate of solving individual types of tasks

Year\Ex	01	02	03	04	05	06	07	08	09	10	11	12	13	14	15	16	17	18	19	20	NA
2010-M	17	11	12	21	21	19	16	16	10	21	8	7	10	3	18	15	9	14	13	19	21
2011-C	25	17	18	35	30	31	27	19	22	33	8	8	9	6	32	26	18	17	14	33	43
2011-M	28	21	24	36	30	33	28	23	19	34	15	13	15	2	31	25	19	20	17	33	38
2012-C	46	32	41	39	39	40	39	28	16	38	12	21	27	28	36	26	43	16	11	29	66
2012-M	20	20	22	25	24	20	15	17	6	20	7	10	12	13	8	11	19	6	5	14	27
2013-C	31	22	8	5	21	29	28	31	26	23	6	9	15	6	25	18	22	17	5	26	42
2013-M	15	13	6	8	12	20	13	18	11	16	2	5	6	6	12	8	14	14	4	14	23
2014-C	15	10	7	15	11	13	12	2	13	14	2	2	6	5	6	5	8	5	5	12	21
2014-M	20	16	14	21	16	19	20	3	9	18	7	2	7	13	15	10	4	12	16	22	33
2015-C	11	12	6	13	5	8	12	2	6	9	4	5	4	3	3	0	11	7	2	17	18
2015-M	18	14	13	19	15	16	14	3	11	13	6	0	7	5	9	3	17	8	5	20	21
SE-M	118	95	91	130	118	127	106	80	66	122	45	37	57	42	75	72	82	74	60	122	163
SE-C	128	93	80	107	106	121	118	82	83	117	32	45	61	48	102	75	102	62	37	117	190
SE	246	188	171	237	224	248	224	162	149	239	77	82	118	90	177	147	184	136	97	239	353
R-M [%]	72	58	56	80	72	78	65	49	40	75	28	23	35	26	46	45	50	45	37	75	53
R-C[%]	67	49	42	56	56	64	62	43	44	62	17	24	32	25	54	40	54	33	19	62	45
R [%]	70	53	48	67	63	71	63	46	42	68	22	23	33	25	50	42	52	39	27	68	49

MSS and SCS students reached the highest average success rate when solving tasks to calculate the discriminant of a quadratic equation (task no. 06, the success rate of 71%), to simplify an expression with powers (task no. 01, the success rate of 70%), to solve an exponential equation and apply logarithmic rules (task no. 10, the success rate of 68%) and to solve word problems about co-operative work or leading to the system of two linear equations (task no. 20, the success rate of 68% as well). The reason of a higher success rate of solving these tasks can be caused by the fact that tasks to define the discriminant of a quadratic equation as well as tasks to solve an exponential equation belong to not complicated algorithmic secondary school tasks.

On the contrary, MSS and SCS students reached the lowest success rate when determining relations between goniometric functions (task no. 11, the success rate of 22%), defining a period of a goniometric function (task no. 12, the success rate of 23%), and when solving a task to calculate with complex numbers (task no. 14, the success rate of 25%). Goniometry tasks cause troubles to a significant number of students. The failure when calculating with complex numbers is probably caused by the fact that students have not had enough experience within the "Complex numbers" area at the time of taking draft entrance tests in mathematics.

Based on the data given in Table 3.1 it can be stated that the FMT UD applicants coming from a military secondary school were better prepared for the draft entrance test in mathematics than students from secondary civilian schools; they reached an average success rate of 53% as opposed to 45% for secondary civilian school students.

A well-arranged survey of the success rate at solving individual types of twenty tasks at draft entrance tests is represented in the following Chart 3.1. Resulting from the abbreviations used in Table 3.1, R-M stands for the success rate of MSS

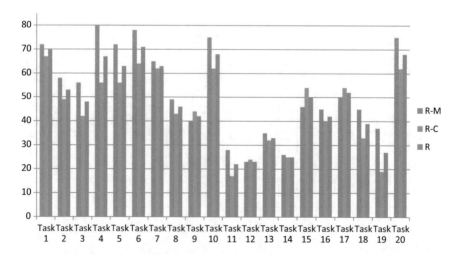

Chart 3.1 Graphical representation of the success rate at solving individual types of tasks reached by MSS, SCS and MSS & SCS applicants at draft entrance tests [in %]

applicants, R-C stands for the success rate of SCS applicants and R stands for the weighted mean of the success rate of MSS and SCS applicants altogether. The highest success rate of MSS applicants is clearly visible from Chart 3.1.

3.3 Qualitative and Quantitative Comparison of the Entrance Draft Tests and the Entrance Tests Results in Particular Years

In the following subsections, there is given a survey of the success rate of the individuals applying for the study at the FMT UD in particular years 2010–2015, a comparison of the results reached by SCS applicants, MSS applicants and all these applicants at the entrance draft tests compared with all the 2010–2015 applicants. There is also provided an overview of the FMT UD admission success rate of SCS applicants, MSS applicants compared with all these SCS and MSS applicants.

3.3.1 Comparison of the Results of the Entrance Draft Tests and the Entrance Tests in 2010

The data presented in Table 3.2 are related to the success rate of applicants achieved at entrance draft tests (hereinafter referred to as EDT), at entrance tests (hereinafter referred to as ET) and at the FMT UD admission procedure. Further, PET stands for Present at ET, PEDT stands for Present at EDT and AFMT stands for Admitted to FMT UD. From this table it can be stated as follows: MSS applicants who had taken part in EDT were very successful because they achieved the average score of 85.00 points at ET in comparison with all the other applicants who achieved the average score of only 62.39 points; another reason of a higher success rate was the fact that 100.00% of MSS applicants were admitted to the FMT UD study compared to only 52.09% of the total of all the other applicants.

Table 3.2 Survey of the success rate of MSS applicants and all the applicants in 2010

2010	Present at EDT	Average score (max. 100 pts) of EDT applicants who took part in ET	Present at ET	PET/PEDT [in %]	Average score at ET (max. 100 pts)	Admitted to FMT UD	AFMT/PET [in %]
MSS	21	72.50	12	57.14	85.00	12	100.00
All applicants	×	×	597	×	62.39	311	52.09

3.3.2 Comparison of the Entrance Draft Tests and the Entrance Tests Results in 2011

From the data presented in Table 3.3 it can be stated that SCS and MSS applicants who had taken part in EDT were more successful than all the applicants since they achieved the average score of 67.16 and 79.41 points, altogether 73.03 points at ET, in comparison with all the applicants who achieved the average score of only 61.16 points. SCS applicants reached 59.46% of the admission success rate, MSS applicants reached 94.12% and altogether 76.06% of all these applicants were admitted to the FMT UD compared to 53.31% of all the other applicants.

3.3.3 Comparison of the Entrance Draft Tests and the Entrance Tests Results in 2012

The data presented in Table 3.4 offer the information as follows: SCS applicants and MSS applicants who had taken in EDT were more successful than all the applicants since they achieved the average score of 62.37 points and 77.35 points, altogether 67.00 points at ET, in comparison with all the applicants who achieved the average score of only 60.11 points. As for the success rate of admission, SCS

Table 3.3 Survey of the success rate of MSS, SCS, MSS & SCS applicants and all the applicants in 2011

2011	Present at EDT	Average score (max. 100 pts) of EDT applicants who took part in ET	Present at ET	PET/PEDT [in %]	Average score at ET (max. 100 pts)	Admitted to FMT UD	AFMT/PET [in %]
MSS	38	61.03	34	89.47	79.41	32	94.12
SCS	43	49.73	37	86.05	67.16	22	59.46
MSS & SCS	81	55.14	71	87.65	73.03	54	76.06
All applicants	×	×	559	×	61.16	298	53.31

Table 3.4 Survey of the success rate of MSS, SCS, MSS & SCS applicants and all the applicants in 2012

2012	Present at EDT	Average score (max. 100 pts) of EDT applicants who took part in ET	Present at ET	PET/PEDT [in %]	Average score at ET (max. 100 pts)	Admitted to FMT UD	AFMT/PET [in %]
MSS	22	58.24	17	77.27	77.35	17	100.00
SCS	44	39.74	38	86.36	62.37	32	84.21
MSS & SCS	66	45.45	55	83.33	67.00	49	89.09
All applicants	×	×	541	×	60.11	374	69.13

applicants reached 84.21% and MSS applicants 100.00%. Altogether 89.09% of all these applicants were admitted to the FMT UD compared to 69.13% of all the other applicants.

3.3.4   Comparison of the Entrance Draft Tests and the Entrance Tests Results in 2013

From the data presented in Table 3.5 can be stated state that SCS and MSS applicants who had taken part in EDT were more successful than all the applicants since they achieved the average score of 69.00 points and 71.47 points, altogether 69.89 points at ET, in comparison with all the applicants who achieved the average score of only 57.83 points. SCS applicants reached 83.33% of the admission success rate similarly to MSS applicants with 82.35%. Altogether 82.98% of all these applicants were admitted to the FMT UD compared to 70.89% of all the other applicants.

3.3.5   Comparison of the Entrance Draft Tests and the Entrance Tests Results in 2014

The data presented in Table 3.6 give the following information: SCS applicants and MSS applicants who had taken part in EDT were more successful than all the applicants because they achieved almost the same average score of 72.00 points and 71.94 points, altogether 71.97 points at ET, in comparison with all the applicants

Table 3.5 Survey of the success rate of MSS, SCS, MSS & SCS applicants and all the applicants in 2013

2013	Present at EDT	Average score (max. 100 pts) of EDT applicants who took part in ET	Present at ET	PET/PEDT [in %]	Average score at ET (max. 100 pts)	Admitted to FMT UD	AFMT/PET [in %]
MSS	23	48.82	17	73.91	71.47	14	82.35
SCS	39	43.50	30	76.92	69.00	25	83.33
MSS & SCS	62	45.43	47	75.81	69.89	39	82.98
All applicants	×	×	426	×	57.83	302	70.89

Table 3.6 Survey of the success rate of MSS, SCS, MSS & SCS applicants and all the applicants in 2014

2014	Present at EDT	Average score (max. 100 pts) of EDT applicants who took part in ET	Present at ET	PET/PEDT [in %]	Average score at ET (max. 100 pts)	Admitted to FMT UD	AFMT/PET [in %]
MSS	33	41.39	18	54.54	71.94	16	88.89
SCS	21	45.33	15	71.43	72.00	14	93.33
MSS & SCS	54	43.18	33	61.11	71.97	30	90.91
All applicants	×	×	375	×	60.13	283	75.47

Table 3.7 Survey of the success rate of MSS, SCS, MSS & SCS applicants and all the applicants in 2015

2015	Present at EDT	Average score (max. 100 pts) of EDT applicants who took part in ET	Present at ET	PET/PEDT [in %]	Average score at ET (max. 100 pts)	Admitted to FMT UD	AFMT/PET [in %]
MSS	21	48.33	9	42.86	71.67	7	77.78
SCS	18	38.67	15	83.33	65.33	12	80.00
MSS & SCS	39	42.29	24	61.54	67.71	19	79.17
All applicants	×	×	294	×	59.64	222	75.51

who achieved the average score of only 60.13 points. SCS applicants reached 93.33%, MSS applicants reached 88.89% and altogether 90.91% of all these applicants were admitted to the FMT UD compared to 75.47% of all the other applicants.

3.3.6 Comparison of the Entrance Draft Tests and the Entrance Tests Results in 2015

From the data presented in Table 3.7 it can be stated that SCS and MSS applicants who had taken part in EDT were more successful than all the applicants because they achieved the average score of 65.33 and 71.67 points, altogether 67.71 points at ET, in comparison with all the applicants who achieved the average score of only 59.64 points. SCS applicants reached 80.00% of the admission success rate similarly to MSS applicants with 77.78%. Altogether 79.17% of all these applicants were admitted to the FMT UD compared to 75.51% of all the other applicants.

3.4 Total Qualitative and Quantitative Comparison of the Success Rate of the SCS, MSS and All These Applicants in Particular Years Compared with All the Other Applicants

The following Table 3.8 summarizing Tables 3.1 up to 7 contains a brief survey of percentage increases reached at the entrance tests by MSS applicants, SCS applicants and altogether MSS and SCS applicants compared to all the other applicants. The Table 3.8 also includes a brief survey of percentage differences between MSS applicants, SCS applicants and altogether MSS and SCS applicants admitted to the FMT UD compared to the all the other applicants admitted to the FMT UD.

Table 3.8 Total survey of the success rate of applicants from MSS, SCS and all the applicants in 2010–2015

Year(s)	Applicants	Average score of EDT for ET applicants	AS at ET	Increase in AS at ET compared to EDT [in %]	Admitted to FMT/present at ET [in %]	Difference between AFMT applicants and all the AFMT applicants [in %]
2010	MSS	72.50	85.00	17.24	100.00	48.91
	All the applicants	×	62.39	×	52.09	×
2011	MSS	61.03	79.41	30.12	94.12	40.81
	SCS	49.73	67.16	35.93	59.46	6.15
	MSS & SCS	55.14	73.03	32.44	76.06	22.75
	All the applicants	×	61.16	×	53.31	×
2012	MSS	58.24	77.35	32.81	100.00	30.87
	SCS	39.74	62.37	56.95	84.21	15.08
	MSS & SCS	45.45	67.00	47.41	89.09	19.96
	All the applicants	×	60.11	×	69.13	×
2013	MSS	48.82	71.47	46.39	82.35	11.46
	SCS	43.50	69.00	58.62	83.33	12.44
	MSS & SCS	45.43	69.89	53.84	82.98	12.09
	All the applicants	×	57.83	×	70.89	×
2014	MSS	41.39	71.94	73.81	88.89	13.42
	SCS	45.33	72.00	58.84	93.33	17.86
	MSS & SCS	43.18	71.97	66.67	90.91	15.44
	All the applicants	×	60.13	×	75.47	×
2015	MSS	48.33	71.67	48.29	77.78	2.27
	SCS	38.67	65.33	68.94	80.00	4.49
	MSS & SCS	42.29	67.71	60.11	79.17	3.66
	All the applicants	×	59.64	×	75.51	×
2010–2015 (arithmetic means)	MSS	55.05	76.14	38.31	90.52	24.45
	SCS	43.39	67.17	54.81	80.07	14.00
	MSS & SCS	46.30	69.92	51.02	83.64	17.57
	All the applicants	×	60.21	×	66.07	×

The data included in the fifth column of the Table 3.8 are graphically represented as a well arranged bar chart in Chart 3.2, which shows the percentage increases in average score at the entrance tests compared to the entrance draft tests of MSS applicants, SCS applicants and altogether MSS and SCS applicants.

The data included in the seventh column of the Table 3.8 are graphically represented in a well arranged form of a bar chart in Chart 3.3, which shows the percentage differences between MSS applicants, SCS applicants and altogether MSS and SCS applicants admitted to the study at FMT UD compared to the all the other applicants admitted to the FMT UD.

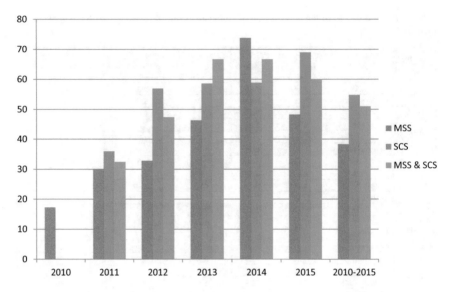

Chart 3.2 Graphical representation of the increase in average score at the entrance tests compared to the entrance draft tests [in %]

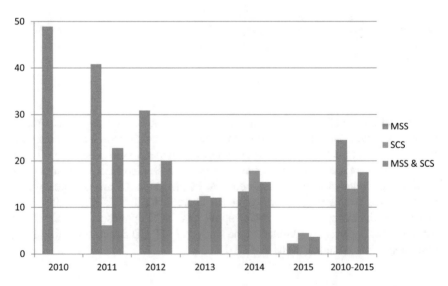

Chart 3.3 Graphical representation of the difference between MSS, SCS and MSS & SCS applicants admitted to the FMT UD study at and all the other applicants admitted to the FMT UD study [in %]

3.5 Conclusion—Evaluation of the Role and Importance of the Entrance Draft Tests in Mathematics at the Faculty of Military Technology at the University of Defence in Brno

Based on the data, tables and charts presented above there can be stated that the draft entrance tests in mathematics at the Faculty of Military Technology at the University of Defence in Brno are a suitable tool for preparation of secondary school students interested in studies at the faculty. From the data included in Table 3.8 and Charts 3.2 and 3.3 (there are summarized results of the draft entrance tests and draft tests in the particular years 2010–2015 and in 2010–2015 as a whole), becomes evident that military and secondary civilian school applicants who had taken part in the draft entrance tests achieved, on average, about 51% better results at entrance tests compared to SCS applicants. Military secondary school applicants reached over 38% better results and secondary civilian school applicants reached almost 55% better results than the all the other applicants who had taken part in the entrance tests in mathematics. Further, it can be stated that differences between the success admission rate to the FMT UD study of a military secondary school and the civilian secondary school applicants who had taken part in the draft entrance tests compared to all the other applicants was almost about 18% in favour of those who had already passed the draft entrance tests before. Military secondary school applicants have, on average, over 24% higher chances and civilian secondary school applicants have 14% higher chances to be admitted at the FMT UD study than the all the other applicants.

Better results of military secondary school applicants can be explained by a higher portion of classes of mathematics, which is taught 5 lessons a week, comparing to secondary civilian schools where there are 3–4 classes a week. Another reason for higher success rate of military secondary school applicants might also be the systematic guidance of military secondary school students by their teachers and purposeful direction of military secondary school students to studies at the FMT UD and the career of a professional soldier. The increasing trend in number of applicants taking part in the draft entrance tests and the entrance tests and also the increasing number of admitted applicants at the FMT UD is in all likelihood caused by the real demographic trends, which might improve in the coming years.

In conclusion, we can state that the draft entrance tests in mathematics make the applicants get better ready for real entrance exams; both the tasks and solutions of the past years can be found on the website (http://www.unob.cz/fvt/studium/Stranky/zkusebni_otazky.aspx). On the Open day, in addition to the draft entrance tests in mathematics, applicants can see the military quarters Šumavská in Brno and visit the lecture hall where the exams in mathematics and the English language take place. The applicants can also get rid of their exam fever and worries of the entrance exams. Resulting from the applicants´ lasting interest in the Open day, which is arranged regularly twice a year at the University of Defence (winter/summer semesters), as well as the interest in passing the draft entrance tests in mathematics (the evaluation can be found on the website http://www.unob.cz/fvt/struktura/k215/ Documents/Výsledky%20PZ%20z%20MN%202015.pdf), it can be expected that the interest in passing this specific type of entrance exams "at no risk" will be permanent despite the demographic trends. Draft entrance tests in mathematics contribute undoubtedly to better preparation of students and increase their chances to be admitted at the FMT UD and be successful in their follow-up military career.

Let us finish with Norman K. Denzin and Yvonna S. Lincoln generic definition of qualitative research and its sense from their *Handbook of qualitative research* published in 2000: *"Qualitative research is a situated activity that locates the observer in the world. It consists of a set of interpretive, material practices that make the world visible. These practices transform the world. They turn the world into a series of representations, including field notes, interviews, conversations, photographs, recordings, and memos to the self. At this level, qualitative research involves an interpretive, naturalistic approach to the world. This means that qualitative researchers study things in their natural settings, attempting to make sense of, or to interpret, phenomena in terms of the meanings people bring to them."*

Acknowledgements The author was supported within the project for "Development of basic and applied research" developed in the long term by the departments of theoretical and applied bases of the FMT UD (Project code: "VYZKUMFVT" (DZRO K-217)) supported by the Ministry of Defence of the Czech Republic.

References

Baysen, E., Hošková-Mayerová, Š., Cakmak, N., Baysen, F. Misconceptions of Czech and Turkısh Unıversity Students in providing Citations. Switzerland: Springer International Publishing, 2016, p. 183–191. *Antonio Maturo et al. (Eds): Recent Trends in Social Systems: Quantitative Theories and Quantitative Models.* Studies in System, Decision and Control 66, 183-190, ISBN 978-3-319-40583-4.

Cvachovec, F., et al. (2011) The results of the current project "Innovation of study programme military technology" at the Department of Mathematics and Physics. In: Distance Learning, Simulation and Communication 2011. Univerzita obrany: Univerzita obrany, 2011, p. 54–59. ISBN 978-80-7231-695-3.

Dny otevřených dveří Fakulty vojenského leadershipu a Fakulty vojenských technologií [online]. Web page of the University of Defence. Available from http://www.unob.cz/sluzby_zarizeni/stranky/den_otevrenych_dveri_fvl_fvt.aspx.

Hošková-Mayerová, Š. (2010) Alexithymia among students of different disciplines. *Procedia–Social and Behavioral Sciences*, 2011, Vol. 9, p. 33–37. ISSN 1877-0428.

Hošková-Mayerová, Š. (2011) Operational program"Education for Competitive Advantage", preparation of Study Materials for Teaching in English. *Procedia-Social and Behavioral Sciences*, 2011, Vol. 15, p. 3800–3804. ISSN 1877-0428.

Hošková-Mayerová, Š. (2014) The Effect of Language Preparation on Communication Skills and Growth of Students' self-Confidence, The Procedia: Procedia - Social and Behavioral Sciences, 2014, Vol. 114, p. 644–648. ISSN 1877-0428.

Hošková-Mayerová, Š., Potůček, R. (2017) Qualitative and quantitative evaluation of the entrance draft tests from mathematics. *Studies in Systems, Decision and Control, Vol. 66, Antonio Maturo et al. (Eds): Recent Trends in Social Systems: Quantitative Theories and Quantitative Models,* p. 53–64, Switzerland: Springer International Publishing, ISBN 978-3-319-40583-4.

Hošková-Mayerová, Š., Rosická, Z. (2012) Programmed learning. Procedia - Social and Behavioral Sciences, 2012, Vol. 31, 782–787. ISSN 1877-0428.

Hošková, Š., Račková, P. (2010) Problematic Examples at the Entrance Exams of Mathematics at the University of Defence, Procedia–Social and Behavioral Sciences, 2010, Vol. 9, p. 348–352. ISSN 1877-0428.

Potůček, R., Račková, P. (2008) Entrance examination from mathematics on the University of Defence (in Czech). 7th International Conference Aplimat 2008, Bratislava, Slovak Republic, 2008. p. 1007–1012. ISBN 80-89313-03-7.

Přijímací zkoušky „NANEČISTO" [online]. Web page of the Department of Mathematics of The Faculty of Civil Engineering of CTU in Prague. Available from https://mat.fsv.cvut.cz/kurzy/nanecisto.

Přijímací zkoušky nanečisto na FFÚ VŠE v Praze [online]. Web page of the Faculty of Finance and Accounting of UE in Prague. Available from http://kbp.vse.cz/testy/.

Račková, P., Hošková-Mayerová, Š. (2013a) Current Approaches to Teaching Specialized Subjects in a Foreign Language. In: ICERI2013 Proceedings. Seville, Spain: IATED DIGITAL LIBRARY, 2013, p. 4775–4783. ISSN 2340-1095. ISBN 978-84-616-3847-5.

Račková, P., Hošková-Mayerová, Š. (2013b) Investigation of Problems Leading to Global Extrema. In: ICERI2013 Proceedings. Seville, Spain: IATED Digital Library, 2013, p. 4745–4750. ISSN 2340-1095. ISBN 978-84-616-3847-5.

Rosická, Z., Hošková-Mayerová, Š. (2014) Motivation to study and work with talented students, The Procedia: Procedia - Social and Behavioral Sciences, 2014, Vol. 114, p. 234–238. ISSN 1877-0428.

Výsledky přijímacích zkoušek z matematiky „nanečisto" [online]. Web page of the University of Defence. Available from http://www.unob.cz/fvt/struktura/k215/Documents/Výsledky%20PZ %20z%20MN%202015.pdf.

Úplné zadání zkušebních otázek a příkladů a jejich správné řešení [online]. Web page of the Faculty of Military Technology of the University of Defence in Brno. Available from http://www.unob.cz/fvt/studium/Stranky/zkusebni_otazky.aspx.

Chapter 4
Digital Ethnography Theories, Models and Case Studies

Vanessa Russo

Abstract The ethnography is a non-standard methodology utilized to study the culture and the interactions of social actors in a cultural context. With the development of the Web were born the first ethnographic applications to the cyberspace and currently this methodology has not yet defined a methodological framework and has taken different labels: Virtual Ethnography, Ethnography Network, Ethnography digital (Murthy, Sociology 42:837–855, 2008) and Netnography (Kosinets, Netnography: The Marketer's Secret Weapon. How Social Media Understanding Drives Innovation, 2010). The aim of this chapter is to define the boundaries of digital ethnography. Finally, with the help of the comparison between empirical cases we will try to understand which are critical points, limits and research prospects for digital ethnographers.

Keywords Digital ethnography · Cyberspace · Social network analysis · Fuzzy system · Big data

4.1 Background: What's Cyberspace?

The aim of this section is to understand dimension, ontological, methodological and applicative of digital ethnography. The starting point is the study of the cyberspace. The cyberspace is the space between two or more technological tools while a conversation is taking place. It began as an electric wire and evolves into a world.

However cyberspace cannot be considered a world unto itself, because it is in constant interrelation with the real world. The process of interaction between the dimension *on* and *off line* is the heart of the Network Society (Castells 2002).

V. Russo (✉)
Department of Business Administration, Università G. d'Annunzio Chieti - Pescara,
Via dei Vestini, 66100 Chieti, Italy
e-mail: russov1983@gmail.com

© Springer International Publishing AG 2017
Š. Hošková-Mayerová et al., *Mathematical-Statistical Models and Qualitative Theories for Economic and Social Sciences*, Studies in Systems, Decision and Control 104, DOI 10.1007/978-3-319-54819-7_4

This information age has never been a technological matter. It has always been a matter of social transformation, a process of social change in which technology. And it's an element that is inseparable from social, economic, cultural and political trends.

The Network Society defined by Castells (2002) is divided according to the interaction of four dominant cultural contexts: techno-elites, virtual communities, hacker and businessmen (Castells 2001).

Tecno-elites is the culture of academia classics, within which the goodness of technological artefacts is placed as the supreme value and the relevance of the discovery evaluated in terms of its social utility.

This form of culture is essential for the birth of the web. In fact, the Internet is a product of the American scientific circles.

The hacker culture (Hošková-Mayerová and Rosická 2015) is supported by aggregate values, and is characterized by a strong sense of community is evident in the establishment of informal institutions. The community of Linux developers is one concrete example. In fact, it appears as a well-defined circle of users that, in addition to cooperate spontaneously to the Linux projects, share the system of values that constitutes the world of Free Software.

Internet entrepreneurs were the first to understand the economic potential of the Web since the early investment in dotcom to Big Data's boom. But by themselves, from the culture that distinguishes them, they could never create a basic support for networking and communication. However, their contribution was, and is essential for the many cultural dynamics that gave rise to the Internet (Castells 2006).

Social groups that generate in the network when some people take part constantly in public debates and start interpersonal relationships in the cyberspace, which is considered the theoretical space where words, human relationships, data, wealth and power are expressed through the use of Internet.

These four cultural layers are arranged hierarchically: the techno-meritocratic culture is structured as a hacker culture by building the network rules and habits of cooperation on technological projects. The virtual community culture adds a social dimension to sharing technology, making the Internet a means of social interaction and selective membership symbolic. The entrepreneurial culture works on top of the Hacker culture and community culture, to spread the use of the Internet in all areas of the company as a means to make profits.

Without the techno-meritocratic culture, hackers would simply be a counter-cultural community specific computer experts a little 'set. Without the hacker culture, community networks on the Internet would not be different from many other common alternatives. Moreover, without the hacker culture and community values, entrepreneurial culture could not be characterized as a specific Internet.

Therefore, in the case of Web-mediated communication it is necessary to make some clarifications:

(1) is an act participatory; (Rainie and Wellman 2012)
(2) gives rise to connections on the basis of related content are also called "Network of Affinity";

(3) It leads to effective mobilization of expertise in terms of new ways of social action and community projects;
(4) It generates temporary zones of consensus based on unexpected alliances between users and concepts. (Lovink 2002)

In the light of these considerations about the complexity of the system and the dynamics of cyberspace, in the following sections will be explored the methodologies, skills and applications useful for study some digital social phenomena.

4.2 Methodology: How Can We Study the Cyberspace?

This field of research methodologically lends an application to ethnographic method. The ethnographic approach is a non-standard methodology that allows you to analyse the culture and the interactions of social actors in a given context. It is based on three basic actions of the human being: to observe, question and read that are accomplished through a set of tools based on direct observation of the phenomenon, in-depth interviews and use of documents (Corbetta 1999; Baysen et al. 2016).

The search field in cyberspace is suitable for application of ethnographic technique. In fact, it consists of virtual social interactions that unfold through different tools and different network but which, in fact, generate communication structures, representations of identity and culture real and shared (Murthy 2008).

In cyberspace, the search field is not clear but it must be investigated through a study of "digital tracks" of the research topic through all the structures of the Network (Social Media, Social Networks, Blogs, Wikis, web sites, search keywords)

Stages of research are divided into six phases:

- hidden observation, individuals in the community are observed and the researcher does not actively involved in the relationship dynamics of the network. This type of observation is also called *passive lurking* (Russo 2015; Russo 2016a, b, c);
- Detecting the presence of content involved;
- Crawling content: with crawling means a computer process by which you can download the contents of a computer database;
- Reconstruction of the relationship network: using open source software Gephi (Bastian et al. 2009) it's possible rebuild the network of relationships and analyse its contents;
- Analysis of "the traces of growth" (Corbetta 1999):
- Classification and ideal types.

These stages are not strictly defined but they must be designed ad hoc time after time. Besides the covert observation it's a very whole skill because it must carefully plan and it's not possible use throughout. Indeed in some social fields the researcher needs a gatekeeper for to join in it (for example in Whatsapp's groups or in Facebook secret groups or in a closed forum). Instead in other cases the limits derives from legally protected areas and penetrable only by a legal agreement, which invalidates the observation covert.

Finally the method of passive lurking is only once of the social research tool for studying digital social realities.

The others phases of research, instead, are characterized by a multidisciplinary approach. In fact, digital ethnographic method get some skill from informatics and mathematical models. Crawling is an informatics skill that enables downloads of web contents from the Net (sites, forum, virtual communities, social network sites) to personal computer. Moreover, the network analysis needs software computer support[1] for data and graph processing.

4.2.1 Mathematical Models

From the mathematical point of view to study the relationship network use the structures and instruments of the Social Network Analysis, the theorems which make up the "Complexity Theory" and Fuzzy Systems.

The SNA is a theoretical and methodological perspective that analyzes the social reality from its reticular structure.

This method puts social relationship as a minimum unit of observation at the expense of individual attributes (e.g., gender, age, education, status, socioeconomic etc.), which are not excluded from the analysis, but traced to one of three possible levels of interdependence of social phenomena: that of the actors, the connecting relations and networks that make up the structure of the whole.

In this way it becomes possible a fundamental objective: analyze the complex system of interdependencies and multiple interconnections within society. In particular, the SNA is important for to explain the dynamics of contamination of interdependencies between systems and social behaviour of individual actors (Trobia and Milia 2011).

The structure at the base of the Social Network Analysis is the graph.

The graph, or network, is a particular type of pattern used for represent the relationships between elements, composed by points called nodes connected to each other through links (Easley and Kleinberg 2010).

Data and their graphical location within the network are described in the adjacency matrix.

---

[1]In this case I've been used Gephi (Bastian et al. 2009) but there are some other tools, for example Node Excel.

The adjacency matrix, also called social matrix, is a square table, composed by a number of rows and columns equal to the number network's nodes. In the matrix, actors are indicated in the same sequence both in line and in column. Each cell reports the binding information between the node of a row and the node of the others So if $x\_ij = 0$ the link is not there; $x\_ij = 1$ if the link is there (Trobia and Milia 2011).

The most important values for analyzing the nodes of a graph are: Centrality. Modularity and Cluster Coefficient (Chiesi 1999).

The analysis of the centrality of a node, compared to the network describes "prestige" node on the entire network and allows to detect also the weight and the level of influence within the other network elements.

The main measures of centrality are three: the centralization Freemans degree, Closeness and Betweeness.

The first indicates the degree and its distribution in the network. This value is based on the concept that the elements, which maintain a greater number of relationships with other members, enjoy a convenient location.

The Closeness measures the level of close to a node to other network. In formal terms, the closeness of a node is the function of its geodesic distance from all other network nodes.

The third measure is the betweeness and detects the degree of intermediaries present between a node and the other; it indicates how often a node is located in the shortest path between two pairs of nodes in the network.

The degree of modularity, instead, determines the sub-networks (or cliques) within a larger network and the quality of their connection. Need to determine if the sub communities have many or few connections between them.

Mathematically the degree of modularity (Q) is calculated by set of connections $(a\_ (i, j))$ between the nodes (i, j) of a less connections module (m) that would result in a random distribution. The Q value can vary from 1 to -1, and in case of modularity tending to 1, the network is composed of modules low connected, instead a degree with Q tending to -1 indicates cracks highly connected.

The clustering coefficient indicates the aggregation level of a network.

Mathematically, this value is calculated considering all actual connections of a network divided by the potential links.

Therefore, a highly structured network presents a clustering coefficient tends to 1 while, on the contrary, a DC tending to zero defines a graph characterized by a strong element of randomness and a low aggregation between the elements that compose it.

The network's studies, along with other disciplines, in a field of study called "complexity theory".

Complexity theory starts the application of the basic rules of Social Network Analysis, for analyzing complex components of network types called small worlds.

The little worlds have a hybrid form than the classical networks, and must be interpreted by using the multi-field paradigms between sociology, biology, physics and the theory of communication.

The fundamental laws that determine the study of complex systems include:

- Six Degrees of Separation; about this theorem in society individuals are separated from any other person in the world by six degrees of knowledge. In other words, society as a whole can be viewed as a huge network of about six million nodes and with a diameter equal to six.
- Strength of Weak Ties. The network structure of a social actor consists of a strongly connected group that represents the strong ties (friends and family) and a circle of acquaintances (weak ties) who have no contact with each other; but each to his acquaintance once a circle of strong and weak ties. In this context, Granovetter (1973) identifies the weak links as the most functional for the circulation of the flow of information.
- Models of small world graph; the theories of Granovetter (1973) and Buchanan (2004), argue that networks, composed of strong and weak ties have the distinction of being part of graph structured and partly random graphs. Watts and Strogatz define these networks "small world" because have some characters of a graph structure and some the characteristics of a random graph.
- Networks to scale invariance; Barabasi (2002) processes the Watt model and Strogatz (1998) and applies it to the study of the structure of digital networks detect a particular form of small world called scale-free network. The two characteristics that make the structure of the scale-free network a network Small Mode "different" from the one analyzed by Watt and Strogatz (1998), are the presence of centres of aggregation of multiple connections (hub) subjected to the Power Law. The scaling law, can be translated in the principle of "the rich who get richer", best described as a preferred combination. According to this theory hubs inside the network will naturally tend to acquire a greater extent than links to nodes with lower grade. For this reason the most popular sites are the most visited of the least popular and tend to acquire more links than others; here is because blogs with more followers are subjects to acquire more contacts than those minors. According to this principle it is built the system in Google. In fact, when you do a search on Google (beyond the sponsored link) the list of links that we get is in order to click (Barabasi 2002).

The set of laws governing the theory of complexity along with the basic techniques of social network analysis help from interpreting mathematically the dynamics of virtual communities. In fact, these types of networks reflect the structure of the model developed by Barabasi (2002) and through the parameters of the SNA are possible to study the position of the nodes and the diversity of the types of links between nodes.

Finally, for describe mathematically the quality of the ties of users within a network it need to invoke the application of fuzzy logic.

In particular Fuzzy logic is important for define varieties of the intensity of the network and the type of bond.

Previously I defined the clustering coefficient as a ranges from 0 (Random graph) to 0.5 (small Word) to 1 (graph-structured) and allows you to study the intensity of the links of a network. Through fuzzy logic, you can define additional subcategories of analysis (Fig. 4.1).

The same reasoning it is possible to apply it to the study of the types of bond. This makes it possible to detect an intermediate range of links between the strong and the weak (Fig. 4.2).

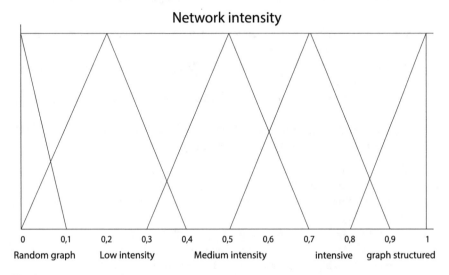

Fig. 4.1 Network intensity

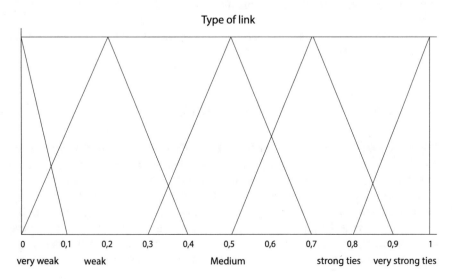

Fig. 4.2 Type of link

4.3 Applications

During my research I applied the techniques and the digital ethnography models in four empirical studies.

4.3.1 The Expression of Sexuality Through Social Network

The aim of the research was to understand how human sexuality is expressed through multimedia networks (Russo 2015).

Again, I have been taken a bearing of affinity networks arisen about sexual inclination. Finally,

I have found the social function of the various social networks about sexuality language.

In order to seek an interpretation of the collected questions, through the digital ethnography tool, the following areas were analyzed:

(1) Affinity networks for sharing a particular conception of sexuality;
(2) Uploading of content related to amateur pornography;
(3) Sharing of symbols that acquire a meaning of sexual type (for example the food, the body or the objects);
(4) The fourth, and last segment of analysis refers to the analysis of time-lapse, a photographic technique that is becoming viral in the social media and has the distinction of being used by users to tell fragments of life.

Research has revealed a universe of shared meanings among users; the Net have also detected forms of virtual community where users share values and social relations.

The analysis shows that each social platform in the field of sexuality has a specific characterization.

4.3.2 The Civic Mediactivism in the City of Chieti

The work, carried out between January 2014 and April 2015, had as its aim to visualize the networks of kinship born around the theme "City of Chieti" and understand the contents and forms of mediactivism (Russo 2016a, b, c).

In a preliminary step, I identified the digital frames in which the traces of the city are present. This initially made it possible to understand the organization and the contents characteristic related to the topic of research and reconstruct the reference social operating system. The second phase, more specific, aimed to the identification of tactical media discussion, in order to isolate the number of media-elements.

In this regard, it highlighted the presence of forms of civic media activism in 21 Facebook[2] virtual groups. The study of tactical networks reported on research has identified:

(1) The rules of social behavior within these communities and, in particular, the status of the group's activities;
(2) Forms of communication and social control exercised by directors and group leaders;
(3) The most frequent discussion topic, namely, the recurring themes within posts and discussions between network users;
(4) The structure; at this stage we used the mathematical models of the social network analysis. In particular, for each structure the graph consideration of relationships has been detected and has been analyzed in relation to the connection medium, in the presence of connector nodes, the diameter, the clustering coefficient and modularity;
(5) the influential social actors: this definition refers to the most active elements in the group. Fingerman and Blau (2009 in Rainie and Wellman 2012) define them consequential strangers (strangers relevant) or elements of a network with which there aren't necessarily a strong bond but acquaintances are always present.

4.3.3   Conflict Mediation in the Institutional Forum on Line[3]

The aim of this work was to identify the types of on line forums in the Institutional Network and related to the Abruzzo area, analyze the relational dimension and forms of conflict and, finally, define the paths of mediation made and achievable.

Specifically I have been detected:

(1) Contents of the forums and discussion groups;
(2) Conflicts and possible conciliation strategies;
(3) User behaviour in relation to situations of conflict and requests of information;

Finally, I have analyzed some interactions related to individual posts for understand: the processes of mediation and the formal and informal role of the social actors.

---

[2]The data were extracted by Netvizz application (Rieder 2013).

[3]This research was presented during the conference *Communication, Mediation, Evalutation and Social Intervention,* Università G. d'Annunzio Chieti—Pescara, 15 Settembre 2015.

4.3.4 Social Network Analysis as a Backup to Community Interactions Online Learning[4]

The Aim of the research project was to determine the relational dimension of a virtual learning community,[5] by applying mathematical models of Social Network Analysis (SNA) outlined in the previous paragraph.

Through the analysis of formal and informal interactions into on line forum community learning we have been detected the social dimension, formal and informal rules between the community users and signs of building links in the group.

In this case SNA became a useful tool for to identify and address the problems within a community of online learning and for facilitate the participation and collaboration among the students.

Through the analysis of graphs it was possible to identify and monitor the informal roles born in the community, some friendly able to contribute positively to the learning process and other antagonists capable of generate dissent and confusion. Also I have detected and monitored subgroups born within the community and studied the behaviour and possible forms of exclusion from the other forum members.

4.3.5 Comparative Analysis

In the four examples described in the preceding paragraphs I applied, to the methodological level, the complete digital ethnography method in all its steps. The covert observation has been implemented in all four examples, however, to get into some communities on the topic of sexuality is needed the help of a gatekeeper:

---

[4]This research, in collaboration with PhD Roberta Di Risio, was presented during the conference *Fuzzy Set, Multivaluted Operations and Application to Social Science,* Università G. d'Annunzio Chieti—Pescara, 18 Settembre 2015.

[5]The learning community detected is part of the project Master *"Koinè. Professione formatore per la didattica della comunicazione"* (Koinè. Profession trainer for teaching communication). The Master was promoted and funded by the Ministry of Education and was activated for the academic years 2013–2014 and 2014/2015 at the "G. d'Annunzio ", with the Scientific Coordination and Teaching Prof. Giselda Antonelli. It covers four Italian universities: Università "G. d'Annunzio "of Chieti-Pescara, Università of Genoa, Università of Salerno and Università of Palermo Studies and is connected to the National Plan Logos.

The Master took two years and was disbursed in blended learning mode. The meetings held in the presence in the classrooms of the University Campus of Chieti, and for the management of on-line teaching using the training platform operated by the University Telematics "Leonardo da Vinci", based in Torrevecchia Teatina (CH). The objectives of this training event were as follows: to train the professional figure of the trainer for teaching communication and enrich their professional profile with knowledge about the use of the most advanced communication technologies in educational settings.

while the observation of the learning community online of Koinè master was not completely hidden because users knew that the course tutor had access to the contents of the forum.

However, in all cases it was protected the privacy and anonymity of users.

The construction of the affinity networks was carried out through Gephi software, in the case of the learning community's forum the construction of the graph was made manually through the establishment of the adjacency matrix, which has automatically generated the graph of relationships. In all other cases I used crawling tools that automatically imported all the required relational data.

In relation to the mathematical model I applied the techniques and paradigms of the SNA as described in the preceding paragraphs.

This mathematical application has been very helpful especially in research on civic mediactivism in the city of Chieti and in the intervention in the case of learning community online of Koinè master.

In the case of civic mediactivism structural analysis of the detected networks showed 18 graphs presenting the typical structure of "networks to scale invariance" (Barabasi 2002): media not equally distributed connections; among those elements were present connector nodes with a preferential amount of connections; mean diameter of the network equal to 5.12; the mean clustering coefficient equal to 0.22; in each graph they were present of the sub-components (cracks) high density and with a clustering coefficient approaches 1.

Otherwise in the case of learning community online of Koinè master view the structural characteristics of the online forum the network of relationships was presented as a graph structured. In this case the application of SNA was useful for detecting the existence of a clique of relations. This subgroup was dominant within the Network but puts in place behaviors in favor to the process of learning lessons.

4.4 Discussion and Conclusion

The digital ethnography, as new methodological syncretism, offered an useful tools for reading and understanding the digital social field.

However, the practical application reveals some critical issues that mentioning.

It is clear consider digital ethnography as an *ex novo* tool, or is a new application of an epistemological process already exists?

As claimed by Bryman and Burgess (Bryman and Burgess 1994 in Corbetta 1999) qualitative research can not be reduced to particular techniques or to a succession of stages but rather it consists of a dynamic process that ties together issues, theories and methods; accordingly the search process is not a well-defined sequence of steps that follow a clear pattern, but a confusing interaction between the conceptual world and empirical, where deduction and induction are realized at the same time (Bryman and Burgess 1994 in Corbetta 1999). Therefore—as mentioned in the previous paragraph—the Digital Ethnography remains a non-standard tool, while maintaining the epistemological structure classic ethnography, can make use of information systems and mathematical models considered as ad hoc tools to activate the application process to a new field of research.

From a methodological point of view, however, it should be clarified what are the critical elements:

(1) the blacks numbers: there are of difficult access Web areas which make necessary the intervention of a gatekeeper thus changing the type of observation;
(2) the difficulty in not pollute, in any way, the scope of the survey, especially in the study of the structure of the network: you must not have ties to members of the group otherwise you may appear as a "hub" thus changing the structure of the network;
(3) the quality of the data, or the adhesion between the digital identity and real identity of the subjects under analysis. Moreover, the understanding of the origin of the data is fundamental to the definition of the type of communicative behavior. Is, therefore, a clear need to contextualize well the scenery of intervention;
(4) Finally, when studying a phenomenon in the network is necessary to consider some aspects that concern the real life and that may affect the search field (Sade-Beck 2004).

For this reason there is no "universal" rules for the application of the ethnographic method, but it is structured in relation to the reference context.

The last point concerns the ethics and ethical dimension. In fact, using the Big Data for the search implies, at times, the presence of sensitive data. In this regard, the AOIR Ethics Working Committee (2012) has developed a document entitled Ethical Decision-Making and Internet Research Recommendations contains some interesting guidelines for social researchers of the Network:

• The greater the vulnerability of communities/author/participant, the greater the requirement for the researcher to protect her;
• When making ethical decisions, researchers need to balance the rights of the subjects participating in the research with the social benefits of research;
• Ethical issues arise and need to be addressed during all phases of the research process, from planning, to research, to the publication and dissemination of results;
• The ethical puzzles are complex and rarely have a binary pattern: there is often a gray area that forces the researcher to self-regulate.

In conclusion, digital ethnography is a hybrid method not only quantitative, but especially qualitative. By its application for Social Research in digital fields it is possible to investigate a multitude of virtual worlds and "subworlds" that compose cyberspace defined as a system halfway between order and chaos.

References

AoIR Ethics Working Committee (2012), *Ethical Decision-Making and Internet Research Recommendations (Version 2.0),* http://aoir.org/reports/ethics2.pdf.

Barabasi A. L. (2002), *Link, La scienza delle reti,* Torino, Einaudi.

Bastian M., Heymann S., Jacomy M., (2009), Gephi: an open source software for exploring and manipulating networks. *International AAAI Conference on Weblogs and Social Media,* San Jose, 17–20 Maggio 2009 (Poster).

Baysen, E., Hošková-Mayerová, Š., Cakmak, N., Baysen, F. (2016) Misconceptions of Czech and Turkish University Students in providing Citations. Switzerland: Springer International Publishing, pp. 183–191. *Antonio Maturo et al. (Eds): Recent Trends in Social Systems: Quantitative Theories and Quantitative Models.* Studies in System, Decision and Control 66.

Buchanan M. (2004), *Nexus. Perché la natura, la società, l'economia, la comunicazione funzionano allo stesso modo,* Arnoldo Mondadori Editore.

Castells M. e AAVV, (2006), *Mobile Communication and Society: A Global Perspective.* Cambridge, MIT Press.

Castells M., (2001), *The Internet Galaxy: Reflections on the Internet, Business and Society,* Oxford, Oxford University Press.

Castells M., Himanen P., (2002), *The Information Society and the Welfare State: The Finnish Model.* Oxford, Oxford University Press.

Chiesi A. M., (1999), *L'analisi dei reticoli*, Milano, Franco Angeli.

Corbetta P. (1999), *Metodologia e tecniche della ricerca sociale*, Bologna, Il Mulino.

Easley D., Kleinberg J., (2010), *Networks Crowds and Markets Reasoning about a Higly Connected World*, Cambridge University Press, New York.

Granovetter M.S. (1973), The Strength of Weak Ties. *American Journal of Sociology*, (vol.) 78-6, pp. 1360–1380.

Hošková-Mayerová, Š; Rosická, Z. (2015) E-Learning Pros And Cons: Active Learning Culture?. *Procedia - Social and Behavioral Sciences,*, vol. 191, no. June 2015, pp. 958–962.

Kozinets R.V. (2010), *Netnography: The Marketer's Secret Weapon. How Social Media Understanding Drives Innovation*, London, Sage.

Lovink G. (2002), *Dark Fiber: Tracking Critical Internet Culture*, Cambridge Mass, MIT Press.

Murthy D., (2008), Digital Ethnography: An Examination of the Use of New Technologies for Social Research. *Sociology*, (vol.) 42-5 pp. 837–855, Doi:10.1177/0038038508094565.

Rainie L., Wellman B. (2012), *Networked: The New Social Operating System,* Cambridge Mass.

Rieder B., (2013) *Studying Facebook via data extraction: the Netvizz application.* In WebSci '13 Proceedings of the 5th Annual ACM Web Science Conference, New York,: ACM, 2013, pp. 346–355.

Russo V. (2015), *L'espressione della sessualità attraverso l'uso dei social network*, in Cipolla C., (a cura di) (2015), *La rivoluzione digitale della sessualità umana*, FrancoAngeli, Milano.

Russo V. (2016a), *Modelli matematici per Comunità Virtuali*, in Maturo A., Tofan I., Fuzziness. *Teorie e applicazioni*, Aracne, Ariccia (Rm).

Russo V. (2016b), *Un'applicazione empirica dei big data allo studio del mediattivismo civico*, in Agnoli S. e Parra Saiani P. (a cura di, 2016), *Sulle tracce di big data. Questioni di metodo e percorsi di ricerca*, numero monografico di Sociologia e ricerca sociale, anno XXXVII, n. 109.

Russo V. (2016c), Urban Mediactivism in Web 3.0. Case Analysis: The City of Chieti, *Recent Trends in Social Systems: Quantitative Theories and Quantitative Models.* Studies in System, Decision and Control 66. *Antonio Maturo et al. (Eds),* Springer International Publishing, pp. 303–316.

Sade-Beck L., (2004). Internet ethnography: Online and offline. *International Journal of Qualitative Methods*, 3(2). http://www.ualberta.ca/~iiqm/backissues/3_2/pdf/sadebeck.pdf.

Trobia A. Milia V., (2011), *Social Network analysis, Approcci, tecniche e nuove applicazioni*, Roma, Carrocci.

Watts D.J., Strogatz S.H. (1998). Collective dynamics of 'small-world' networks. *Nature*, 393 (6684): 440–442.

Chapter 5
The Demand Side of Negativity and Privatization in News: Experimental Study of News Consumer Habits

Lenka Hrbková, Jozef Zagrapan and Roman Chytilek

Abstract Negativity in media and emphasis on personal side of politics are often cited as a common journalist practice which is harmful to democratic processes. Journalists and media houses are often held accountable for these phenomena because they prioritize profit over the quality of content. However, we offer an analysis focused on demand side of both negativity and privatization of political news. Using the Dynamic Process Tracing Environment (DPTE), we test the assumption that both of these features of political media coverage may be driven by audience demand for negative and personal news. According to the available literature, personal news can serve as a useful heuristics in citizens' political judgment. We have confirmed a negativity bias; however, a tendency to select news about politicians' private affairs was not confirmed. Even thought respondents preferred political news, personal news showed to be more memorable, which might have further implications for formation of political attitudes.

Keywords Negative news · Privatization of news · Consumer demand · Experiment · Heuristics

L. Hrbková (✉) · R. Chytilek
Faculty of Social Studies, Department of Political Science, Masaryk University, Joštova 10, 602 00 Brno, Czech Republic
e-mail: hrbkova@mail.muni.cz

R. Chytilek
e-mail: chytilek@fss.muni.cz

J. Zagrapan
Institute for Sociology, Slovak Academy of Sciences, Klemensova 19, 813 64 Bratislava, Slovakia
e-mail: zagrapan@gmail.com

© Springer International Publishing AG 2017
Š. Hošková-Mayerová et al., *Mathematical-Statistical Models and Qualitative Theories for Economic and Social Sciences*, Studies in Systems, Decision and Control 104, DOI 10.1007/978-3-319-54819-7_5

5.1   Introduction

Scholars of political communication are often concerned with two important characteristics of media coverage of politics. First, politics is often captured in a negative light, and second, there is a shift of focus from political parties to individual politicians and their private lives. Both of these features—negativity and focus on personal stories—of media coverage of politics may be understood as parts of a more general process of strategic media frames (Capella and Jamieson 1997; Patterson 1994). But regardless their mutual relation, both concepts are common phenomena of current media practice, both are also considered to be negative and harmful and a result of journalistic malpractice. Growing negative coverage of public affairs, as a result of the shift from descriptive to interpretative journalism (Patterson 1994), together with greater attention to personal affairs and the private lives of political actors are conceptually linked to strategic news coverage (Trussler and Soroka 2014; de Vreese and Semetko 2002; Capella and Jamieson 1997) and are both seen as a part of a media malaise contributing to general political cynicism and distrust (Jerbil et al. 2013; Kleinnijenhuis et al. 2006; Calepella and Jamieson 1997; Patterson 1994). Our research examines these traditional views of negativity and privatization of political news as a result of solely journalists' preferences. We also focus on how these types of news can influence citizen information processing and whether citizens use these types of information as cognitive shortcuts. We base our experimental study on two assumptions. First, media content is a result of an interplay between journalist supply and reader demand. Second, even if media consumers prefer a type of news which is normatively considered to be inferior to hard news, they are still able to use acquired information to form political attitudes. While most of the research of media coverage of politics is focused on the character of the news, we focus on the audience and on their preferences in order to clarify whether the current state of media content is a result of any demand-driven processes. We believe that this perspective has been rather unexplored and our research may contribute to the empirical knowledge of how citizens select news and how they use the acquired information in their political reasoning. Our basic research questions ask what type of news do people really prefer. Do negative and personal stories about politicians really sell better than serious policy oriented news? Moreover, if this assumption holds true, we are interested in its' implication for political attitudes of those who consume this type of news. Do those people who prefer to read negative and personal news about politicians systematically differ in updating their political attitudes compared to those who prefer policy oriented and more positive news?

Both negative and personal news may have an increased information value, which plays an important part in evaluation of politicians and generating political attitudes. We base our theoretical assumptions on research of cognitive shortcuts in a low-rationality environment. Citizens rely on various heuristics to compensate for knowing very little about politics (Lau and Redlawsk 2001). Popkin shows that citizens are able to assimilate various cues about politicians and identifies media as the key source of these shortcuts (Popkin 1991). We study the demand-side of media usage and the way citizens use personal and negative news about politicians as effective heuristics. We combine interest in citizens' approach to personal news about public figures with the overall tendency of media towards negativity (Chytilek and Tóth 2016).

5.2 A Supply-Side Perspective

Media play a crucial role within the democratic process, since democracy requires journalists to perform important functions in relation to political information (Strömbäck 2005). Regarding these normative assumptions about work of journalists, it is often the case that media practice is criticized for prioritizing other aspects of news. As media is the watchdog of democracy, the journalistic approach towards political actors is supposed to be naturally critical. However, the overall high levels of negative and strategic frames in the media are often regarded as a result of a general journalistic cynicism towards politicians and public officials (e.g. Trussler and Soroka 2014; de Vreese and Semetko 2002; Capella and Jamieson 1997). Additionally, negative frames and a focus on conflict often mean more exciting and sensational stories, and both journalists and their editors understand that bad news sells better (Zaller 1999; Allport and Lepkin 1943; Diamond 1978; Patterson 1994; Niven 2000).

Apart from more attention paid to bad news, there has also been a shift in the media focus towards individual politicians as the key political actors (personalization of politics and political news). Together with a shift towards personally-oriented political coverage of politics there has also been an evident shift from perception of politicians as public office-holders to reporting about politicians as private individuals, their personal lives and characters. This change of discourse from political to non-political characteristics of politicians is called *privatization* (e.g. Van Aelst et al. 2011; Rahat and Sheafer 2007; Holtz-Bacha 2004). Stanyer writes about *intimization*, defined as a process of information circulation in which information flows between the private sphere (personal lives and relationships) and the mediated public sphere of politics. Intimate politics consists of publicizing information and imagery of personal character, public scrutiny of personal relationships, and family life (Stanyer 2013: pp. 12–15). This phenomenon is facilitated by the willingness of politicians across countries to disclose aspects of their personal lives (e.g. Dakhlia 2008; Kuhn 2010; Langer 2012; Holtz-Bacha 2004). Although it has been established that increase of privatization is affected by structural factors, such as the media environment the degree of legal protection for public figures etc. (Stanyer 2013), there is still a surprisingly large gap in the state of knowledge about the role of the public in this process. We believe the increasing negativity and privatization of political news to be a two-way road, which depends not only on the intention of journalists but also on the habits and preferences of consumers.

5.3   A Demand-Side Perspective

The presence of both negativity and privatization of news has been documented, as well as their pervasively negative effects on the citizens (Jerbil et al. 2013; Kleinnijenhuis et al. 2006; Calepella and Jamieson 1997; Patterson 1994). We, on the other hand, are interested in these phenomena from the consumer-side perspective. We believe that audience habits in media consumption is a rather underexplored topic, although it is highly relevant in respect to the perception of political communication. Research (Lichter and Noyes 1995; West 2001) shows that citizens express discontent with negative news frames. Unfavourable attitude towards negativity in the news was expressed by experimental subjects in Trusser's and Soroka's study (2014). The same subjects, however, showed a gap between attitude towards such news and their actual behaviour. There is a reason to believe in a discrepancy between citizens' expectations about media and the type of news they actually prefer (Graber 1984).

Media negativity bias seems to be a function of a natural human inclination towards negativity. Evidence shows that negative information simply matters more and invokes certain cognitive reactions compared to positive information. Research in psychology has documented negativity bias in information processing (Ito et al. 1998; Pratto and John 1991) and impression formation (Hamilton and Huffman 1971; Rozin and Royzman 2001). Bad things in general consume more thinking than good things (Abele 1985; Fiske 1980) and recipients show more cognitive processing when they are in a bad mood (Isen 1987; Schwarz 1990).

The prevailing emphasis on negativity also applies in politics. Citizens prefer to receive information about politicians' personal failures, scandals, and gaffes to information about their performance (Ryan and Brader 2013). There has been a large body of research on negative campaigning, showing that negative ads are more powerful than positive (for a review see Lau and Rovner 2009). Klein has proved that political impression formation is vulnerable to negativity bias, negatively perceived traits of candidates matter more in the overall evaluation of the candidates (Klein 1991, 1996).

This negativity bias can be explained by the utility of negative information, which is considered to be more perceptually salient and informative (Skowronski and Carlston 1989). Atkin emphasized the need for surveillance that produces specific information-seeking behaviors. All pieces of information pose possible threats or opportunities. Negative information evokes more intense inspection because of the higher level of possible danger or negative consequences (Atkin 1973). Negativity bias was demonstrated in relation to readers' selection of political messages (Trussler and Soroka 2014; Soroka and McAdams 2015; Meffert et al. 2006; Donsbach 1991). Meffert also shows that voters read more negative messages about their preferred candidate compared to the less preferred candidate (Meffert et al. 2006), which supports the assumption that negativity requires more scrutiny, since a negative story about one's own favourite politician may pose more of a threat to one's attitudes and evoke more rigorous inspection.

If the negativity bias—based on a greater information value of negative news—applies, then we predict that:

H1a: Participants will be more likely to select negative news compared to positive news about politicians.

H1b: Participants will pay more attention to negative news and therefore spend more time on reading negative news.

H1c: Negative news will provide subjects with information of better accessibility, and therefore subjects will be able to recall more negative information compared to positive information gained from positive news.

The demand side of the privatization of political news is a much less explored. It is possible that reporting about politicians' private lives is also a demand-driven media practice stemming from citizens' bias towards personal information about politicians. Newspaper editors themselves agree that however much the private affairs of political candidates are overrepresented in the news, the situation is driven simply by public interest in this type of information because people want to hear gossip (Splichal and Garrison 2003). The argument that people actually prefer personal stories about political actors to more substantial political news has, however, not been empirically tested so far.

There are assumptions that would suggest higher attractiveness of this kind of news leading to a privatization bias in news selection. Personal information can serve the public as a cue to evaluate a politician. It was already noted by Sennet (1974) that politicians are continually being scrutinized and any misconduct in one sphere of their lives is automatically equated with their capability and competence in all other spheres. Personality politics is a deflection of public interest away from measuring personal character in terms of effective public action and it makes personal character symbolic in sense that any flaw can become an instrument of self-destruction (Sennet 1974: 286). Personal information about politicians thus serve citizens, whose cognitive capacities and interest in politics are rather limited, as a useful source of information. People usually don't follow political news, even when they report paying attention to public affairs, they actually prefer more entertaining stories. Still, voters are informed enough to make sense of the political world. Popkin argues that they use their "gut rationality" to gain information quite effortlessly in everyday life. Cognitive shortcuts enable them to evaluate the acquired information and maintain running tallies about political actors (Popkin 1991: 44). Personalities of politicians work as heuristics for voters to form attitudes and make political decisions. Popkin summed up the importance of politicians' personality in Gresham's law of political information: a small amount of personal information about a politician can drive out a large amount of previous impersonal information. Personal news help voters generate narratives about politicians; it is easier to take personal data and fill in the political facts and policies than vice versa. These narratives are easily compiled and stored in memory longer than hard facts (Popkin 1991: 79). As other research shows, voters often use a candidates and their reputation as a source of their policy evaluations and as a means to connect with political issues (Mondak 1993; Capelos 2010).

Based on the theory of privatization of news, we predict that if privatization of political news applies:

H2a: Participants will be more likely to select personally focused news compared to political news about politicians.

H2b: Participants will pay more attention to personally focused news and spend more time on reading personal news about politicians.

H2c: Personally focused news will provide subjects with information of better accessibility and subjects will be able to recall more personal information compared to political information gained from politically focused news.

Theoretical assumptions suggest that personal news about politicians is not only attractive because it is more entertaining than strictly political news, but also it may have a specific information value for political judgment. Therefore, if personal information about politicians works as a heuristic, those who are interested predominantly in the privatized news would be able to make inferences similar to those who prefer the hard news. The assumption of political heuristic as a useful decision making tool has been accepted in political science literature (Popkin 1991; Lupia 1994; Lau and Redlawsk 2001). However, there is not a consensus on the effectiveness of heuristic reasoning. The original heuristics research in psychology (Tversky and Kahneman 1974) understands heuristics as a biased way of thinking which leads to errors in inferences. Part of the political science scholars also challenge the assumption of heuristics as a tool to make "as if" fully informed decisions by uninformed voters (Bartels 1996; Kuklinski and Quirk 2000). Therefore, it is necessary, to test the potential of heuristics to overcome cognitive limits of the public and to contribute to political reasoning. Even though the issue of heuristic inferences in politics has been studied for some time, we suggest that it is still useful to test and retest heuristics-based hypothesis in new contexts and settings. Based on the theory of personality as heuristics in political judgment we predict that:

H3: Reading personal news affects the ability of the subjects to update their evaluation of persons as politicians in the same way as reading political news does.

5.4   Experimental Design and Procedure

We test which kind of information voters deliberately search for and how they use the information in the evaluation of political leaders. We designed an experiment to see how people process information and what kind of impact various types of political information have. Our design involved no obvious "control group" (people who would receive no treatment) but we still report it as an "experiment" as we have a rather strong "theoretical baseline" (c. Morton and Williams 2010: 311) about how the subjects' information seeking strategies should look like, which has never been properly tested at the individual level. We also use a common practice for framing experiments (Kinder and Sanders 1990; Nelson and Kinder 1996) that also mostly do not conflate experimentation and necessary and sufficient conditions for causal inference; comparing alternative treatments that we describe in this and following section of the article. Still, it wouldn't be meaningful to define one of them as "baseline".

For our purpose, we used information about the previous presidents of the Czech Republic, Václav Havel and Václav Klaus. We have intentionally used politicians who were not politically active at the time of the experimental sessions.[1] Using contemporary political leaders would pose an ethical threat to the integrity of the experiment, since we believe that experimental reality should in no possible way intersect with subjects' real lives (Gadarian and Lau 2011).

Havel and Klaus, although not personally present in the political life of the Czech Republic, still represent very important symbolic values. Their political fates were largely similar: in every election, the parliament voted them into office by very narrow margins (Havel in 1993 and 1998, and Klaus in 2003 and 2008, with the last election requiring a full seven rounds of voting). The trajectories of their presidencies also developed comparably, and were marked by frequent disputes with the political elites, especially over the formation of governments. In a post-communist country where democracy consolidated itself slowly and hesitantly, their long periods in office led to the gradual erosion of their images, so that at the end of their second terms they were leaving the office as relatively controversial public figures, particularly criticized for their remoteness from the problems of ordinary citizens.

Both Havel and Klaus also had to deal with rather detailed public examinations —not limited to the tabloids—of their personal lives. Speculations about Havel's health, connected with his proneness to alcohol and tobacco consumption attracted significant attention and the public long found it difficult to relate to his second wife Dagmar, with whom Havel already kept a close relationship while his first wife Olga was still alive; he married her soon after Olga's death. With Klaus, there was a string of affairs with younger women, but speculations also surfaced as to whether he might have been a homosexual. Although the intellectual impact of Havel and Klaus in the world has been of unequal influence, all of the above makes them comparable research subjects for a study which focuses on the Czech population and poses research questions about perception, reception, and evaluation of personal and political messages about politicians.

Only part of the Czech public considers the personal lives of politicians to be relevant to how they evaluate them. In 2002, 36% of respondents believed that politicians ought to be rated solely on the basis of their discharge of duties, whereas 59% said that they would include private life into the evaluation. In 2013 46% indicated that they would judge politicians solely on the basis of performance in office, and 50% would consider private life as well (Tuček 2013). The importance of personal life for the evaluation of politicians remained fairly stable throughout the period under consideration, fluctuating between 49% and 61%. Yet almost two-thirds of respondents (62% in 2002 and 64% in 2013) demanded the morality of

---

[1]Václav Klaus was six months out of office at the time of experiment. Václav Havel's last term ended in 2003; he died in 2012.

politicians to be judged more stringently than that of ordinary citizens. This number too remained extraordinarily stable throughout the period. All of this suggests that personal stories, especially negative ones, have a potentially significant impact on the public's evaluation of politicians, and it is the mechanisms of these evaluations that we sought better to understand in our experiment.

5.4.1 Procedure

For the purpose of this study, we used part of the data we collected in experimental sessions held in the computer facilities of Masaryk University in Brno, the Czech Republic. Part of the subjects were undergraduate students at the university and part of the subjects were recruited from general public via online advertising to achieve a greater variation in political sophistication. A total of 186 participants (120 women, 66 men; 112 student subjects and 71 non student subjects; mean age 25) took part in the study. All participants were paid 150 CZK (approx. €5.50).

Subjects were randomly divided into experimental groups. The experiment was performed in the Dynamic Process Tracing Environment created by David Redlawsk and Richard Lau—a computer-based dynamic information board designed to simulate the information environment of political campaigns (e.g. Lau and Redlawsk 2001; Redlawsk 2002, 2004). In DPTE, subjects can see a flow of newspaper-style headlines that are directly linked to articles. After reading the articles, subjects return to the headline flow. The system tracks the information search by each of the subjects (including the sequence of opened articles and the time of processing of each article).

Subjects filled in a questionnaire of political sophistication consisting of questions about their interest in politics, political participation, and political knowledge. Basic demographic questions followed and then subjects had to evaluate political leaders. According to the assigned experimental group participants evaluated one of the former presidents, either Václav Havel, Václav Klaus, or both presidents. Evaluation focused both on personal and political profiles of the politicians.

After this introductory questionnaire, the dynamic information board followed. Subjects could click on any of the headlines in the information flow on the computer screens and open a full article to read. Depending on the experimental group, some subjects read articles about one of the former presidents and others were exposed to articles about both presidents. The main reason for creating these experimental groups was to have greater control over the procedure and to see whether participants' strategy of selecting articles to read differs when they have the chance to read about one or two politicians.

Subjects were informed that the articles were fully authentic. The articles were based on real various sources—(1) media, (2) blogs, (3) official presidential websites —and were edited so that they were similar in length and complexity. There were 38 texts about each politician; 18 about their personal lives, 20 about their political activities. One half (of both personal and political) articles presented positive information about the president and the other half was negative. Political articles focused on the same policy issues for Havel and for Klaus. For each issue there was one positive and one negative article about both Havel and Klaus. Positive and negative impressions of the articles and their headlines were tested in a pre-test.

Subjects could open any of the articles that were flowing down their screens. In the meantime, the headlines kept flowing in the background. Whenever they wanted, subjects could close the story and return to the information flow. This part of the experiment lasted for 13 min in one-politician condition; the subjects in the two-politicians condition had 20 min to search for information. Each article showed up three times. After the information flow, the subjects were asked to evaluate the presidents once again with questions identical with those at the beginning of the session. Finally, they were invited to state everything they could remember about the politician or politicians based on the previously read information. The session was ended with a short debriefing.

5.4.2   Variables

In our analysis we work with following variables:

Evaluation of Václav Havel and Václav Klaus as a politician and as a personality: We asked participants "What was the overall contribution of Václav Havel/Václav Klaus as a politician for the Czech Republic since 1989 to today like?" (0 = very negative, 100 = very positive)" and "How do you rate Václav Havel/Václav Klaus as a personality? (0 = very negatively, 100 = very positively)" to find evaluations of both of them.

Change of evaluation We asked participants to evaluate politicians before and also after they read the articles about them, and we measured the difference in the evaluation. We used absolute values to determine how much each participant changed his or her evaluation.

Read articles: In our experiment, we used four types of articles: negative personal, negative political, positive personal and positive political. DPTE allowed us to track which articles participants read and also how much time (in seconds) they spent reading them.

As a control variable, we used *political sophistication*: We calculated how politically sophisticated a participant was by adding the answers from the questionnaire about political participation and the test of political knowledge. We also controlled for *gender* and *age*.

5.5   Results

We divided the results by groups and politicians. Since some participants read articles only about one politician and others about two, we report them as Havel 1 (results of group that read articles only about Vaclav Havel), Klaus 1 (participants

Table 5.1 Total number of read articles, how much time participants spent reading them and total number of recalled articles about each politician

	Total read	Time	Total recalled
Havel 1 (N = 58)	12.36 (3.22)	50.18 (14.05)	7.34 (3.94)
Klaus 1 (N = 45)	12.91 (6.44)	49.54 (20.49)	6.72 (2.89)
Havel 2 (N = 83)	9.13 (4.35)	49.04 (16.85)	7.81 (3.29)
Klaus 2 (N = 83)	9.65 (4.11)	48.43 (15.14)	6.67 (2.32)

Note Table entries are means (with standard deviation in parentheses), time measured in seconds

that read articles about Vaclav Klaus), Havel 2 and Klaus 2 (results from a group that read about both Havel and Klaus) (Table 5.1).

There was not much difference in the amount of read articles and the time spent reading them in different groups. Since the participants in a group with two politicians had more time to read, they read about 18 articles, comparing to less than 13 in a groups with one politician, but they read approximately the same number about Havel and Klaus. People could recall more information about Havel than about Klaus, but the difference is small.

We can observe similar behaviour even when we take a closer look at the articles that people selected to read. In each group the highest number of the articles read were negative articles about presidents' personal lives. Participants also recalled these articles the most. On the other hand, they spent most time reading negative stories which focused on political issues (Table 5.2).

Table 5.2 Number of specific articles read, time spent reading them a number of specific articles recalled about each politician

	Read N Per	Read N Pol	Read P Per	Read P Pol
Havel 1	3.57 (1.98)	2.86 (1.56)	3.09 (1.83)	2.84 (1.47)
Klaus 1	4.02 (2.27)	3.65 (2.10)	2.27 (2.35)	3.00 (1.79)
Havel 2	3.07 (1.92)	2.04 (1.40)	1.93 (1.57)	2.10 (1.81)
Klaus 2	3.25 (1.79)	2.70 (1.81)	1.81 (1.72)	1.89 (1.48)
	Time N Per	Time N Pol	Time P Per	Time P Pol
Havel 1	53.86 (14.51)	59.83 (18.90)	43.66 (12.74)	53.71 (17.36)
Klaus 1	50.52 (18.17)	58.77 (24.16)	42.12 (20.13)	54.66 (20.09)
Havel 2	53.61 (19.52)	60.52 (20.38)	45.73 (16.45)	55.83 (23.22)
Klaus 2	50.51 (16.92)	57.54 (21.21)	44.86 (15.54)	55.18 (24.12)
	Recalled N Per	Recalled N Pol	Recalled P Per	Recalled P Pol
Havel 1	2.80 (1.52)	1.86 (1.02)	2.10 (1.57)	2.00 (1.07)
Klaus 1	2.56 (1.46)	1.83 (0.86)	1.75 (0.91)	2.06 (0.99)
Havel 2	2.55 (1.29)	1.52 (0.74)	1.66 (0.89)	1.55 (0.94)
Klaus 2	2.01 (1.18)	1.56 (0.99)	1.57 (0.83)	1.40 (0.63)

Notes Table entries are means (with standard deviation in parentheses), time measured in seconds
Abbreviations: N Per/N Pol/P Per/P Pol—read negative personal, negative political, positive personal, positive political

To determine whether participants preferred negative over positive news and personal over political, we combine the types of articles together (it means, e.g. negative news = negative personal + negative political, personal news = negative personal + positive personal). While we can confirm that participants read more negative articles than positive ones (H1a), they spent more time reading them (H1b) and there is also statistically significant difference between recalled number of negative and positive articles (H1c).

However, results concerning the personal-political character of news aren't obvious. Even though there was statistically significant difference between recall of personal and political stories (H2c), we cannot say that participants in every group preferred personal over political news (H2a). This is evident in the group that read articles only about Klaus, when people had preferred more political articles. What is more, subjects spent more time reading political stories than personal ones (H2b). Nevertheless, the results show that even though participants spent more time cognitively processing political articles, they could recall more of the personal ones. We may also conclude that behaviour of participants across groups was very similar.

Read articles and change of evaluation
Our results confirm that the exposure to the information flow led subjects to update their evaluations of the presidents. In around 80% of cases participants changed their initial evaluation of Havel and Klaus as politicians and in 86% of cases evaluation of them as personalities. To measure the change of evaluation, we subtracted initial evaluations from the evaluations that participants made after the information flow.

To test our third hypotheses, we used an OLS regression where the dependent variable was the size of change in evaluations and independent variables were the numbers of read of articles. In addition, we assessed political sophistication, gender and age as control variables in the model (Table 5.3).

We found no evidence that any of the various types of articles influenced the change of evaluation of Havel and Klaus as politicians (H3). Even though change in some cases certainly happened, we are not able to conclude that it was the news that participants chose to read to be the reason. Similar result can be drawn about the change in evaluation of politician as personalities. The only exception is group Havel 1. If subjects read more negative personal stories, their evaluation of Havel decreased, on the other hand, if they read more of positive personal articles, their evaluation of the former president increased. However, we did not find this effect for any other group. Type of articles selected by subject does not to seem to affect change of evaluation. This also applies to control variables—political sophistication, gender and age. We can conclude that even thought people were changing their initial evaluations, we cannot identify the reason for this changes.

Table 5.3 The effect of articles on change of evaluation

	Havel				Klaus			
	Group 1		Group 2		Group 1		Group 2	
	Eval. Pol	Eval. Per	Eval. Pol	Eval. Per	Eval. Pol	Eval. Per	Eval. Pol	Eval. Per
Read N Per	0.134 (0.841)	−2.406* (0.902)	−0.74 (0.680)	−0.891 (0.983)	0.169 (0.999)	0.597 (1.354)	−0.876 (1.088)	0.021 (0.976)
Read N Pol	−2.400 (1.214)	0.619 (1.302)	−0.157 (1.035)	0.514 (1.495)	−0.194 (0.876)	−0.327 (1.188)	0.18 (1.031)	0.598 (0.924)
Read P Per	−0.259 (1.082)	2.572* (1.103)	−0.651 (0.849)	1.367 (1.226)	−0.059 (1.044)	1.518 (1.415)	1.204 (1.138)	1.745 (1.021)
Read P Pol	−1.154 (1.186)	0.891 (1.272)	1.201 (0.761)	2.022 (1.100)	−1.344 (1.056)	−0.263 (1.432)	1.264 (1.216)	0.110 (1.090)
Pol sophisti	0.482 (0.581)	−0.397 (0.623)	0.694 (0.374)	0.327 (0.541)	1.890 (0.603)	0.440 (.817)	−0.287 (0.540)	−0.361 (0.484)
Const	5.141 (7.925)	−3.523 (8.500)	−10.476 (4.958)	−19.806 (7.166)	−15.215 (6.363)	−16.003 (8.625)	−0.588 (6.977)	5.221 (6.254)
R2	0.091	0.202	0.095	0.098	0.245	0.190	0.048	0.063

$*p < 0.05$

Note Table entries are unstandardized regression weights (with standard errors in parentheses)

All regressions are controlled for gender and age

Abbreviations: Eval Pol—change of evaluation of politician as a politician, Eval Per—change of evaluation of politician as a personality, Read N Per/N Pol/P Per/P Pol—read negative personal articles, negative political, positive personal, positive political, Pol sophistic—political sophistication, Const—constant

5.6 Conclusion

We focused on the demand side of media coverage of politics. Our goal was to test what kind of information concerning political personalities people tend to select, and whether the selection of such information influenced their political opinions. We confirmed negativity bias (Meffert et al. 2006) meaning that when free to choose people prefer to read negative news to positive ones. Not only did participants show a tendency to naturally select negativity over positivity, they also paid more attention to negative news and thus spent more time processing it. Higher levels of negativity in the media thus might mirror higher demand for this type of news. Moreover, our subjects were also able to recall more negative information, which shows that negativity relates to better accessibility of information in memory and this way could have important implications for attitude formation (Zaller 1992). Negativity bias has been a topic of a scientific inquiry for some time; our experimental study has confirmed its presence in the way subjects select information about real-life political leaders. The general tendency of the news to report negatively about politicians might not be solely a manifestation of journalists' cynicism towards political elites but also a reflection of the natural interest of public in this type of news.

Our research also asked whether a similar type of bias applies for privatized news, another important feature in the media political coverage. In this case we were not able to confirm that people prefer tabloid-style personal stories about politicians over the ones of substantial political matter. Surprisingly, we found that when people can choose, they read more personal news compared to political news, but simultaneously they tend to spend more time reading political articles than personal ones. This indicates that purely political news requires more attention and that it is more cognitively demanding than information about politicians' private affairs. In the light of this finding, it is quite interesting that in spite of more attention paid to political news, subjects still recalled more personal than political articles. Therefore, we can conclude that personal news is easily accessible in the memory despite the lack of higher interest or cognitive effort paid to it by readers. People do not have to actively search for personal information about politicians. Also, we found out that the complexity of the information environment does not affect our findings, because there was no difference between a complexity of information environment in which subjects were exposed to stories about one politician and a situation where there was a possibility to choose between the two presidents.

We were also interested in implication of personalized news (which supplies people with more accessible information than politically focused news) for citizens' political attitudes. We did not find any variable that would influence the change of evaluation of the presidents and our participants' opinions. Our results suggest that people do update their opinions after being exposed to the news. However, there is no evidence that different types of news content (either personal or political) have different effects on the degree of change in the evaluation of politicians. We haven't found evidence that people use personal stories about politicians heuristically to asses their political profiles. Based on our data we cannot either confirm that political news has a different effect on political judgement than personal news. Further research focused on the effects of various types of information on political attitudes is necessary.

Acknowledgements We gratefully acknowledge the support of the Czech Science Foundation-Research Project "Experimental research of electoral behaviour and decision-making in highly personalised elections" (GA13-20548S).

References

Abele A (1985). Thinking about thinking: Causal, evaluative, and finalistic cognitions about social situations. *European Journal of Social Psychology* 15(3): 315–332.

Allport FH and Lepkin M (1943). Building War Morale with News-Headlines. *The Public Opinion Quarterly* 7(2): 211–221.

Atkin C (1973). Instrumental Utilities and Information Seeking. In: Clark P (ed) *New Models of Communication Research*. Newbury Park: Sage, pp. 205–242.

Bartels L (1996). Uninformed Votes: Information Effects in Presidential Elections. *American Journal of Political Science* 40(1): 194–230.

Capella JN and Jamieson KH (1997). *Spiral of Cynicism: The Press and the Public Good.* New York: Oxford University Press.

Capelos T (2010). Feeling the Issue: How Citizens' Affective Reactions and Leadership Perceptions Shape Policy Evaluations. *Journal of Political Marketing* 9(1): 9–33.

Chytilek, R. and Tóth, M. (2016) Study of Costly Voting with Negative Payoffs in a TRS Electoral System, Recent Trends in Social Systems: Quantitative Theories and Quantitative Models. Studies in System, Decision and Control 66. Antonio Maturo et al. (Eds), Springer International Publishing, pp. 89–106.

Dakhlia J (2008). *Politique People*. Paris: Edition Breal.

Diamond E (1978). *Good News. Bad News*. Cambridge: MIT Press.

Donsbach W (1991). Exposure to political content in newspapers: The impact of cognitive dissonance on readers' selectivity. *European Journal of Communication* 6(2): 155–186.

Fiske ST (1980). Attention and weight in person perception: The impact of negative and extreme behavior. *Journal of Personality and Social Psychology* 38(6): 889–906.

Gadarian SK and Lau RR (2011). Candidate advertisements. In: Druckman JN, Green DP, Kuklinski JH and Lupia A (eds) *Cambridge Handbook of Experimental Political Science*. New York: Cambridge University Press, pp: 214–227.

Graber D (1984) *Processing the News*. New York: Longman.

Hamilton DL and Huffman LJ (1971). Generality of impression formation processes for evaluative and non-evaluative judgments. *Journal of Personality and Social Psychology* 20(2): 200–207.

Holtz-Bacha C (2004). Germany: How private life of politicians got into the media. *Parliamentary Affairs* 57 (1): 41–52.

Isen AM (1987). Positive affect, cognitive processes, and social behavior. In: Berkowits L (ed) *Advances in Experimental Psychology, Vol. 20*. San Diego: Academic Press, pp: 03–253.

Ito TA. et al. (1998). Negative Information Weights More Heavily on the Brain: The Negativity Bias in Evaluative Categorizations. *Journal of Personality and Social Psychology* 75(4): 887–900.

Jerbil N, Albaek E and de Vreese CH (2013). Infotainment, cynicism and democracy: The effects of privatization vs. personalization in the news. *European Journal of Communication* 28(2): 105–121.

Kuklinski JH and Quirk PJ (2000). Reconsidering the Rational Public: Cognition, heuristics, and Mass Opinion. In Lupia A, McCubbins MD, and Popkin SL (eds) *Elements of Reason. Cognition, Choice, and the Bounds of Rationality*. Cambridge: Cambridge University Press, pp: 153–182.

Tversky A and Kahneman D (1974). Judgment under Uncertainty: Heuristics and Biases. *Science* 185 (4157): 1124–1131.

Kinder DR and Sanders LM (1990). Mimicking political debate with survey questions: the case of white opinion on affirmative action for blacks. *Social Cognition* 8: 73–103.

Klein JG (1991). Negativity Effects in Impression Formation: A Test in the Political Arena. *Personality and Social Psychology Bulletin* 17(4): 412–418.

Klein JG (1996). Negativity in Impressions of Presidential Candidate Revisited: The 1992 Election. *Personality and Social Psychology Bulletin* 22(3): 288–295.

Kleinnijenhuis J, Van Hoof MJ and Oegema D (2006). Negative news and the sleeper effect of distrust. *Press/Politics* 11(2): 86–104.

Kuhn R (2010). President Sarkozy and news media management. *French Politics* 8(4): 355–376.

Langer AI (2012). *The Personalization of Politics in the UK: Mediated Leadership from Atlee to Cameron*. Manchester: Manchester University Press.

Lau, RR and Rovner IB (2009). Negative Campaigning. *Annual Review of Political Science* 12: 285–306

Lau RR and Redlawsk DP (2001). Advantages and Disadvantages of Cognitive Heuristics in Political Decision Making. *American Journal of Political Science* 34(4): 91–71.

Lichter RS and Noyes RE (1995). *Good Intentions Make Bad News: Why Americans Hate Campaign Journalism*. Lanham: Rowman & Litterfield.

Lupia, A (1994). Shortcuts versus Encyclopaedias: Information and Voting Behavior in California Insurance Reform Elections. *The American Political Science Review* 88(1): 63–76.

Meffert MF. et al. (2006). The Effects of Negativity and Motivated Information Processing During a Political Campaign. *Journal of Communication* 56(1): 27–51.

Mondak JJ (1993). Source cues and policy approval: The cognitive dynamics of public support for the Reagan agenda. *American Journal of Political Science* 37(1): 186–212.

Morton RB and Williams KC (2010). *Experimental Political Science and the Study of Causality: From Nature to the Lab*. Cambridge: Cambridge University Press.

Nelson TE and Kinder DR (1996). Issue frames and group-centrismin American public opinion. *Journal of Politics* 58: 1055–1078.

Niven D (2000). The Other Side of Optimism: High Expectations and the Rejection of Status Quo. *Political Behavior* 22(1): 71–88.

Patterson TE (1994). *Out of Order*. New York: Vintage Books.

Popkin SL (1991). *The reasoning voter: communication and persuasion in presidential campaigns*. Chicago: The University of Chicago Press.

Pratto F and John OP (1991). Automatic vigilance: The attention-grabbing power of negative social information. *Journal of Personality and Social Psychology* 61(3): 380–391.

Rahat G and Sheaffer T (2007). The Personalization(s) of Politics: Israel, 1949–2003. *Political Communication* 24(1): 65–80.

Redlawsk DP (2002). Hot Cognition or Cool Consideration? Testing the Effects of Motivated Reasoning on Political Decision Making. *The Journal of Politics* 64(4): 1021–1044.

Redlawsk DP. (2004). What Voters Do: Information Search during Election Campaigns. *Political Psychology* 25(4): 595–610.

Rozin P and Royzman EB (2001). Negativity bias, negativity dominance, and contagion. *Personality and Social Psychology Review* 5(4): 296–320.

Ryan TJ and Brader T (2013). Gaffe Appeal: A Field Experiment on Partisan Selective Exposure in a Presidential Election. Paper presented at the annual APSA meeting, August 28–31, Chicago, IL.

Schwarz N (1990). Feelings as Information: Informational and Motivational Function of Affective States. In: Higgins ET and Sorrentino RM (eds) *Handbook of Motivation and Cogniton: Foundation of Social Bahavior, Vol. 2*. New York: Guiltford Press, pp: 527–561.

Sennet R (1974). *Fall of Public Man*. Cambridge: Cambridge University Press.

Skowronski JJ and Carlston DE (1989). Negativity and extremity biases in impression formation: A review of explanations. *Psychological Bulletin* 105(1): 131–142.

Soroka S and McAdams S (2015). News, politics, and negativity. *Political Communication* 32(1):1–22.

Splichal S and Garrison B (2003). News Editors Show Concern For Privacy of Public Officials. *Newspaper Research Journal* 24(4): 77–87.

Stanyer J (2013). *Intimate Politics. Publicity, Privacy and the Personal Lives of Politicians in Media-Saturated Democracies*. Cambridge: Polity Press.

Strömbäck J (2005). In Search of a Standard: four models of democracy and their normative implications for journalism. *Journalism Studies* 6(3): 331–345.

Trussler M and Soroka M (2014). Consumer Demand for Cynical and Negative News Frames. *The International Journal of Press/Politics*: 1–20.

Tuček M (2013). Morálka politiků a důvody pro jejich odstoupení z funkce – Březen 2013. Centrum pro výzkum veřejného mínění, Sociologický ústav AV ČR. Tisková zpráva. http://cvvm.soc.cas.cz/media/com_form2content/documents/c1/a6984/f3/pd130408.pdf.

Van Aelst P, Sheaferand T and Stanyer J (2011). The personalization of mediated political communication: A review of concepts, operationalizations and key findings. *Journalism* 13(2): 1–18.

de Vreese, CH and Semetko HA (2002). Cynical and engaged: strategic campaign coverage, public opinion and mobilization in a referendum. *Communication Research* 29 (6): 615–641.

West DM (2001). *The Rise and Fall of the Media Establishment*. New York: Nedford/St. Martin's.

Zaller JR (1992). *The Nature and Origin of Pass Opinion. Cambridge*: Cambridge University Press.

Zaller JR (1999). *A Theory of Media Politics*. Unpublished manuscript. Stopping the Evil or Settling for the Lesser Evil: An Experimental.

Chapter 6
Cost Assessment of Training Using Constructive Simulation

Martin Hubacek and Vladimir Vrab

Abstract Modeling and simulation represent a common part of most human activities. Development of computer technology causes a massive advancement of computer simulation. Computer simulation offers many new views on the modeling and simulation while allowing penetration of simulation into other disciplines. Simulation has an irreplaceable role in the fields of training and education for centuries. Its application and development are largely associated with its use in the military. There is an analogous situation with constructive simulation, which is used as a tool for training of commanders and staffs of military units. The benefits of simulation for a higher quality of training are beyond doubt. Therefore, constructive simulation gradually penetrates into other spheres such as the training of emergency staff. However, relevant studies about the economic benefits of the use of constructive simulation for training are relatively rare. The presented cost comparison of the exercises is based on the authors' experience gained during the implementation of various types of exercise at the Center of Simulation and Training Technologies Brno with the use of constructive simulation OneSAF.

Keywords CAX (computer assisted exercise) · Training · Constructive simulation · Calculation of coasts

M. Hubacek (✉)
Faculty of Military Technology, Department of Military Geography
and Meteorology, University of Defence, Kounicova 65, 662 10 Brno,
Czech Republic
e-mail: martin.hubacek@unob.cz

V. Vrab
Center of Simulation and Training Technologies, Kounicova 65, 662 00 Brno,
Czech Republic
e-mail: vladimir.vrab@unob.cz

© Springer International Publishing AG 2017   71
Š. Hošková-Mayerová et al., *Mathematical-Statistical Models and Qualitative Theories for Economic and Social Sciences*, Studies in Systems, Decision and Control 104, DOI 10.1007/978-3-319-54819-7_6

6.1 Introduction

Simulation of human activity is a phenomenon which is exploited for centuries. Its use is very varied and has many forms. Children perform unconsciously simulation during their games when they imitate adults and their activities. Various games were used to simulate the military conflict since the ancient times. Since the industrial revolution machine models were used to determine the behaviour of the real machine (Bennis 1966; Hudson 2014). Development of computer technology has caused a great progress in the field of modelling and simulation in the 20th century. This was also evident in the field of training and education of people. This part of the simulation has always been primarily developed especially in the military, whose approaches and technologies were later generally applied in other domains of human activities (Hofmann et al. 2013).

Generally three basic types of simulation are used in training. These are live, virtual and constructive simulations. Live simulation is a classic training in the field, which is carried out from the time when the first military units started to be organized. It was perfected through the use of a variety of sophisticated systems for simulation of effects of lethal and non-lethal weapons. Its main purpose is to conduct training with real equipment like during a real deployment. In contrast, virtual simulations use models of equipment (vehicles, weapon systems, …) for training people and thus saving real equipment and the cost of operating machines. Training of individuals, crews and small units is the primary objective of these kinds of simulation. The last type of simulation (constructive) is designed for training commanders and staffs of larger units. This kind of simulation was fully developed after the emergence of computer technology. It is therefore the youngest simulation type, although the staff exercises on maps and various war games can be considered as its forerunner.

Development of virtual reality, more accurate mapping of terrain and new information technologies enabled significant development of virtual simulation, however, constructive simulation still has its justification and it is irreplaceable in training of staffs. This article will be devoted to constructive simulation and evaluation of cost training of staffs in different types of operations. The results are based on the authors' experience gained from more than fifteen years of action on the Center of Simulation and Training Technologies (CSTT) in Brno, which is a training facility of the Army of the Czech Republic.

6.2 The Used Simulation Tools and Their Utilization

The CSTT uses constructive simulation system based on semi-automatic behavior SAF (Semi-Automated Forces) for training of commanders and staffs since its inception in 1999. ModSAF (Modular Semi-Automated Forces) was the first one and it was gradually replaced by a system OTB (OneSAF Testbed Baseline).

Currently OTB is used along with the new system OneSAF (One Semi-Automated Forces). This system is used in other countries especially in the United States, Australia, Slovakia and others (Grega and Bucka 2013; Lui and Watson 2002; Macedonia 2002; Prochazka et al. 2002; Wittman and Courtemanche 2002). The main principles of simulation and simulation environment are preserved in all development versions. It was always system with minimal aggregation and semi-automatic behaviour, which can be influenced by the system operator according to the order of commander of simulated unit. The staff participating in the training is physically separated from the simulation and its members can communicate with subordinate units using only radio, telephone and data communications. If necessary, it is possible to transmit the 3D images of simulated situation to the command post. Simulation system is also connected to the command and control system. This connection allows to transmit some information from the simulation to command and control system. The information is similar to those in the real command and management systems. These include mainly data about the position and status of vehicles and other weapon systems.

As stated above, CSTT had been providing training for several years and it extended the range of operations from the primary capability to conduct primarily combat operations to training units for missions abroad, training in the deployment of army units in favour of the Integrated Rescue System (IRS) and the preparation of specific units such as military engineers, military police etc.

Expanding the capabilities of the simulated activities brought greater possibility of using the simulation center. Combat units are not currently the only exercisers on CSTT. Units designed for use in foreign operations, engineer and rescue units, military police, air defense units, units of the integrated rescue system (especially firefighters and police) use constructive simulation for their training today. Table 6.1 shows the relationship between combat exercises and non-combat exercises over the past 10 years. Qualitative benefits in the training of commanders and staffs of units are unambiguous and indisputable. It is mainly due to continuous capacity utilization of the center. The question then arises, whether such center

Table 6.1 The number of exercises in the individual categories over the past 10 years

	Battle operation	Rescue operation	Stabilization operation	Other
2006	9	1	1	0
2007	7	3	1	0
2008	6	1	1	2
2009	8	2	3	2
2010	5	1	3	3
2011	7	3	3	3
2012	4	2	3	0
2013	8	4	4	0
2014	6	9	4	0
2015	2	8	2	5

provides any economic benefits in comparison with field training (Maturo and Hoskova-Mayerova 2017).

In general, the widespread opinion prevails that training using simulators is significantly cheaper than field training using real machinery and equipment. Literature and other available sources concerned with training and simulations do not provide enough evidence for this claim or instructions for comparing the costs of comparable field training and training using solely simulation technologies. One can readily agree with many conclusions about the benefits of training with simulation technologies (Smith 2009; Rybar et al. 2000; Vrab 1998), which mainly apply to:

- possibility of repeating the same training session;
- training staff in safe conditions;
- saving costs for fuel and ammunition;
- causing no damage to the environment and countryside;
- reduction of the impact on the environment (no disturbance of everyday life of the population);
- identifying weaknesses in the decision-making process.

Even though these benefits are undeniable, and many of them would also offset any potential increased costs of training using simulation technologies because the possible losses of life and damage to environment are hard to quantify.

The following chapters outline the way compared to the cost of training at the CSTT compared to the cost of making the same field exercises using real techniques.

6.3 Evaluation Model of the Training Efficiency

The basis of the model for calculating the efficiency (or rather costs) of an exercise stems from the premise that in an analytic formula for calculating the costs, identical variables are accepted that characterize a given phenomenon, process and behaviour as a process of constructive, virtual, and live simulations (training in a real environment). The relationship for calculating the costs associated with exercise N_{cv} can be expressed by the following equation:

$$N_{cv} = (N_a + N_b + N_t + N_p + N_m + N_s + N_h + N_u + N_n) \cdot k \qquad (6.1)$$

where

- N_a refers to the costs of machinery employment during the exercise [CZK],
- N_b refers to the rental costs of the area for the exercise [CZK],
- Nt refers to the costs of technical equipment [CZK]
- N_p refers to the costs of support of the exercise [CZK],
- N_m refers to the costs of consumed ammunition [CZK],

- N_s refers to the costs of boarding of the participants of the exercise [CZK],
- N_h refers to the costs of accommodation of the participants of the exercise [CZK],
- N_u refers to the costs related to maintenance of the equipment, depreciation [CZK],
- N_n refers to the costs that have not been specified yet [CZK],
- k is a tolerance coefficient.

The variable N_a represents the cost of using all types of machinery during the exercise (passenger cars, transport vehicles, trucks, special vehicles, aircrafts, reconnaissance aircraft etc.). Its value can be expressed as:

$$N_a = \sum_{j=1}^{n} a_j \cdot l_j \cdot c_j \tag{6.2}$$

where

- a_j refers to the number of pieces of j-th piece of machinery employed in the exercise,
- l_j refers to the average distance covered by j-th piece of machinery during the exercise in [km],
- c_j refers to the costs of j-th piece of machinery spending on 1 km ride [CZK/km].

The actual amount of mileage in real training can be easily determined with the odometer on used vehicles. In the case of constructive simulation it is only possible in systems with a low degree of aggregation. OTB and its successor OneSAF have a small degree of aggregation. Used tools of AAR (After Action Review) allow to obtain the actual mileage for each simulated vehicle.

Live simulation (real exercise in the field—LIVEX) can take place in areas whose use is a subject to payment for the rental. Quantifying its value is based on prices per m^2 of j-th area (c_j) multiplied by the overall use of the j-th area (P_j) or the total invoiced cost for rental may be put into Eq. 6.1.

Calculation of N_b can be performed as follows:

$$N_b = \sum_{j=1}^{n} P_j \cdot c_j \tag{6.3}$$

where

- P_j refers to the area of j-th sector used during the exercise [m^2],
- c_j refers to the costs of 1 m^2 of j-th sector [CZK/m^2].

In the case of training using virtual or constructive simulation, the cost for renting space equals zero. Exercise may in fact take place in any area. The spatial location of the exercise is possible in any territory which is mapped in a terrain database (TDB) or where the data for building the TDB exist. The cost of TDB is not insignificant, ranging in the hundreds of thousands of crowns, depending on the area and its details. The costs can be potentially increased by the cost of collecting digital geographic data required for the creation of TDB. Because of possible multiple repetitive use of a given TDB, the potential increased costs are considerably reduced with each conducted exercise; unlike with the exercise in real terrain, where it is necessary that new resources are devoted to every other exercise again. Conducting exercises on simulators also allows training in areas that are not available for traditional training for various reasons (territory of another state, different climate area, considerable distance, region with ongoing combat operations, national park or otherwise valuable territory etc.). TDBs of these areas can be created from various data sources (national, allied, international) (Hubacek 2012) or it is feasible to use non-contact methods of data collection for creating of geographic data (Robinson 1995; Kovarik 2011) which ought to be supplemented by information obtained through GEOINT, IMINT methods and geographic analysis.

Calculation of the remaining coefficients is defined in a similar way. A more detailed description can be found for example in Hubacek and Vrab (2012a, b). The proposed procedure was applied to selected exercises, which took place at the CSTT in the recent years. Three representative exercises were selected as representative of the main types of training.

6.4 Selected Exercises and Calculation of Their Costs

Three typical exercises were selected for comparison of the costs of field training and simulator training. These exercises have been regularly practiced by the Czech Army units on the CSTT for more than 5 years. The topics of these exercises are:

- battalion task force in attack
- deployment of military troops on disposal of consequences of natural disasters
- task force in stabilization operation.

Selected exercises were evaluated according to the methodology described above. The calculation was based on parameters that could be quantified. During the training with constructive simulation only some of the expense issues arise. They are mainly:

- service for data processing (partial edits in data files)
- operators of the simulator
- consumption of electricity by the simulator
- accommodation and catering for the participants of the training.

Data extracted from the simulation were used to quantify the cost of the same exercise carried out by in the field. When using the constructive simulation, it is possible to use the data used for presenting entities (models) in the simulator. These data are stated in the Protocol Data Units (PDU) (Pavlu and Vrab 2007). Currently,

Table 6.2 Selected input data entering the calculation of the exercise price

	Battle operation	Rescue operation	Stabilization operation
Number of trainees	55	42	40
Service personnel	30	35	30
Simulation stations	42	49	36
Other used equipment	56	29	62
Number of simulated people	1511	975	527
Number of simulated vehicle	375	183	94
Total mileage covered terrain	10951	2405	862
Total mileage on the road	4661	7283	5429

it is not possible to calculate and quantify all the items according to the methodology described above. For this reason, only the familiar costs and the costs of each item that can be computed from available information are stated. The basic parameters that were used primarily are the following (Table 6.2):

- number of simulated people
- number of simulated vehicles
- mileage
- quantity of consumed ammunition and other material

Values in the table represent the total amount. Vehicles, ammunition and other items were categorized in the calculation by:

- the type of chassis
- the movement in the terrain and on roads
- the type of ammunition and other consumed material

Categorical data and publicly known information on prices were used to calculate the costs of field exercises (Hubacek and Vrab 2012a). Ammunition is an exception, and its cost is based on military tariffs. Some costs may change over time. However, this change is not usually significant and it does not affect the overall relation between exercise in the field and exercises on the simulator. The individual items and the total cost of execution of exercises on the simulator and in the field are summarized in Table 6.3.

The proposed solution procedure is based on publicly known prices in the Czech Republic. The final rating may be different in other countries, or using a different simulation system, or use of other vehicles, weapons and equipment. Some expenses in the calculation of live training is not possible to quantify reliably, therefore it is not included in the calculation. Likewise, the cost of the center construction is not included in the cost of the simulator exercises, while operating costs are included in the price of training. Similarly, the cost of procurement of

Table 6.3 The result of comparing costs between the different kinds of exercises in variations CAX and LIVEX. Prices are in CZK

	Battle operation		Rescue operation		Stabilization operation	
	CAX	LIVEX	CAX	LIVEX	CAX	LIVEX
N_{a1} (vehicle)	5 650	595 000	7 480	215 000	7 480	111 000
N_{a2} (plane)	0	1 600 000	0	1 800 000	0	600 000
N_b	0	0	0	NA	0	0
N_t	1 580	10 560	1 560	4 500	1 460	8980
N_p	320 000	0	150 000	750 000	150 000	300 000
N_m	0	7 830 000	0	0	0	150 000
N_s	42 620	793 270	32 550	377 000	31 000	222 040
N_h	44 000	0	33 600	NA	32 000	105 400
N_u	NA	NA	NA	NA	NA	NA
N_n	NA	NA	NA	500 000	NA	100 000
k	1	1	1	1	1	1
N_{cv}	**413 850**	**10 828 830**	**224 540**	**3 646 500**	**221 940**	**1 597 420**

Note NA—not available in this time

equipment and its amortization are not included in case of a live training. Again, the costs for the exercises are only counted.

6.5 Conclusion

Simulation technologies are increasingly being applied in the area of staff training. Areas of their use focuses not only on the military, but also on other security forces, rescue units and crisis staffs. Their use is advantageous for many reasons, such as the possibility of a repetition of situations, training in geographically varying environment, simulations of all sorts of meteorological phenomena, preventing damage to nature, reducing the impact on the population, and so on. Despite this fact the economic contribution of simulation technologies is debated quite often due to the relatively high costs of acquisition of simulation technique.

CSTT has been existing in the Czech Army for more than 15 years. This unique center was primarily designed only for training military units. Throughout the years, the center has been adapted for conducting of training in non-combat operations and in rescue operations. Free capacities of the center are used by units of the integrated rescue system. The implementation of diverse training can determine the advantages of training in different types of operations. Three typical examples were chosen from more than two hundred realized exercise. Costs for selected exercises have been calculated according to the proposed methodology for both variants of exercise CAX and LIVEX. These cost calculations show a clear economic advantageousness of exercise on a simulator compared to the same exercises in the field. Although all costs are not included in the calculation, calculations show a much higher costs of exercises in the field.

The most economically advantageous is the use of constructive simulation for training of staff in command and control of combat operations. It is mainly due to significantly greater deployment of personnel and equipment in this type of training. Ammunition is another important issue which generates important savings. On the other hand, the number of trained persons is much higher in case of LIVEX than during CAX. Even though, calculations have shown greater profitability of combat units training, simulator training in emergency and non-combat operations is still about ten times cheaper than the same exercise carried in the field. It is necessary to point out the fact that a number of exercise topics can not be carried in the field without restrictions of everyday life. The second advantage is possibility of training of crisis staffs that were never in a control of any operation. Training using constructive simulation can prepare crisis staffs to the real situation, although the most important benefit is the fact that members of crisis staffs can practice in command, control and communication during crisis situations.

The calculations clearly demonstrated the benefits of using constructive simulation for training of staffs. Generally, it can be declared that the economic advantage of using this technology grows with the size of the simulated situation. Costs could be reduced also by the use of model scenarios or parallel training of staff of small units.

Acknowledgements This chapter is a particular result of the defence research project DZRO K-210 NATURENVI managed by the University of Defence in Brno.

References

Bennis, W. G. (1966). Changing organizations. *The Journal of Applied Behavioral Science*, 2(3), 247–263.

Grega, M. and Bucka, P. (2013). Exercise of crisis management of non-military character in practice (In Slovak). *In Proceedings of Riesenie krizovych situacii prostrednictvom simulacnych technologii.*

Hofmann, A., Hoskova-Mayerova, S., & Talhofer, V. (2013). Usage of fuzzy spatial theory for modelling of terrain passability. *Advances in Fuzzy Systems, 2013*, p. 13.

Hubacek, M. (2012) Constructive simulation and GIS. *In Proceedings of the International Conference on Military Technologies and Special Technologies 2012.*

Hubacek, M., and Vrab, V. (2012a). Cost and efficiency evaluation model of training methods. Science & Military Journal, 7(2), 5.

Hubacek, M., and Vrab, V. (2012b). The use of constructive simulation for policemen training (In Czech). The science for population protection 2012, 3.

Hudson, P. (2014). *The industrial revolution*. Bloomsbury Publishing.

Kovarik, V. (2011) Possibilities of Geospatial Data Analysis using Spatial Modeling in ERDAS IMAGINE. *In Proceedings of the International Conference on Military Technologies 2011 - ICMT'11.*

Lui, F., and Watson, M. (2002). Mapping Cognitive Work Analysis (CWA) to an intelligent agents software architecture: *Command agents. Proceedings of the Defence Human Factors Special Interest Group.*

Macedonia, M. (2002). Games soldiers play. *IEEE Spectrum*, 39(3), 32–37.

Maturo, F., Hošková-Mayerová, Š. (2017) Fuzzy Regression Models and Alternative Operations for Economic and Social Sciences Recent Trends in Social Systems: Quantitative Theories and Quantitative Models, Decision and Control, Vol. 66, Maturo (Eds.), 235–248.

Pavlu, P. and Vrab, V. (2007). Interconnection OTBSAF and NS-2. *Improving M&S Interoperability, Reuse and Efficiency in Support of Current and Future Forces, 19-1.*

Prochazka, D., Hubacek, M. and Rapcan, V. (2002) Using ModSAF in Czech Army: The current status. *In: 6th World multiconference on systemics, cybernetics and informatics, Vol VIII, Proceedings: Concepts and applications of systemics, cybernetics and informatics II.*

Robinson, A. H. (1995) *Elements of cartography.* Wiley.

Rybar, M. et al. (2000). *Modeling and simulation in military (In Slovak).* MO SR.

Smith, R. D. (2009). *Military simulation: where we came from and where we are going.* Modelbenders.

Vrab, V. (1998). *Concept of training officers, staffs and commanders using computer simulations and simulator (In Czech).* Vojenská akademie Brno.

Wittman, R. L., and Courtemanche, A. J. (2002). The OneSAF product line architecture: an overview of the products and process. *In Proceedings of the Simulation Technology and Training Conference (SimTecT'02).*

Chapter 7
Social Problems and Decision Making for Teaching Approaches and Relationship Management in an Elementary School

Luciana Delli Rocili and Antonio Maturo

Abstract The present chapter illustrates specific cases of the teaching experience in a primary school and the application of theories regarding the choice of teaching methods in this particular context. In order to decide the most appropriate intervention strategies, both in terms of teaching and for an efficient management of relationships within the class and with the students' families, the first step was to discover the students' social and environmental background. The experience described herein is at the basis of the final proposal on how to concretely implement some decisional procedures at school, as for instance those linked to the limited rationality and the Analytic Hierarchy Process theorized by Saaty.

Keywords Teaching experiences in primary school · Social problems · Decisional procedures for teaching and relationship management

7.1 A Second-Year Class Composed of Undesired Pupils

In the early stages of her teaching career, the first author was teaching in a difficult and complex situation which seemed to be devoid of solutions, but later on it allowed her to thoroughly understand, numerous social and psychological issues and to learn how to intervene effectively in difficult situations on the field.

In those days, teachers had to report directly to the Health National Service all those children who were considered as 'problem child', because of their behavior deemed as inappropriate to the school environment and discipline in order to

L. Delli Rocili
Istituto Comprensivo Statale Pescara 5, Via Gioberti no 15,
65100 Pescara, Italy
e-mail: lucianadr@live.it

A. Maturo (✉)
Department of Architecture, University of Chieti-Pescara,
Viale Pindaro 42, 65127 Pescara, Italy
e-mail: antmat@libero.it

Š. Hošková-Mayerová et al., *Mathematical-Statistical Models and Qualitative Theories for Economic and Social Sciences*, Studies in Systems, Decision and Control 104, DOI 10.1007/978-3-319-54819-7_7

receive a diagnosis concerning their problems. The child was then screened by a specialized team of the Health National Service composed of doctors, psychologists, and other specialized staff. He was subjected to specific tests in order to evaluate the character and to measure the intelligence quotient (IQ) based on specific formulas. In (Parent and Gonnet 1981: 80) the following Termann classification of the mental status of children according to the value of their IQ was described:

- IQ from 0 to 25 = Idiocy;
- IQ from 25 to 50 = Imbecility;
- IQ from 50 to 60 = Severe debility;
- IQ from 60 to 70 = Slight debility;
- IQ from 70 to 80 = Limit range;
- IQ from 80 to 90 = Field of latency;
- IQ from 90 to 100 = Field of normality.

In addition to their IQ, the medical team assessed various other parameters, such as the space-time orientation, the degree of sensory perception, the ability to report, the curriculum of the child, his/her medical history etc.

As a result, 'problem children' were those who were negatively diagnosed: they were either given an official certificate of their pathology by the National Health Service or they simply had a very low academic performance.

In that school, the teachers in service since several years, and who had taught in a first class in the last school year, had managed to "clean up" their classes forming a second class for the following year in which they gathered all the 'problem children'. This class was assigned to the newly arrived teacher, the first author, who was to perform the trial period before entering as an official state teacher. There were seven children diagnosed by the Health National Service and others with poor school performance. None of the children in the class had learned how to read and write, so she had to teach the program of a first and a second class squeezing the two programs in one year time and with very difficult children.

Further, coming from three different classes, the process of socialization among children who were also aggressive one with the other was still to be built. The time spent in the class was very challenging, with sudden explosions of anger, frequent quarrels, and a general negative attitude towards the school. The children felt rejected and marginalized by their group of origin, and they felt they were being thrown in a group of lower level—a situation perceived as punitive.

There being no classrooms available, this class was located in a corner of the corridor closed by two makeshift plywood walls supported by wooden slats.

Given the proximity of the exit, the first author was forced very often to chase a restless child who suddenly would run out from the class, and outside the school very fast. At the end, for security reasons, she was forced to lock the classroom during lessons.

Without any support from the school or from other entities, the first author started a period of self-directed and self-financed study of volumes of psychology, education, medicine, teaching practices, in order to deal with this specific situation. (Hoskova-Mayerova and Mokra 2010).

7.2 The Socialization, Educational and Teaching Path Under Emergency Conditions

The situation had to be tackled with energy and solicitude.

Petracchi (1976: 16) stated that: the intentionality in education is basically conditioning: the point is to condition for deconditioning. That is, you have to help the child to adapt himself or herself to certain situations in an intelligent manner, so that he or she can develop through his or her commitment to learning creative and productive thinking.

However, it was not possible to expect to overcome all issues, but action was needed to produce a significant change in the relational skills within the class in which there were children who were demonstrating extreme aggression, overreacting in any frustrating situation or who showed emotional unstable and violent emotions.

Some pupils showed character disorders attributable to family environment (as proved by medical, psychological, and pedagogical diagnosis) that, even if unintentionally, had made them feel a sense of 'abandonment', but the school had not done much to help them, on the contrary it had probably increased in these children a sense of inadequacy and the certainty of being rejected once again.

Children who live in a situation of social maladjustment in the family and who do not feel accepted experience major difficulties in adapting to the needs of the school, a structure which requires compliance with specific rules and in which positive results in learning are expected; the lack of affection that the child experiences generates insecurity resulting in regressive or aggressive attitudes, the latter is typical when it is seen as the only way to defend himself or herself against a world perceived as hostile.

For these reasons, it was necessary to provide the children with a school environment which had a more 'human' approach and was more attentive to social relations. It was not considered useful to point exclusively to learning and content related to the curriculum, because it would be unproductive; to make the school environment livable, it was more appropriate to aim first of all to make it a social space. In the first period of the school year, therefore, it was critical to dissolve all those conflict situations among some children, encouraging moments of dialogue, conversation, analysis of situations, search for overcoming misunderstandings, and, above all, occasions to reflect and communicate their thoughts and feelings. Some of these aspects are well analyzed in (Canevaro 1976; Spinelli 2005; Delli Rocili and Maturo 2013, Osterrieth 1963; Winnicott 1979). Many research has considered further aspects on specialized teaching during the last decade (Hoskova-Mayerova 2011a, b; Rackova and Hoskova-Mayerova 2013; Ceccatelli et al. 2013a, b).

Fig. 7.1 Circular
relationships

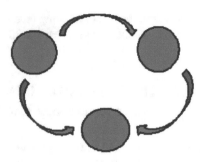

In this way, the children have learned to know each other better, to accept and to welcome suggestions. They were encouraged to play in pairs or small groups to expand their social skills and to facilitate cooperative activities. Each child had to restructure himself or herself and his or her "living space" building relationships which were no longer *unidirectional* but *circular*, in which they could meet and get to know other points of view (Fig. 7.1).

At the same time, games for the recognition of the letters of the alphabet in words and sentences were inserted, as well as games for the construction of new words. The progress shown in children was evident, both in terms of learning and on a relational level. By the Christmas' holiday break, the children were able to focus on a task proposed to them, they could read and write sentences, they acquainted with numerical skills by using additions and subtractions to solve problematic situations. Further, it was possible to introduce gradually teaching units with a higher level of difficulty and, by the end of the school year, pupils reached a level higher than that of children attending a first class in school (for example, they were able to build stories starting from comics and to use the direct and indirect speech). In the end, the results were very satisfactory and the effort was paid off.

7.3 An Analytical Hierarchical Process for Selecting the Best Teaching Strategy

We are looking for a formalization, i.e. a mathematical model, to arrive at a choice of the strategies to be followed in circumstances such as those described in the first two paragraphs.

In general, if you are in a difficult situation, the first step is to try to clarify what are the objectives to be achieved and which are the feasible strategies. It is essential to understand how you can make a 'rational' decision which brings to acceptable results.

The concept of rationality of the decision is understood not in an absolute sense, but as a decision consistent with the objectives (March 1994; Luce and Raiffa 1957; Von Neumann and Morgenstern 1944; Maturo et al. 2010).

Perfect rationality cannot be reached, because of the incompleteness of the information, the social and psychological constraints, and of unclear goals or alternatives. That is the reason why we should settle for a "bounded rationality" (March 1994).

The methods of probability theory and the theory of fuzzy sets may help to control this uncertainty. Very importantly, it is a procedure that is able to identify and describe analytically the objectives and the extent to which the various alternatives meet these objectives.

Such a procedure, called Analytical Hierarchy Process (AHP), was developed in (Saaty 1980, 2008) and taken up by various authors for different types of decision problems. In particular, in (Maturo and Ventre 2009a, b) the AHP algorithm is applied for the purpose of seeking consensus in decision-making problems with multiple decision makers.

We believe that the AHP procedure can be particularly effective for the decisions in schools, for its careful analysis of goals and sub-goals and the effect of possible strategies for effective teaching, for an optimal system of relations in the classroom, and in interactions between school, family, and society.

Another possible method to control the uncertainty due to the joint action of more decision makers is the Game Theory (Luce and Raiffa 1957; Von Neumann and Morgenstern 1944). This theory can be applied in schools, in which students, teachers, and families are considered stakeholders with different interests and opinions, partly conflicting, and each stakeholder has to decide his or her own strategy, which, combined with that of others, allows to obtain the maximum profit for everybody. Alternative decision making models are considered in (Delli Rocili and Maturo 2013; Maturo et al. 2010).

We believe that, at least in a first stage, it is preferable to start from Saaty AHP procedure, because, in fact, the teacher ends up being the only decision-maker (after hearing the everybody's views, taking note of their needs, and trying to mediate between conflicting interests) (Hoskova-Mayerova and Rosicka 2012, 2015).

We briefly recall the theory upon which the procedure AHP is based.

We wish to recall that a *directed graph* or *digraph* is a pair $G = (V, A)$, where V is a non-empty set, said *vertex set* and A is a subset of $V \times V$, whose elements are said *arcs*. The vertices will be named with Latin letters. An ordered pair (u, v), of vertices, belonging to A is said arc with initial vertex u and final vertex v.

A n-tuple of vertices $(v_1, v_2, ..., v_n)$, $n > 1$, where (v_i, v_{i+1}) is an arc, for $i = 1, 2, ..., n - 1$, is said *path* of length $n - 1$.

The first step of the AHP procedure is to represent the decision problem with a directed graph $G = (V, A)$, said AHP graph, with the following properties:

1. The vertices are distributed in a given number $n > 2$ of *levels*, numbered from 1 to n;
2. There is only one vertex of level 1, called the *root* of the graph;
3. For each vertex v different from the root, there is at least one path with initial vertex the root and final vertex v;

Fig. 7.2 AHP graph for n = 3

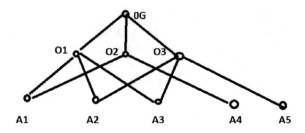

4. Each vertex of level i < n is the initial vertex of at least one arc and there are no outgoing arcs from the vertices of level n;
5. If an arc has the initial vertex of level i, then his final vertex is at level i + 1.

The vertex of level 1 is said *overall objective* (OG), the vertices of level 2 are said *objectives* (or sometimes, more accurately, *specific objectives*). The vertices of level n are said *alternatives* or *strategies*. The vertices of level k < n are said also *objectives of level k*.

A particular case, for n = 3, with 3 goals and 5 alternatives is represented by the following Fig. 7.2. The arcs are intended to be oriented from top to bottom.

It can be observed that, for n = 3 we are facing a usual problem of multi-objective decision, in which the analysis process is reduced simply to identify objectives and alternatives, so that, for the Saaty method to actually represent an analytical procedure must be n ≥ 4.

In many problems, a good compromise between simplicity and analyticity may be the choice of n = 4. We believe that, for school problems, which require an accurate analysis of the objectives, the best choice may be n = 5.

For n ≥ 5, the vertices of level 3 are said *sub-objectives* and have the function to specify the various aspects of the objectives. Furthermore, the vertices of level n − 1 are said *criteria* and represent, to some extent, the tests to be applied to the alternatives. For each strategy and each criterion, it must be possible to measure the degree to which the strategy meets the criterion.

In Fig. 7.3, we have a AHP graph with n = 5. There are 3 objectives O1, O2, O3; 5 sub-objectives S1, S2, S3, S4, S5; 4 criteria C1, C2, C3, C4, and 3 alternatives A1, A2, A3.

The second stage of the AHP procedure consists in assigning scores to the arcs of the graph that do not have as second extreme an alternative, i.e. to assign scores to the objectives of any level k with 1 < k < n with respect to the objectives of level k − 1. These scores, which involve a hierarchy of objectives of different levels, are usually called "weights".

The third phase of the AHP process is assigning scores to the arcs of the graph that have as final extreme an alternative. It is evaluating alternatives or strategies against criteria assigned. In certain appropriate conditions, the scores of the arcs that have as final extreme an alternative are called "utilities".

Fig. 7.3 AHP graph for n = 5

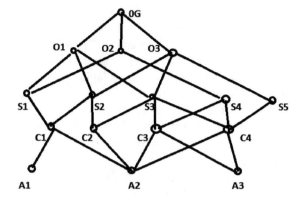

The fourth step of the AHP procedure consists in calculating the scores of the paths and vertices of the graph with the help of matrix algebra.

The fifth and the last stage is the analysis of the results obtained in order to decide a choice or a ranking among the alternatives.

The analysis of each of the various steps may be performed by different makers or committees of experts, because, in general, require different skills. For example, in the school system the first two phases can be decided at ministerial level, the third by the teachers, the fourth by a computer, and the fifth by politicians.

7.4 Assigning Scoring and Selection or Arrangement of the Alternatives

We define the criteria and the algorithms on which the decision makers (or commissions) assign a score to each arc. For each arc (u, v) the score assigned is a number belonging to the interval [0, 1]. Such a score, if it is a weight, measures the degree of importance of v in defining u. If v is an alternative and u is a criterion, the score measures the degree to which the alternative v meets the criterion u.

There is a constraint: for each vertex u from which comes out at least one arc (i.e. for each objective of any level) the sum of the weights or scores of the outgoing arcs from u must be equal to 1.

In (Saaty 1980, 2008) the following procedure for assigning scores is proposed. Let $x_1, x_2, \ldots, x_p$ be the final vertices of the outgoing arcs from a vertex v.

The first step is the pairwise comparison between these vertices using linguistic judgments. If the decision maker D believes that the vertex x_r is more important that x_s, or equally important, he estimated the importance of x_r with respect to x_s with one of the following judgments: *indifference, weak preference, preference, strong preference, absolute preference*.

Subsequently, judgments are transformed into numerical values according to the following numerical scale: indifference = 1, weak preference = 3, preference = 5, strong preference $= 7$, absolute preference $= 9$. Scores 2, 4, 6, 8 are used for intermediate evaluations.

If the vertex x_r score is a number k from 1 to 9 when x_r is compared with x_s then x_s is attributed to the score $1/k$ when it is compared with x_r.

It is thus associated with the p-tuple $(x_1, x_2, \ldots, x_p)$ a matrix $A = (a_{rs})$ with p rows and p columns, where a_{rs} is the score of x_r when compared with x_s. The elements of this matrix are all positive numbers satisfying the condition $a_{rs} \, a_{sr} = 1$.

Afterwards, you calculate the principal eigenvalue λ_1 of the matrix A and among the eigenvectors associated with λ_1, you choose $w = (w_1, w_2, \ldots, w_p)$ having all of the components nonnegative and such that the sum $w_1, w_2, \ldots, w_p$ is equal to 1. The real number w_r is the score that the procedure AHP assigns to the arc (v, x_r).

Before accepting the final score w_i it is necessary, however, to check the accuracy or rationality of the opinions expressed, i.e. the consistency of the opinions expressed. For example, it must be respected the transitivity of preferences.

Saaty suggests to check the consistency of the opinions expressed by the number:

$$\mu = (\lambda_1 - p)/(p - 1).$$

If $\mu \leq 0.1$ then the consistency is considered acceptable, otherwise the decision maker must revise his or her judgments. In practice, however, there may be valid reasons, in particular circumstances, to raise the threshold of 0.1. Often you can accept a higher level of inconsistency and, for example, $\mu \leq 0.2$ or $\mu \leq 0.3$ can also fit.

Once the scores assigned to the arcs, the scores of paths and then that of the vertices are therefore calculated. The score of a path is defined as the product of the scores of the arcs forming the path. The score of a vertex v different from the root is calculated as the sum of the scores of the paths starting from the root and arriving in v.

In particular, we calculate the scores of the alternatives and the decision problem is solved: between the two alternatives, the one with the highest score is preferable.

The presence of errors, incomplete information, constraints of any kind can, however, lead to different evaluations if the difference between the scores of two alternatives is not sufficiently high. In that case, assessments of economic, political, social, religious or other type can also lead to prefer an alternative with a score lower than another.

7.5   A Hierarchical Representation of the General Objective "The Good School"

Given the current debate on education reforms (Hoskova-Mayerova 2011a; Rackova and Hoskova-Mayerova 2013), we take as general objective "the Good School", which can be defined according to the following purposes: "The school aims at the harmonious and integral development of the person, within the principles of the Italian Constitution and the European cultural tradition, at promoting awareness and respect and appreciation of individual differences, with the active involvement of students and their families."

The particular objectives would be:

O1 = *Learning*. Promote cultural and social literacy through the acquisition of languages and codes.

O2 = *Processing experience*. To become aware of themselves and the world; develop the analytical and critical thinking.

O3 = *Conscious citizenship*. Foster conscious attitudes of solidarity, cooperation and respect.

The sub-objectives can be the following:

S1 = *Language and communicative mastery*. Listen, speak, read, understand, think about the use of language.

S2 = *Mathematical—scientific mastery*. Know, represent, understand, use numbers, spaces, shapes, relationships, functions, data, forecasts; make assumptions and use the scientific method to discover the reality, reflect on living and not, etc.

S3 = *Historical mastery*. Reconstruct historical facts, reflect critically on facts and consequences, on mankind and his life on the planet, on the use of renewable energy.

S4 = *Geographical mastery*. Orient themselves in space, analyze, represent, locate regions, learn the elements of the landscape.

S5 = *Respect for cultural heritage*. Appreciate the value of cultural heritage, identifying the problems of protection of the area and possible solutions and to promote cultural heritage.

S6 = *Education to legality*. Acquire a sense of responsibility, legality, and the ability to recognize and respect the values of the Constitution.

S7 = *Education to affectivity*. Acquiring the ability to respect themselves, others, according to the values of acceptance, acceptance, sharing.

As criteria we propose the ability to:

C1 = Participation in communicative exchanges.
C2 = Listening, reading, reading comprehension.
C3 = Production of written texts respecting the rules of spelling, grammar, syntax.
C4 = Master calculations.
C5 = Recognize, represent, classify shapes and geometric figures.
C6 = Solve problems.
C7 = Identify elements and historical traces.
C8 = Use the timeline to organize information.
C9 = Understanding historical texts even with the use of geo-historical papers.
C10 = Telling the facts investigated.
C11 = Finding space, using the language of geo-graphic.
C12 = Identify elements and characters that characterize the landscapes.

C13 = Understanding the changes made by man on the natural environment.
C14 = Taking care of yourself and others.
C15 = Taking care of the environment.
C16 = Cooperation in the group.
C17 = Participation in collective decisions and the life of the school on the basis of shared rules.
C18 = Capacity to respect moral and religious values.
C19 = Ability to respect the laws and values enshrined in the Italian Constitution.

A proposed commission should make a judgment:

(a) on the pairwise comparison of the objectives to the overall objective;
(b) for each objective Oi, the pairwise comparison of sub-objectives that define the objective Oi;
(c) for each sub-objective Sj, the pairwise comparison of criteria defining Sj.

Subsequently, according to the Saaty method, scores are calculated for each arch, path and vertex of the graph. In particular, we obtain the weights of the criteria Ck with respect to the overall objective.

7.6   Some Possible Alternatives When Teaching

The hierarchical representation of the general objective "the Good School" could be made by a ministerial commission or appointed by a regional government. Instead it can be the task of a council of teachers of a whole school or group of schools in a particular territory to see what alternatives may be feasible in the context in which they operate.

For example, let us consider the following alternatives:

A1 = Lectures.
A2 = Teaching workshop.
A3 = Teaching with the game.
A4 = Individualized teaching.
A5 = Personalized teaching.

For each criterion Ck an appropriate commission (usually different from that which defined the criteria weights) should give an opinion for each pairwise comparison between alternatives, evaluating which one best meets, and to what extent, the criterion.

According to the Saaty method, a matrix $M = (m_{ij})$ is finally obtained, with 5 rows representing the alternatives, and 19 columns, each of which represents a criterion. The element m_{ii} measures the extent in which the alternative Ai satisfies the criterion Cj.

Let K be the column vector with 19 elements which represents the weights of the criteria. The product rows by columns P = MK provides the scores of alternatives in relation to the overall objective.

These scores are provided to politicians and school administrators as aid in decision making processes.

The model is dynamic. Taking into account the views of politicians, students, families and the results of the experiences that gradually obtained, the list of alternatives and the matrix M may be updated periodically, and can therefore be evaluated improvements over time.

7.7 Conclusions and Perspectives of Research

The Saaty method, thanks to the hierarchical analytical procedure that allows us to analyze and evaluate every aspect of objectives, sub-objectives, criteria, and alternatives, seems to be particularly well adapted to addressing the most appropriate intervention strategies in primary school.

Indeed the problem of choosing a teaching method in this type of school should take into account the complexity of the context in which teachers and staff work.

There is a need not only to provide effective teaching, but above all to adapt it to situations as they arise, to discover the students social and environmental background and apply an efficient management of relationships within the class and with the students families.

As well described in the book by March, the rationality of decisions is limited because the decision makers have limited knowledge, often rooted prejudices and opinions, and they cannot consider all the variables and possible action alternatives, many of which are unknown or hardly relevant.

However a hierarchical analytical method provides a procedure for ordering of alternatives that results to be consistent with the views and objectives of decision makers. Also it reveals aspects and relationships, not recognizable by other methods.

As a research perspective it can be expected a more in-depth analysis using fuzzy operations and algebraic fuzzy numbers that take into account the uncertainty in the allocation of scores such as fuzzy regression models (Maturo and Maturo 2013, 2014, 2017; Maturo and Fortuna 2016; Maturo 2016; Maturo and Hošková-Mayerová 2017).

References

Canevaro, A. (1976). I bambini che si perdono nel bosco. La Nuova Italia, Firenze.
Ceccatelli C., Di Battista T., Fortuna F., Maturo F. (2013a) Best Practices to Improve The Learning of Statistics: The Case of the National Olympics of Statistic in Italy. Procedia. Social and Behavioral Sciences. Elsevier Ltd. (2013). Vol. 93. pp. 2194–2199.

Ceccatelli C., Di Battista T., Fortuna F., Maturo F. (2013b) L'Item Response Theory come Strumento di Valutazione delle Eccellenze nella Scuola. Science & Philosophy. Telematica Multiversum Editrice, 1(1), (2013).

Delli Rocili L., Maturo A., (2013). Teaching mathematics to children: social aspects, psychological problems and decision making models, in Interdisciplinary approaches in social sciences, Soitu, Gavriluta, Maturo Eds, Editura Universitatii A.I. Cuza, Iasi, Romania.

Hoskova-Mayerova, S., Mokra, T. (2010). Alexithymia among students of different disciplines. Procedia–Social and Behavioral Sciences, 9, 33–37, doi:10.1016/j.sbspro.2010.12.111.

Hoskova-Mayerova, S. (2011a). Operational program "Education for Competitive Advantage", preparation of Study Materials for Teaching in English. Procedia-Social and Behavioral Sciences, 2011, 15, 3800–3804. doi:10.1016/j.sbspro.2011.04.376.

Hoskova-Mayerova, S., (2011b). Operational programm "Education for Competitive Advantage", preparation of Study Materials for Teaching in English. Procedia-Social and Behavioral Sciences, Vol. 15, no. 2011, p. 3800–3804.

Hoskova-Mayerova, S., Rosicka, Z. (2012). Programmed learning. Procedia - Social and Behavioral Sciences, 2012, 31, 782–787. doi:10.1016/j.sbspro.2011.12.141.

Hoskova-Mayerova, S., Rosicka, Z. (2015). E-Learning Pros and Cons: Active Learning Culture? Procedia - Social and Behavioral Sciences, 191, 958–962. doi:10.1016/j.sbspro.2015.04.702.

Luce, R.D. and Raiffa, H. (1957). Games and Decisions. Wiley. New York.

March J.G. (1994). A primer on decision making. How decisions happen. The Free Press, New York.

Maturo, A., Maturo, F. (2013). Research in Social Sciences: Fuzzy Regression and Causal Complexity. Springer Berlin Heidelberg, Berlin, Heidelberg. pp. 237–249, doi:10.1007/978-3-642-35635-3_18.

Maturo A., Maturo F. (2014). Finite Geometric Spaces, Steiner Systems and Cooperative Games. Analele Universitatii "Ovidius" Constanta. Seria Matematica. Vol. 22(1), pp. 189–205 ISSN: Online 1844-0835. doi:10.2478/auom-2014-0015.

Maturo, A., Maturo, F. (2017). Fuzzy Events, Fuzzy Probability and Applications in Economic and Social Sciences. Springer International Publishing, Cham. pp. 223–233, doi:10.1007/978-3-319-40585-8_20.

Maturo, A., Ventre, A.G.S. (2009a). An Application of the Analytic Hierarchy Process to Enhancing Consensus in Multiagent Decision Making, Proceeding of the International Symposium on the Analytic Hierarchy Process for Multicriteria Decision Making, July 29-August 1, 2009, paper 48, 1–12. Pittsburgh: University of Pittsburg.

Maturo, A., Ventre, A.G.S. (2009b). Aggregation and consensus in multi objective and multi person decision making. International Journal of Uncertainty, Fuzziness and Knowledge-Based Systems vol. 17, no. 4, 491–499.

Maturo A., Squillante, M., Ventre. A.G.S., (2010). Coherence for Fuzzy Measures and Applications to Decision Making, S. Greco et al. (Eds.): Preferences and Decisions, STUDFUZZ 257, Springer-Verlag Berlin Heidelberg, 2010, 291–304.

Maturo, F. (2016). Dealing with randomness and vagueness in business and management sciences: the fuzzy probabilistic approach as a tool for the study of statistical relationships between imprecise variables. Ratio Mathematica 30, 45–58.

Maturo, F., Hošková-Mayerová, S. (2017). Fuzzy Regression Models and Alternative Operations for Economic and Social Sciences. Springer International Publishing, Cham. pp. 235–247, doi:10.1007/978-3-319-40585-8_21.

Maturo, F., Fortuna, F. (2016). Bell-Shaped Fuzzy Numbers Associated with the Normal Curve. Springer International Publishing, Cham. pp. 131–144, doi:10.1007/978-3-319-44093-4_13.

Osterrieth, P. A. (1963). L'enfant et la famille. Editions du Scarabée, Paris.

Parent P., Gonnet C. (1981). Problemi del disadattamento scolastico, Armando Editore, Roma.

Petracchi, G. (1976), Decondizionamento. Editrice La Scuola. Brescia.

Rackova, P., Hoskova-Mayerova, S., (2013), Current Approaches to Teaching Specialized Subjects in a Foreign Language. In: ICERI2013 Proceedings. Sevilla, Spain: IATED DIGITAL LIBRARY, p. 4775–4783.

Saaty, T.L. (1980). The Analytic Hierarchy Process, New York: McGraw-Hill.
Saaty, T.L., (2008). Relative Measurement and Its Generalization in Decision Making, Why Pairwise Comparisons are Central in Mathematics for the Measurement of Intangible Factors, The Analytic Hierarchy/Network Process, Rev. R. Acad. Cien. Serie A. Mat., Vol. 102 (2), 251–318.
Spinelli, A.S. (2005). Baro romano drom, Melteni Editore, Roma.
Von Neumann, J. and Morgenstern, O. (1944). Theory of Games and Economic Behavior, Princeton University Press, Princeton, New Jersey.
Winnicott D. W. (1979). La famiglia e lo sviluppo dell'individuo, Armando Editore, Roma.

Chapter 8
From the Pictorial Art to the Linear Transformations

Ferdinando Casolaro and Alessandra Rotunno

Abstract This chapter proposes a path for the teaching of geometry in Italy that reflects the development, which took place over the past two centuries; specifically, it emphasizes the expansion of the Euclidean plane to the projective plane. Moreover, we highlight the social aspects of a teaching based on the graphical visualization as required by the projective geometry.

Keywords Art · Geometry · Education · Social problem

8.1 Introduction

In this article, after a brief historical overview, we present the first elementary concept of projective geometry in a context that does not neglect the fundamental Euclidean properties, such as similarities and isometric, but also takes into account the contribution of Art and Architecture.

In the late '80s, a project on the interrelations between the teaching of Mathematics and Design was added to the model M.P.I. Experimentation "Brocca" (for the introduction of Computer Science in the School and the University). The goals were to introduce the use of new information technologies in representation and combine the requirement of educating students with an understanding of geometry that was behind these processes, namely the projective geometry (Cundari 1997).

This article highlights one of the fundamental aspects of the project which, from an educational point of view, led to fresh the approaches in teaching the expansion of the Euclidean model to projective geometry and, through the knowledge of vector spaces, to Affine geometry.

F. Casolaro (✉)
Department of Economics and Business, University of Sannio, Benevento, Italy
e-mail: ferdinando.casolaro@unina.it

A. Rotunno
High School "Genovesi", Naples, Italy
e-mail: alerotu@tin.it

© Springer International Publishing AG 2017   95
Š. Hošková-Mayerová et al., *Mathematical-Statistical Models and Qualitative Theories for Economic and Social Sciences*, Studies in Systems, Decision and Control 104,
DOI 10.1007/978-3-319-54819-7_8

The project aimed:

(1) For Design, to use computer techniques (CAD, GET, CABRI ', etc.) which replaced representation by ruler and compass;
(2) For Mathematics, to educate teachers (and then students) with a basic knowledge of geometry which supports the new techniques, that is, projective geometry, which in recent decades has been virtually (though not officially) excluded from teaching programs in universities.

The authors of this study have been commissioned to undertake the drafting of the Mathematics course, whose results are in the bibliography (Casolaro 1997).

We believe that, in relation to Mathematics, good results have been achieved; however, they should be excellent if we were not subject to limitations of political institutions imposing authorizations for the general activities. Indeed, the dissemination of the issues took place mainly proposing local seminars in various schools, sending materials through the World Wide Web, and with a great spirit of collaboration of young teachers who, after testing this path, communicated their findings, also allowing continuous corrections.

Topics, such as projective geometry, and outlines about non-Euclidean geometries are the subjects of the 2012 Ministerial Guidelines for second grade Secondary Schools; therefore, in the following paragraphs, a teaching route that was demonstrated as part of the accreditation courses (SSIS, TFA, PAS, etc.) and trialed first in the training, and then in the teaching activities, of the candidates for the teaching accreditation, is presented. On this topic, significant tests were also proposed abroad in various contexts (Hošková-Mayerová and Potůček 2017).

8.2 Evolution of Geometry Through the Arts: Historical Overview

The first demonstrations of geometry were established by the school of Thales during the sixth century BC. Previously, few fragments had been found, in particular on spherical geometry, with Chaldean-Babylonians who, attracted by the charm of the dome of the sky, tried to learn the properties of space, being the sphere the field of study on which they operated.

Also the Greeks were interested in plane geometry but also in spherical geometry; in 1885, two texts were translated (Casolaro and Pisano 2011, 2012): "On mobile spheres" and "The rising and setting"; these works had been written in the third century BC by a contemporary of Euclid, Autolycus Pitane (these are the two most ancient texts that have been found intact).

In his research "On mobile spheres", Autolycus discussed meridian, maximum and parallel circles; this book assumes spherical geometry theorems, thus these concepts were well known to the Greeks in that period.

Moreover, in this work, the propositions are arranged in a logical order; indeed, each proposition is firstly enunciated in a general form, and then repeated, with explicit reference to the figure, and finally a demonstration is given. Euclid uses the same style in his work "Elements" which summarized and reorganized the main researches of prior mathematicians.

Euclid dealt with some issues of Spherical Geometry; indeed, in his study "The phenomena", for the first time, he defined the spherical surface as a surface of rotation of a circumference around its own diameter. Furthermore, he introduced, even if not explicitly, the concept of geometric transformation (in consideration of the various positions that the circumference assumes by rotating around its own diameter).

The type of geometry described by Euclid in his seminal work "Elements", which is based on the methods of Greek's geometry, is called "elementary geometry". This definition is not severe because there are no clear boundaries between elementary geometry and other geometries that have emerged in recent centuries.

Following modern perspectives, elementary geometry involves, almost exclusively, properties of equalities or similarities of figures; indeed, it deals with properties such that, if they are valid for one figure, they are also valid for equal or similar figures.

The analysis of the geometry of the structures, which has led to this point of view, was performed by the German mathematician Felix Klein in the second half of the nineteenth century, after disputes over the crisis of the fundamentals and discussions on the parallel postulate. This lead to the development of the so-called "non-Euclidean geometry" and the projective geometry which presented itself as a more general geometry than the Euclidean one (Casolaro and Paladino 2012a,b). Klein's results are the consequence of research for geometries different from Euclidean one, mainly due to architects and lovers of pictorial art.

After several attempts to consider art as a science, in the fifteenth century, Filippo Brunelleschi set out the rules of perspective and gave a scientific identity to the works of architects and painters. Perspective is universally accepted as the theoretical foundation of pictorial representation and art; indeed, in geometry, it is one of the methods used to represent spatial figures on a plane.

The representation of the technical details of a figure takes place with the methods of descriptive geometry. However, the representation of the transformations it undergoes with the operations of projection and section is the task of projective geometry.

Those who attempted to solve the issues of representation were almost certainly the Greek artists, and we cannot exclude that in ancient epoch there was a system of representation not very dissimilar to that developed during the Renaissance. Apparently, however, we can rule out any research into spatial, mathematical, and unitary definition in the Greek figurations.

It was in the Hellenistic period, in fact, that there originated a research group whose results are "the optics of the ancients", which arose from the desire to study luminous phenomena in order to distinguish what is "appearance" from what is reality.

Starting from the premise that "the light propagates in a straight line" many theorems were established (mostly by Euclid who was a member of the group) that are still considered among the foundations of the mathematical treatment of light.

In addition, Marcus Vitruvius Pollone, who probably lived in the first century BC and can be considered the most significant treatise writer about Architecture in the Latin world, highlighted the interest of the Greeks for representation and art of painting.

We know little about Vitruvius but the originality of his work "De Architectura" (27 BC) in which he describes the Basilica of Fano (he is the builder), is widely recognized; indeed, "De Architectura" was taken as a model in the Renaissance. However, in the XII century, with the Gothic architecture, we observe a principle of representation more strictly rational. The main aim of the Gothic architecture was to achieve the highest possible brightness and the greatest possible amplitude of the rooms with the smallest inconvenience of the building walls and structures.

The first works of Brunelleschi, does not allow us to predict the revolution that would then be put in place, as these works are still in keeping with the Gothic tradition.

It is with the Crucifix of Santa Maria Novella (1409) that we begin to glimpse a new artifact with perfect proportions and a symmetrical distribution of the parts because, at that time, Brunelleschi was seeking the means to make bodies in space objectively representational. He found this means in the perspective of which he determined the rules.

In fifteenth century, Leon Battista Alberti was working along the same lines. He also dedicated to Brunelleschi his treatise "De Pictura" in which he wrote that in the Florentine art of those years the surpassing of the works of antiquity was apparent.

The ingenuity and culture of Alberti is also manifested in his literary and ped-agogical work; he defined the literature of the art of the fifteenth century, almost in opposition to another great artist of the same period, Piero della Francesca, who represents the mathematics of the art. It is in the wake of Piero della Francesca that in the sixteenth century, the prospective has passed from the hands of the artists to those of the scientists, mainly thanks to the work of an eminent commentator, Federico Commandino. In his treatise on linear perspective, he established the principles on which is based the method of projection, originated by the Greek astronomer Claudius Ptolemy (II century AD). Nowadays, it is called "stereo-graphic projection"; this reports all the figures considered at two levels orthogonal to each other, the horizontal and vertical (in the language of the architects the base and the upright) and to depict a plane perpendicular to both and from a point of view above the vertical plane. Then, he imagines that the framework is overturned above the vertical plane by rotation around its intersection with the same plane, to fix also the united elements. This is a first idea of the composition of two operations of perspective that, with the development of projective geometry leads to the concept of homology.

Meanwhile, the teaching of the great Italian artists, from Brunelleschi to Alberti, Piero della Francesca and Leonardo da Vinci, was also assimilated by other non-Italian painters. A prominent among these, is the German artist Albrecht to whom a particularly important place in the history of mathematics must be assigned. He was a mathematician, painter, and treatise writer about architecture and, in his country; he taught many things related to perspective, probably learned from Piero della Francesca during a long stay he made in Italy. However, the key step for the expansion of the Euclidean model was made in France where, some decades later, new horizons for geometry were outlined.

In particular, the French mathematician Girard Desargues is considered the father of the general method for descriptive geometry and the precursor of projective geometry in a work, dated 1639, in which he inaugurated the method of central projection, and introduced for the first time the concept of the point at infinity, laying the foundations for the development of projective geometry, that is, the discipline that studies the properties of figures that do not change by means of projection and section.

Projective geometry was developed in the next century, first with Gaspard Monge and then with his student, Jean-Victor Poncelet to find its precise rigor in 1872 with "The Erlangen program" of Felix Klein who was appointed full professor at the University of Erlangen, and published his "Program" in which he considered the geometrical properties of figures compared to groups of transformations. This involved the application of the "Theory of Groups" (in the organization provided by Jordan), to geometric theories. In this way, Klein presented a unifying theory, which allowed him to classify the various geometries, which progressed independently from one another.

Klein observed that there are in the space transformations that do not alter the geometrical properties of the figures. For geometric properties, Klein intended those properties independent of the position of the figure in space, of its absolute magnitude, and the order of its parts (Casolaro 1997).

The set of transformations that leaves intact such properties is said to be the principal group of transformations, in that:

(1) The composition of two or more transformations is still a transformation;
(2) There is the transformation that transforms the figures into themselves (identity);
(3) There is the inverse transformation;
(4) The composition of transformation is associative.

8.3 Invariants of Linear Transformations

Within the groups of transformations, some properties are preserved. With regard to the transformation of straight lines (invariants expressed by equations of the first degree) and the transformations of conical lines (invariants expressed by quadratic equations), a homography changes a straight line into another straight line and a circle into a real non-degenerate conic; also it preserves the cross ratio between quadruples of points on the corresponding straight lines (Casolaro and Eugeni 1996).

If the homography transforms parallel lines into parallel lines, it is called an "Affinity". The affinities form a group (a subgroup of the homography) under the operation of composition. The affinity group characterizes affine geometry (the first extension of Euclidean geometry).

With regard to the properties expressed by equations of the first degree, since an affine transformation preserves the parallelism, an improper point is changed into another improper point, for which reason the improper straight line is united; relative to the transformations expressed by quadratic equations, a circle is changed into a closed conic (an ellipse). It also preserves the simple ratio of triads of points on corresponding straight lines, and the ratio between the measurements of the areas of the corresponding figures is constant.

An affinity, which preserves the amplitudes of the angles, is called a "similitude"; similitude forms a group (a subgroup of the affinity) under the operation of composition. The similitude of the group characterizes Euclidean geometry. With regard to the properties expressed by linear equations, in a similitude, perpendicular lines are transformed into perpendicular lines; relative to the properties expressed by equations of the second degree, a circle is changed into another circle (Casolaro and Miglionico 2009). Moreover, the similitude preserves the relationship between the measurements of the corresponding segments pairs. With regard to the latter property, it is interesting to note that in the transition: "homography → affinity → similitude, in which it retains respectively the "cross ratio → simple ratio → distance ratio", we pass from a relative property:

(1) to a correspondence between two quadruples of points (a cross ratio);
(2) to a correspondence between two triples of points (a simple ratio);
(3) to a correspondence between two couples of points (a distance ratio).

If the similitude preserves distances, it is called an isometry. The isometry preserves all the properties of the similitude (and therefore of the homography and affinity) and transforms congruent segments into congruent segments. The study of the properties of similitudes and isometries is the subject of Euclidean geometry texts (Casolaro and Miglionico 2009).

8.4 The Projective Plane: Didactic Approach to the Introduction of Points at Infinity

The projective plane is an expansion of the Euclidean plane with the addition of points at infinity. It is in the *projective geometry* that is emphasized the importance of the *points at infinity*. It arises, therefore, the problem of the representation of a point and a straight line in the projective plane.

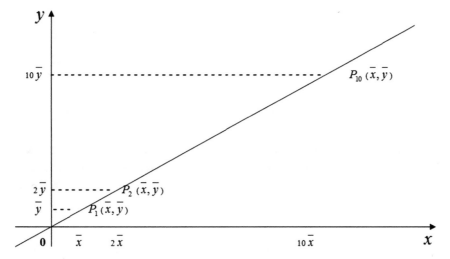

Fig. 8.1 Didactic approach to the introduction of points at infinity

From an educational point of view, we believe it appropriate to present the question starting from the representation of a point $P(\bar{x}, \bar{y})$ in the Cartesian plane, then detecting the improper point with the addition of a third coordinate, by means of the extension to the projective plane.

With reference to the Fig. 8.1, we consider a direction δ of the plane (improper point P_∞); be r the straight line through the origin that identifies P_∞ and be P_1 $(\bar{x}, \bar{y})$ a point of r.

Consider the sequence of points:

$$P_1\,(\bar{x}, \bar{y}) \equiv \left(\frac{\bar{x}}{1}, \frac{\bar{y}}{1}\right);$$

$$P_2\,(2\bar{x}, \overline{2y}) \equiv \left(\frac{\bar{x}}{\frac{1}{2}}, \frac{\bar{y}}{\frac{1}{2}}\right); \ \ldots$$

$$P_{10}\,(10\bar{x}, \overline{10y}) \equiv \left(\frac{\bar{x}}{\frac{1}{10}}, \frac{\bar{y}}{\frac{1}{10}}\right); \ \ldots$$

$$P_{1000}\,(1000\bar{x}, 1000\bar{y}) \equiv \left(\frac{\bar{x}}{\frac{1}{1000}}, \frac{\bar{y}}{\frac{1}{1000}}\right); \ \ldots$$

That, considering the denominator a third coordinate, we can also write:

$$P_1\ (\bar{x}, \bar{y}, 1);$$

$$P_2\ (\bar{x}, \bar{y}, \frac{1}{2}); \ \ldots$$

$$P_{10}\ (\bar{x}, \bar{y}, \frac{1}{10}); \ \ldots$$

$$P_{1000}\ (\bar{x}, \bar{y}, \frac{1}{1000}); \ \ldots$$

The points of this sequence belong to the line r and with the increase of the index k, also increases the distance $\overline{OP_k}$, that is, when the common denominator to the two coordinates (or third coordinate) tends to zero, the point P_k moves away indefinitely.

Let us suppose:

$$x = \frac{x_1}{x_3} \qquad y = \frac{x_2}{x_3}$$

We can define the coordinates of improper points as triples of real numbers (x_1, x_2, x_3) in which is $x_3 = 0$.

Therefore, the expression

$$x_3 = 0$$

identifies the *locus* of improper points of the plane, namely the improper straight line.

8.5   The Projective Geometry and New Geometric Theories

New geometric theories, based on an integrated vision of algebra and geometry, are connected to algebraic hyperstructures (Hošková-Mayerová and Maturo 2017), which constitute a bridge between algebra and geometry, to the fuzzy theories, in which a geometric representation of the uncertainty is connected with new algebraic structures (Maturo and Hošková-Mayerová 2017; Maturo 2016), and finally to subjective probability and its fuzzy extensions, in which the consistent probability assessments form particular geometric structures (Maturo and Maturo 2013, 2017; Maturo and Fortuna 2016).

Teaching geometry through design, in order to avoid the classical Euclidean demonstrations (which in the fast paced world of today are heavy and cumbersome) allows the approach to this discipline even to school children who are left behind;

Fig. 8.2 Homology: composition of two projection operations

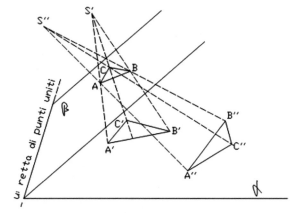

Fig. 8.3 Projection of the window on the *horizontal* plane at 9:00 am

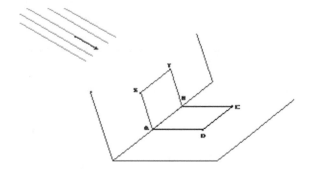

thus, we think that the projective geometry is also a social vehicle that certainly facilitates learning.

In particular, the concept of homology is considered essential since some teachers of the Primary School who, having studied this in training courses for the school districts organized by the Ministry in the school year just finished, where we had the opportunity to be involved, let their students in the fourth and fifth years of the elementary school draw the window and the shadow it produces on the floor at different times of the day (Casolaro and Pisano 2009).

Figure 8.2 shows the homology through the technical aspects of the composition of two projection operations from its S′ and S″ centers.

Figures 8.3, 8.4 and 8.5 show the construction of homology with easy designs proposed to the pupils of the Primary School, where the Center of Projection is the improper point of the direction of the sun's rays (Casolaro 1997). Therefore, the homology is a transformation that takes place in the plan but is generated in three-dimensional space.

The significance of these so-called "little sketches" goes beyond the objectives of representation because it allows the teacher to show specifically the movement of the Earth around the Sun (the interrelation with Science) and the changes of the

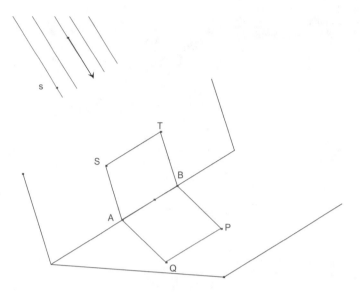

Fig. 8.4 Projection of the window on the *horizontal* plane at 12.00

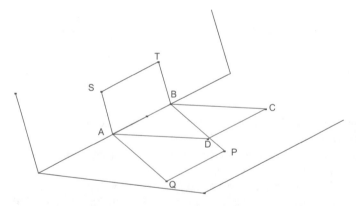

Fig. 8.5 Superposition of the projections: the transformation in the *horizontal* plane that changes in ABPQ ABCD is the homology

figures through the perspective (the interrelation with Art). Therefore, it is clear that to the need to disseminate important issues for the cultural growth of the younger generation must be added also the hope that those who work within the school community are aware of this requirement.

8.6   Conclusion

Teaching Geometry in Italian schools has become a social problem because Euclidean geometry is particularly difficult for students; moreover, they use speeding up any activities with technological resources and Geometry is quite far from the reality of the world today. Unfortunately, 90% of teachers of Mathematics omit teaching Geometry and concentrate their activities exclusively on algebra with some applications of information technology.

In the previous section, we have shown that the presentation through drawing and graphical display allows an immediate understanding of the topics, without neglecting the rigor that requires discipline. In addition, with the introduction of improper elements (point and line at infinity) we have a more complete view of the universe around us. This is the method of projective geometry, already developed in the nineteenth century, but that still finds little application in Italian Schools. For future development of our method we aim to extend it to IRT (Ceccatelli et al. 2013) and finite geometric spaces (Maturo and Maturo 2014).

Unfortunately, resistance to give the weight they deserve to these topics in education is precisely from professionals (school directors, inspectors and more experienced teachers) who are not willing to deal with issues that were not subjects on their own university path of study (Casolaro and Paladino 2012a, b). For the future, however, we must be optimistic because we do not think this resistance to the dissemination of issues that are essential to the training of young people for a correct observation of the physical world can continue.

References

Casolaro, F., Eugeni, F.: Sulle Trasformazioni geometriche che conservano la norma iperbolica e i settori iperbolici. Ratio Mathematica 11, 23–33 (1996).

Casolaro, F.: Il Programma di Erlangen e le Trasformazioni geometriche. Disegno e Matematica, Nuova didattica finalizzata all'uso delle nuove tecnologie – Dipartimento di Rappresentazione e Rilievo dell'Università "La Sapienza" di Roma: 101–118 (1997).

Casolaro, F., Prosperi, R.: Aspetti qualitativi ed interdisciplinari delle funzioni elementari: monotonia e concavità nell'insegnamento della Fisica e delle Scienze economiche. Mathesis' Conference Proceedings "Conoscere attraverso la Matematica: linguaggio, applicazioni e connessioni interdisciplinari" – Nettuno (Rm) 18-21/11/2004 – pp. 296–302 (2005).

Casolaro, F., Iorio, L., Prosperi, R.: Le applicazioni della matematica da Eulero ad oggi. Mathesis' Conference Proceedings "Nel III centenario della nascita di Leonhard Euler (1707–2007)" – Chieti, 1-4/11/2007; pp. 117–127 (2009).

Casolaro, I., Miglionico, C. n. 7: Rappresentazioni grafiche di primo e secondo grado. Dal piano euclideo al piano proiettivo. Epistemologia Didattica, anno IV (luglio-dicembre 2009) - Ed. Laveglia&Carlone. (2009).

Casolaro, F., Pisano, R.: Riflessioni sulla geometria nella Teoria della relatività, SISFA 2009, Facoltà di Architettura dell'Università "La Sapienza" di Roma. 221–231 (2009).

Casolaro, F., Pisano, R.: An Historical Inquiry on Geometry in Relativity: Reflections on Early Relationship Geometry-Physics (Part One). History Research - Vol. 1, Number 1, December 2011, 47–60 (2011).

Casolaro, F., Pisano, R. (part two): An Historical Inquiry on Geometry in Relativity: Reflections on Early Relationship Geometry-Physics - History Research - Vol. 2, Number 1, January 2012, 57–65 (2012).

Casolaro, F., Paladino, L.: Evolution of the geometry through the Arts - APLIMAT 2012, Slovak University of Tecnology in Bratislava, 481–490 (2012a).

Casolaro, F., Paladino, L.: Analisi sociale e rigore scientifico. Scelta di equilibro per l'ottimizzazione dei risultati nell'insegnamento della Matematica. Logica, Linguaggio e Didattica della matematica – Ed. Franco Angeli. 67–82 (2012b).

Ceccatelli, C., Di Battista, T., Fortuna, F. and Maturo, F.: Best practice to improve the learning of statistics: the case of the national olympics of statistics in italy. Procedia: Social & Behavioral Sciences XCIII, 2194–2199. (2013). doi:10.1016/j.sbspro.2013.10.186.

Cundari, C.: Disegno e Matematica, Nuova didattica finalizzata all'uso delle nuove tecnologie. Dipartimento di Rappresentazione e Rilievo dell'Università "La Sapienza" di Roma. Ed. GRAF SRL – TARQUINIA (1997).

Hoskova-Mayerova, Š., Maturo, A.: Fuzzy Sets and Algebraic Hyperoperations to Model Interpersonal Relations, in Studies in Systems, Decision and Control, Vol. 66, Antonio Maturo et al. (Eds): Recent Trends in Social Systems: Quantitative Theories and Quantitative Models. (2017). doi:10.1007/978-3-319-40585-8_19.

Hoskova-Mayerova, Š., Potucek, R.: Qualitative and quantitative evaluation of the entrance draft tests from mathematics, in Studies in Systems, Decision and Control, Vol. 66, Antonio Maturo et al. (Eds): Recent Trends in Social Systems: Quantitative Theories and Quantitative Models. (2017). doi:10.1007/978-3-319-40585-8_6.

Maturo, F.: Dealing with randomness and vagueness in business and management sciences: the fuzzy probabilistic approach as a tool for the study of statistical relationships between imprecise variables. Ratio Mathematica 30, 45–58 (2016).

Maturo, A., Maturo, F.: Research in Social Sciences: Fuzzy Regression and Causal Complexity. Springer Berlin Heidelberg, Berlin, Heidelberg. pp. 237–249 (2013). URL: http://dx.doi.org/10.1007/978-3-642-35635-3 18, doi:10.1007/978-3-642-35635-3 18.

Maturo A., Maturo F.: Finite Geometric Spaces, Steiner Systems and Cooperative Games. Analele Universitatii "Ovidius" Constanta. Seria Matematica. Vol. 22(1), pp. 189–205 (2014). ISSN: Online 1844-0835. doi: 10.2478/auom-2014-0015.

Maturo, A., Maturo, F.: Fuzzy Events, Fuzzy Probability and Applications in Economic and Social Sciences. Springer International Publishing, Cham. pp. 223–233. (2017). URL: http://dx.doi.org/10.1007/978-3-319-40585-8 20, doi:10.1007/978-3-319-40585-8 20.

Maturo, F., Fortuna, F.: Bell-Shaped Fuzzy Numbers Associated with the Normal Curve. Springer International Publishing, Cham. pp. 131–144 (2016). URL: http://dx.doi.org/10.1007/978-3-319-44093-4 13, doi:10.1007/978-3-319-44093-4 13.

Maturo, F., Hošková-Mayerová, S.: Fuzzy Regression Models and Alternative Operations for Economic and Social Sciences. Springer International Publishing, Cham. pp. 235–247 (2017). URL: http://dx.doi.org/10.1007/978-3-319-40585-8 21, doi:10.1007/978-3-319-40585-8 21.

Chapter 9
Identification of Effective Leadership Indicators in the Lithuania Army Forces

S. Bekesiene, Šárka Hošková-Mayerová and P. Diliunas

Abstract Leadership is of overriding importance in the military sphere since the foundation for leading a unit consists in influence, motivation and soldiers' inspiration by the leader's personal example. The Lithuanian Army seeks to develop a military leadership identity as a way to promote mission success. This study is sought to identify the effective leadership style, which is appreciated by soldiers in the Lithuanian Armed Forces. The behavior of the leader was measured using the Leader Behavior Description Questionnaire (LBDQ), which was developed and adapted by Andrew W. Halpin at the Ohio State University. The 204 participants were selected on an easy sample basis. The data collected from military personnel holding different ranks and doing their professional military service of all the units of the Lithuanian Armed Forces were analyzed using the structural equation modelling (SEM). Moreover, the findings from statistically significant modified model showed the strongest leadership style indicators.

Keywords Leadership style · SEM · AMOS

9.1 Introduction

It is difficult to clearly identify what leadership is and provide its accurate definition, for there is no unique approach towards the notion of leadership. While studying the scientific literature, we have found out that there are many definitions of leadership and concepts. After having analyzed the research on leadership and its

S. Bekesiene (✉) · P. Diliunas
The General Jonas Zemaitis Military Academy of Lithuania, Vilnius, Lithuania
e-mail: svajone.bekesiene@lka.lt

P. Diliunas
e-mail: paulius.diliunas@mil.lt

Š. Hošková-Mayerová
University of Defence, FMT, Brno, Czech Republic
e-mail: sarka.mayerova@unob.cz

© Springer International Publishing AG 2017
Š. Hošková-Mayerová et al., *Mathematical-Statistical Models and Qualitative Theories for Economic and Social Sciences*, Studies in Systems, Decision and Control 104, DOI 10.1007/978-3-319-54819-7_9

theories, Stogdill (1974) states that there are just as many different definitions of the concept of leadership as individuals intending to describe it (Stoner et al. 2000). Different scholars put forward different definitions of leadership. Usually, leadership is defined as a behavior that affects the follower's attitude and actions while seeking to achieve the underlined goal.

The leaders in the military organization are usually identified with managerial personnel; so to be a serviceman (commander, officer) leader means to appropriately deal with subordinate soldiers—that is, to know and be able to inspire them to conduct joint activities (sometimes under very difficult environmental conditions) in order to achieve the established objective. The leader's (officer's) behavior is appropriate when he is held up as an example to others, not only just exercising his powers and giving orders. Although leadership in the military organization is usually based on situational leadership and subordinates' motivation, the practical application of leadership theories and leadership itself can be effective only if the chosen leadership style and the ways, forms and means of influence are suitable to subordinates (Military Leadership, TRADOC (LAF) 2016, pp. 2–6) (Lithuanian Military Doctrine 2016).

The Lithuanian Armed Forces as an organization as well as other world armed forces need those who are able to effectively use modern military equipment and combat systems; who are incentivized to take initiatives under rapidly changing and uncertain conditions; and who are capable of making right decisions, even those which might be important to the state. Therefore, our present-day army is in need of those leaders who could take risks and would not be afraid to take an initiative in a difficult combat situation; who would be able to perform multiple tasks simultaneously and, if necessary, act in an unconventional way; who would be able to collect information and process it; who would be loyal, reliable, self-confident, and capable of raising a degree of confidence in those around; and who would seek for continuous personal and professional development, and not avoid taking on new challenges.

The Military Strategy of the Republic of Lithuania (2012) states: "Lithuanian soldiers shall be characterized by a combat spirit and patriotism that would never let them lose their motivation and unquestionable commitment to their nation and the State (The military strategy. Nr. V-1305 2012). Readiness of Lithuanian military personnel to implement collective defense and other military operations, as well as the success of these operations, demands an ability to work in international military structures and act in a multinational and multicultural environment. Therefore, the LAF' personnel must be capable of using official NATO languages and cooperating with representatives of different cultural backgrounds, know the principles and procedures of interoperability of military forces, constantly improve their qualifications in international courses."

The two most important factors contributing to the coherent functionality of modern military and its structural units are leadership and mission command. Leadership in the army is understood not only as an individual act but also as a complex phenomenon embracing organizational, social, and personal processes. Mission command in the army is perceived as a flexible, pragmatic, and decentralized mission (task, function) execution in accordance with clearly formulated and understood senior commander's intent (Lithuanian Military Doctrine 2016, pp. 4–15).

The Military Strategy of the Republic of Lithuania (2012) says: "Leadership is crucial in ensuring the motivation of soldiers to execute the tasks assigned to them and maintaining their morality in difficult combat situations. Therefore, the LAF shall educate professional leaders, capable of managing their subordinates and striving for constant improvement with an aim to execute military tasks in a timely, precise and creative manner" (The military strategy. Nr. V-1305 2012).

It is emphasized in the Lithuanian Military Doctrine (2016) that leadership in the army at a strategic level is based on the theory of transformational leadership, while at a mission level (in armed forces, formations, elements, and other military units) it is supported by the theory of situational leadership (Lithuanian Military Doctrine 2016; Hoskova-Mayerova 2016, 2017). The practical application of certain leadership theories and leadership itself can be efficient only if the chosen leadership style and the ways, forms and means of influence are suitable to those being under supervision.

According to Higgs and Rowland (2005), trying to establish the sequence of events, leaders must unequivocally perceive an influencing environment (Higgs and Rowland 2005).

The processes of globalization affect an economic, political, social, and cultural life of all countries of the world. Globalization is characterized by integration, dynamism, and speed, whereas modern information technology allows one to efficiently process, store, and transmit information. Globalization and information technology have encompassed national defense and security. In the past, to exhibit industrial and military might one had to have mass (weight); now one needs speed (agility). Due to the process of globalization, political, military, economic or other crises have more serious consequences on neighboring and other countries, on the whole region or even several areas.

Rapid processes of social development and globalization affected the army. Certain changes will undoubtedly occur in the military, and in turn affect the army's transformation and staff training.

The dissemination of information and its access create the conditions for the public to take an interest in the Lithuanian military activities in the country and beyond. The public is particularly interested in the allocations earmarked for a national defense system and for the army, in the eligibility of costs, in the professionalism of the activities conducted, as well as in the administration of military resources (financial and human).

Along with the changes in the social environment, human resources, and security there occur some transformations in the nature of Lithuanian military missions and operations, in the organizational arrangement of structural units, as well as in the staff composition. Therefore, the army and its personnel must not only adapt to new operating requirements but also to the new characteristics of young people joining the army. New properties are as follows (to which one should not however confine himself):

1. Liberal views (a liberal attitude);
2. Internalization of values;
3. Awareness.

The above-mentioned phenomena require a new attitude towards the formulation of the requirements for military personnel, towards leadership training and realization. Leadership is a relatively new and widely discussed topic. For most people, leadership is a possibility and (or) a challenge to realize themselves both at work and in personal life, and an officer-leader is the key organizer of activities both in private and public sectors. To this day, many have attempted to answer the question whether one becomes an officer-leader entirely due to his inherent characteristics, or these characteristics can be acquired. Great leaders are thought to possess certain distinctive characteristics (charisma, talent, and so forth) and features (an ability to communicate, to anticipate event, and so forth), which may be innate and (or) acquired during the training.

Leadership can stand for an ability to maintain an equilibrium of attention in defining and implementing a specific task, in designing, developing, and educating a group. If one is intended to achieve the objectives without taking into account the needs of the group, it may fall apart; no attention is paid to the task—it may not be fulfilled; ignoring individual needs—followers can become disloyal. Whatever personal attributes of the leader (commander, officer) are, in order to effectively fulfill a task (mission), he must be able to properly allocate attention to the listed things above.

Successful leadership is inextricably entwined with one's ability to have authority among his subordinates who respect him as a professional; that is, he must have the qualities of both the commander and the leader. The commander's powers to have the right to command are assigned by upper management structure, while the group itself voluntarily acknowledges the leader's possibility to influence its members; so the best situation is when the formal commander is also the informal leader. An officer-leader will achieve much better results at work rather than act as an individual officer or as an individual leader. However, the cases in which one person has all the characteristics are rare. And sometimes instead of waiting for a genius with all these qualities to appear, it is better to think of how to create a professional group wherein different people with different characteristics will be brought together for unified and efficient work.

Leadership is closely connected with the situation around. The specific time, place or situation the group and its leader are faced with determine who will take the leadership in this case and how it will be done. A cultural dimension can be attributed to the situation. All this leads to the formation, emergence, and survival of a true leader. There should be extraordinary and unpredictable conditions for a leader to come into being. During a relatively quiet time of peace it is enough to have a good commander who has excellent knowledge of leadership and management, and who also has definite rules of operation and acts in compliance with them. However, in the constantly changing environment and during natural disasters, accidents or fighting, a commander may be not enough. It is the unpredictable situations that create the need for a leader to emerge. Historical examples confirm that true leaders materialize in difficult conditions. With regard to the situation, it should be emphasized that it is only the situation that changes, comparing leadership in different environments. The principles of leadership in the army have many links with leadership in the civil milieu. And only the situation determines what leadership styles (or methods) will be applicable. Typically, leadership styles (methods) always remain the same.

Not only can a leader influence and lead others to achieve their goals, but he is also capable of assessing his own capabilities and limitations, and consulting with his followers.

There is a lot of research conducted on leadership, leader, his qualities and abilities. The follower's role is often forgotten. It is in the leadership theories the term 'followers' is used, emphasizing a natural choice whether or not to follow the leader and implement his ideas (Military Leadership, TRADOC (LAF) 2016, pp. 2–9) (Lithuanian Military Doctrine 2016). Thus, leadership is a relationship between the leader and his followers. In this interconnection the leader without his followers cannot exist. Followers are not quiet and submissive people who only realize the leader's instructions. They primarily shape the leader's status; contribute to the formation of his behavior; communicate openly with him on the issues of building a team; and, of course, assist him when difficult decisions are to be made. Followers are not obliged to obey the leader. As soon as the follower understands that the leader's acts are illegal, dishonest or are not in tune with common values, the follower may stand aside. In this case, we refer to both informal leaders and formal commanders communicating illegal orders.

In summary, it can be said that leadership is of overriding importance in the military sphere since the foundation for leading a unit consists in influence, motivation and soldiers' inspiration by the leader's personal example. All of these measures should not be coercive or based upon the commander-leader's position. Only then is it possible to talk about leadership, not about formal commanding.

9.2 Theoretical Considerations

The behavior of the leader was measured using the Leader Behavior Description Questionnaire (often referred to as LBDQ), which was developed and adapted by Andrew W. Halpin at the Ohio State University (https://cyfar.org/sites/default/files/LBDQ_1962_Self_Assessment.pdf). It was found in empirical research that a large number of hypothesized dimensions of leader behavior could fall into two strongly defined factors. These were identified by Halpin and Winer and Fleishman as as Consideration and Initiation of Structure (Halpin et al. 1956). The two defined subscales have been widely used in empirical research, particularly in military organizations, industry, and education.

The LBDQ (Form XII) incorporates 100 hypothesized dimensions of leader behavior. These are divided into 12 subscales; each subscale is composed of either

five or ten items Brief definitions of the subscales in the Questionnaire are presented
in Table 9.1 and structured model of relationships between leadership style is
presented in Fig. 9.1 as general structural equation model (GSEM) in (a) and as
theoretical model in (b). GSEM measurement model structure (Fig. 9.1a) can be
described by equations:

$$y = \Lambda_y \eta + \varepsilon, \quad \eta = B_y \eta + \zeta$$

Where y—indicators of η, Λ_y—factor loadings of η on y, δ—measurement error
for x, ε—measurement error for y and ζ—model typically include a structural error
term. In addition for measurement and structural model for SEM analysis, which is
described in Fig. 9.1a we can write equations in matrix like this.

The joint estimate of a subscale is calculated by summing up the scores received
in the evaluation of every statement. Higher scores always or often reflect the way
the leader behaves; on the contrary, lower scores demonstrate that one can rarely
observe such behavior of the leader working in the group.

Table 9.1 Descriptions of the subscales of the leader behavior description questionnaire

Subscale	Description	Label in model
Superior orientation	Maintains cordial relations with superiors; has influence over their decisions; is striving for higher status	AB1
Integration	Maintains a closely knit organization; demonstrates intermember relations	AB2
Predictive accuracy	Displays foresight and ability to predict outcome accurately	AB3
Product emphasis	Formulates the team goals; constantly seeks for better results; applies pressure for productive output	AB4
Consideration	Regards the comfort, well-being, status, and contributions of followers	AB5
Role assumption	Actively exercises the leadership role rather that surrendering leadership to others	AB6
Tolerance and freedom	Allows followers scope for initiative, decision and action	AB7
Initiation of structure	Defines own role, and lets followers know what is expected	AB8
Persuasiveness	Uses persuasion and argument effectively; exhibits strong convictions	AB9
Tolerance of uncertainty	Is able to tolerate uncertainty and postponement without anxiety or upset	AB10
Demand reconciliation	Reconciles conflicting demands	AB11
Representation	Speaks and acts as the representative of the group and indicates its importance in an organization	AB12

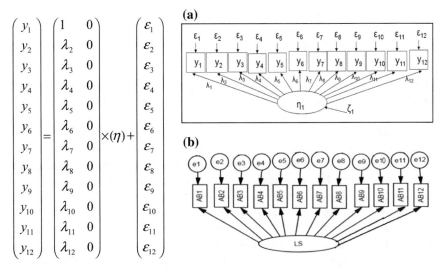

$$
\begin{pmatrix} y_1 \\ y_2 \\ y_3 \\ y_4 \\ y_5 \\ y_6 \\ y_7 \\ y_8 \\ y_9 \\ y_{10} \\ y_{11} \\ y_{12} \end{pmatrix} = \begin{pmatrix} 1 & 0 \\ \lambda_2 & 0 \\ \lambda_3 & 0 \\ \lambda_4 & 0 \\ \lambda_5 & 0 \\ \lambda_6 & 0 \\ \lambda_7 & 0 \\ \lambda_8 & 0 \\ \lambda_9 & 0 \\ \lambda_{10} & 0 \\ \lambda_{11} & 0 \\ \lambda_{12} & 0 \end{pmatrix} \times (\eta) + \begin{pmatrix} \varepsilon_1 \\ \varepsilon_2 \\ \varepsilon_3 \\ \varepsilon_4 \\ \varepsilon_5 \\ \varepsilon_6 \\ \varepsilon_7 \\ \varepsilon_8 \\ \varepsilon_9 \\ \varepsilon_{10} \\ \varepsilon_{11} \\ \varepsilon_{12} \end{pmatrix}
$$

Fig. 9.1 Diagram for predictive effective leadership style in Lithuania armed forces: **a** measurement and structural model for SEM analysis; **b** theoretical model

To mark one's answer there was a five-letter format used in the LBDQ. The Questionnaire asks the respondent to describe how often his leader exhibits (does not exhibit) his behavior while working in the group.

There are presented 100 items, which reflect different work situations. Along with every item the respondent has to mark one of the five letters:

1. A—behavior is always demonstrated (a score of 5)
2. B—behavior is often demonstrated (a score of 4)
3. C—behavior is occasionally demonstrated (a score of 3)
4. D—behavior is seldom demonstrated (a score of 2)
5. E—behavior is never demonstrated (a score of 1)

In order to prevent respondents answering questions automatically and force them to read each item carefully, the following 20 items (6; 12; 16; 26; 36; 42; 46; 53; 56; 57; 61; 62; 65; 66; 68; 71; 87; 91; 92; 97) are scored in the reverse direction:

1. A—behavior is always demonstrated (a score of 1)
2. B—behavior is often demonstrated (a score of 2)
3. C—behavior is occasionally demonstrated (a score of 3)
4. D—behavior is seldom demonstrated (a score of 4)
5. E—behavior is never demonstrated (a score of 5)

9.3 Methodology

9.3.1 Data Collection

The research was carried out in March and April, 2016. Participants filled in the questionnaire on *manoapklausa.lt*, a link was sent personally or through Facebook. They were asked for their verbal consent, and confidentiality was guaranteed. A research group included the professional military service soldiers who were serving in Lithuanian army units during the research. 204 professional military service soldiers with different military ranks and those from all parts of the Lithuanian Armed Forces participated in this research. The participants were selected on an easy sample basis. They had to complete the Leader Behavior Description Questionnaire (LBDQ—Form XII).

The demographic and social characteristics of the respondents, who served as research participants, are summarized in tables below. In Table 9.2, their gender distribution is presented; 168 (82.4%) out of 204 respondents were males, and only 36 (17.6%) were females. This fact corresponds to gender distribution within the entire LAF. The majority have a bachelor's degree −56.9% (N = 116).

In Table 9.3, we can see a summarized and categorized number of years the research participants have been working in military forces. Presented in Table 9.3, the distribution shows that a vast majority of the respondents, comprising about 47.1% (N = 96), have been working in the Lithuania Army for a time period from twelve to seventeen years. As Table 9.3 indicates, 23.5% (N = 48) of the respondents have been working in their forces for a time period of more than eighteen years, while 7.8% (N = 16)—for a period ranging from one year to five years. 21.6% (N = 44) of the respondents spent from one year to five years in the Lithuania army.

The respondents were asked what part of Lithuanian Armed Forces they served in. A summary of their service is presented in Table 9.4. 86% (N = 174) of the respondents stated that they served in the Land Forces; about one tenth of them 12%

Table 9.2 The respondents' distribution within the level of education

Level of education	Frequency (N)	Percentage	Valid percentage	Cumulative percentage
Secondary	8	3.9	3.9	3.9
Bachelor	116	56.9	56.9	60.8
Higher education	4	2.0	2.0	62.8
Unfinished higher education	4	2.0	2.0	64.8
Master's degree	68	33.3	33.3	98.1
Higher than Master's degree	4	2.0	2.0	100.0
Total	204	100.0	100.0	

Table 9.3 A summary of working years in the Lithuanian army

Working years	Frequency (N)	Percentage	Valid percentage	Cumulative percentage
1–5 years	44	21.6	21.6	21.6
6–11 years	16	7.8	7.8	29.4
12–17 years	96	47.1	47.1	76.5
18+ years	48	23.5	23.5	100.0
Total	204	100.0	100.0	

Table 9.4 A summary of service in the Lithuanian armed forces

Military forces	Frequency (N)	Percent	Valid percent	Cumulative percent
Land forces	172	84.3	86.0	86.0
Air force	4	2.0	2.0	88.0
Navy	24	11.8	12.0	100.0
Total	200	98.0	100.0	
Missing system	4	2.0		
Total	204	100.0	100.0	

(N = 24) served in the Navy; and 2% (N = 4) of the respondents did their service in the Air Force.

In this survey, the participants' military degree was also disclosed. More than half of them 68.6% (N = 136) were senior officers; about one fifth of them 21.6% (N = 48) were junior officers; and only 9.8% (N = 20)—enlisted grade. Moreover, 68.6% (N = 136) of the respondents had their subordinates, and 31.4% (N = 68) did not have any. The numbers of descriptive analysis imply that research participants are fairly knowledgeable about managerial and structural characteristics of their specific working environments.

9.3.2 Data Analysis

The SPSS 20 version software and the structural equation modeling (SEM) by using IBM SPSS the analysis of moment structures AMOS 24 program were used (http://www-03.ibm.com/software/products/en/spss-amos). Before SEM analysis all items were tested. Their formed unities and measurement properties (construct validity and reliability) were checked. In this case, for the primary constructs the Principal Component Analysis (PCA) was done (Anderson and Gerbing 1988). As for SEM analysis, the testing of measurement properties of the instruments was done simultaneously with the hypotheses testing.

9.4 Results and Discussions

At the start of the analysis the prominent characteristics of leader behavior among servicemen were examined, and the sum of items of every subscale was calculated. In Table 9.5, we can see the respondents have best evaluated qualities such as Representation (3.79), Initiation of Structure (3.60), Demand Reconciliation (3.55), Role Assumption (3.53), and Tolerance and Freedom (3.50).

The reliability results for the commander-leader behavior description components showed, that Cronbach's alfa coefficients variation interval is from 0.70 till 0.95 for constructed measurable factors scale (Fig. 9.1). These results let us continue with SEM analysis, which was performed by AMOS. The computed analyses and get results are presented in this article below.

A. *Factor analysis of indicators predictive relationship with leadership style in Lithuania armed forces*

For GSEM analysis was included labeled constructs, which were proved by theory. AMOS software test for model fit in confirmatory factor analysis and the unstandardized and standardized model-fits were generated for leadership style (LS). The computed results for indicator variables (measurable factors) and LS as a

Table 9.5 Numerical characteristics and validity of the commander-leader behavior description

Label	Measurable factors	Number of variables	Calculated min	Calculated max	Mean	Standard deviation	Cronbach's alfa for scale
AB1	Superior orientation	10	2.10	4.40	3.46	0.54	0.700
AB2	Integration	5	1.00	4.80	3.31	0.89	0.901
AB3	Predictive accuracy	5	1.40	4.80	3.35	0.74	0.852
AB4	Product emphasis	10	1.80	4.70	3.21	0.63	0.814
AB5	Consideration	10	1.60	4.60	3.41	0.81	0.893
AB6	Role assumption	10	2.20	4.80	3.53	0.68	0.824
AB7	Tolerance and freedom	10	1.20	4.90	3.50	0.82	0.921
AB8	Initiation of structure	10	1.00	4.80	3.60	0.73	0.899
AB9	Persuasiveness	10	1.20	4.90	3.38	0.94	0.949
AB10	Tolerance of uncertainty	10	1.50	4.80	3.24	0.57	0.762
AB11	Demand reconciliation	5	1.40	5.00	3.55	0.84	0.837
AB12	Representation	5	1.00	5.00	3.79	0.77	0.847

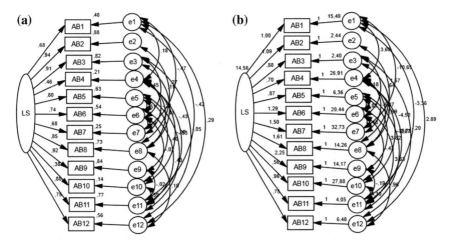

Fig. 9.2 AMOS 24 software generated diagram for predictive effective leadership style in Lithuania armed forces: **a** standardized indicators loadings; **b** unstandardized indicators loadings. Model fit for a predicted model structure was not proved

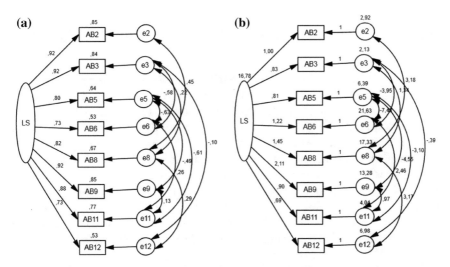

Fig. 9.3 AMOS 24 software generated diagram for predictive effective leadership style in Lithuania armed forces: **a** standardized indicators loadings; **b** unstandardized indicators loadings. Model fit for a predicted model structure was proved (Hasilová 2014)

latent variable are presented in graphs (Figs. 9.2 and 9.3). More over the model evaluation for fit summary is also presented in Table 9.6.

The AMOS software tested twelve predictive indicators by the measurement and structural model, which was presented in Fig. 9.1a above and minimum was achieved for the constructed model. The analysis was based on the goodness of fit

Table 9.6 Estimates from latent variable to indicator variables

Measurable factors	Label in model	Standardized regression weights estimate		Squared multiple correlations	
		Not proved model	Proved model	Not proved model	Proved model
Superior orientation	AB1	0.68	–	0.48	–
Integration	AB2	0.94	0.92	0.88	0.85
Predictive accuracy	AB3	0.91	0.92	0.82	0.84
Product emphasis	AB4	0.46	–	0.21	–
Consideration	AB5	0.80	0.80	0.63	0.64
Role assumption	AB6	0.74	0.73	0.54	0.53
Tolerance and freedom	AB7	0.68	–	0.25	–
Initiation of structure	AB8	0.82	0.82	0.68	0.57
Persuasiveness	AB9	0.92	0.92	0.85	0.85
Tolerance of uncertainty	AB10	0.38	–	0.14	–
Demand reconciliation	AB11	0.88	0.88	0.77	0.77
Representation	AB12	0.75	0.73	0.56	0.53

statistics. It become obvious that the overall model fit does not appear quite good. The estimated $\chi 2$ of 195.629 (df = 33) the null hypothesis of a good fit is rejected at the 0.05 level (p < 0.000). Moreover, the estimated Root Mean Square Error of Approximation (RMSEA) of 0.156 is too large, and it rejects the null hypothesis of a good fit at the 0.05 level (p < 0.000). Also that the model fit to the data is problematic the estimate for the Comparative Fit Index (CFI) of 0.935 indicates.

The poor fit model results showed that model proposed in Fig. 9.1 need to be modified. Therefore to identify the variables (indicators) that have significant relationships with the leadership style latent variable (LS) the factor analysis was carried out. The calculations obtained from AMOS (factor loadings) are associated with arrows from LS (latent variable) to the indicator variables are presented as in Table 9.6 as in Figs. 9.2a and 9.3a.

By well-known rule, the indicators to be significant must have the loadings of 0.7 or higher on the latent variable. The results, which are highlighted in Table 9.6, clear shows who of indicators have to be excluded as not having significant predictive relationship with the leadership style in the Lithuania armed forces: (AB1) Superior Orientation (r = 0.68, α = 0.48); (AB4) Product Emphasis (r = 0.46, α = 0.21), (AB7) Tolerance and Freedom (r = 0.68, α = 0.25); (AB10) Tolerance of Uncertainty (r = 0.38, α = 0.14). On the other hand, the indicator variables, which were identified with loadings of 0.7 or higher, were used to modify the one factor model (Fig. 9.2), which was then retested using Confirmatory Factor Analysis.

B. *Confirmatory factor analysis of indicators predictive relationship with leadership style in Lithuania armed forces*

The confirmatory factor analysis helped to test the importance of latent variable (LS) and its indicators (AB2, AB3, AB5, AB6, AB8, AB9, AB11 and AB12). The analysis by AMOS software generated the unstandardized and standardized model fits. The computed results are shown in the Fig. 9.3a, b. The estimates of standardized regression weights and squared multiple correlations for proved model are in the Table 9.6.

In the modified model eight predictive indicators were tested and minimum was achieved. As it is highlighted in Table 9.7 overall model fit quite well to the data. Based on the goodness of fit statistics.

$\chi 2$ of 6,260 (df = 10) was estimated and the null hypothesis of a good fit at the 0.05 level (p < 0.793) was not rejected. Additionally, the estimated Root Mean Square Error of Approximation (RMSEA) of 0.000 (p < 0.793) and the Comparative Fit Index (CFI) of 1.00 indicate that the modified model fits the data well. The proved model fit calculations are presented in Table 9.7.

The maximum likelihood (ML) estimates entailing the standardized regression estimates and squared multiple correlations are summarized in the Table 9.6. All the estimated loadings for the indicators are significant and have higher of 0.7 loadings on the latent variable (Table 9.6 and Fig. 9.3a). The implied correlations for all the variables obtained from AMOS graphics are presented in Table 9.8.

The standardized regression estimates which represent the factor loadings (i.e. indicator coefficients) associated with the arrows from latent variables to the respective indicator variables in the modified model are highlighted in the Table 9.6 as proved model measures. The standardized regression estimates (indicator loading coefficients) for: AB2 (i.e.—maintains a closely knit organization; demonstrates intermember relations) is 0.92 ($\alpha = 0.85$); AB3 (i.e.—displays foresight and ability to predict outcome accurately) is 0.857 ($\alpha = 0.735$); AB5 (i.e.—regards the comfort, well-being, status, and contributions of followers) is 0.80 ($\alpha = 0.64$); AB6 (i.e.— actively exercises the leadership role rather that surrendering leadership to others) is 0.73 ($\alpha = 0.53$); AB7 (i.e.—allows followers scope for initiative, decision and action) is 0.92 ($\alpha = 0.85$); AB8 (i.e.—defines own role, and lets followers know what is expected) is 0.92 ($\alpha = 0.85$); AB9 (i.e.—uses persuasion and argument effectively; exhibits strong convictions) is 0.92 ($\alpha = 0.85$); AB11 (i.e.—reconciles conflicting demands) is 0.92 ($\alpha = 0.85$); AB12 (i.e.—speaks and acts as the representative of the group and indicates its importance in an organization) is 0.92 ($\alpha = 0.85$).

Table 9.7 Modified model fit summary

CMIN

Model	NPAR	CMIN	DF	P	CMIN/DF
Default model	26	6.260	10	0.793	0.626
Saturated model	36	0.000	0		
Independence model	8	1720.988	28	0.000	61.464

RMR, GFI

Model	RMR	GFI	AGFI	PGFI
Default model	0.306	0.992	0.973	0.276
Saturated model	0.000	1.000		
Independence model	20.439	0.224	0.002	0.174

Baseline comparisons

Model	NFI Delta1	RFI rho1	IFI Delta2	TLI rho2	CFI
Default model	0.996	0.990	1.002	1.006	1.000
Saturated model	1.000		1.000		1.000
Independence model	0.000	0.000	0.000	0.000	0.000

Parsimony-adjusted measures

Model	PRATIO	PNFI	PCFI
Default model	0.357	0.356	0.357
Saturated model	0.000	0.000	0.000
Independence model	1.000	0.000	0.000

NCP

Model	NCP	LO 90	HI 90
Default model	0.000	0.000	5.093
Saturated model	0.000	0.000	0.000
Independence model	1692.988	1560.707	1832.631

FMIN

Model	FMIN	F0	LO 90	HI 90
Default model	0.031	0.000	0.000	0.025
Saturated model	0.000	0.000	0.000	0.000
Independence model	8.478	8.340	7.688	9.028

RMSEA

Model	RMSEA	LO 90	HI 90	PCLOSE
Default model	0.000	0.000	0.050	0.950
Independence model	0.546	0.524	0.568	0.000

(continued)

Table 9.7 (continued)

AIC				
Model	AIC	BCC	BIC	CAIC
Default model	58.260	60.672	144.531	170.531
Saturated model	72.000	75.340	191.452	227.452
Independence model	1736.988	1737.731	1763.533	1771.533

ECVI				
Model	ECVI	LO 90	HI 90	MECVI
Default model	0.287	0.305	0.331	0.299
Saturated model	0.355	0.355	0.355	0.371
Independence model	8.557	7.905	9.244	8.560

HOELTER		
Model	HOELTER 0.05	HOELTER 0.01
Default model	594	753
Independence model	5	6

Table 9.8 Implied correlations for all variables in the modified modell

	LS	AB8	AB9	AB12	AB11	AB6	AB5	AB3	AB2
LS	1.000								
AB8	0.818	1.000							
AB9	0.921	0.754	1.000						
AB12	0.730	0.710	0.672	1.000					
AB11	0.878	0.718	0.833	0.640	1.000				
AB6	0.731	0.599	0.674	0.534	0.728	1.000			
AB5	0.797	0.652	0.619	0.582	0.524	0.323	1.000		
AB3	0.919	0.802	0.847	0.643	0.807	0.516	0.733	1.000	
AB2	0.923	0.854	0.850	0.673	0.810	0.675	0.736	0.848	1.000

9.5   Conclusion

Though this study there were used SEM analysis to test twelve indicators of leadership style in the Lithuanian army forces. The findings from this study have shown that only some of them don't provide the strong influence on analyzed leader style: AB10—is able to tolerate uncertainty and postponement without anxiety or upset ($r = 0.38$, $\alpha = 0.14$); AB4—applies pressure for productive output ($r = 0.46$, $\alpha = 0.21$); AB1—is striving for higher status ($r = 0.68$, $\alpha = 0.48$); AB7—allows followers scope for initiative, decision and action ($r = 0.68$, $\alpha = 0.25$).

The received features obtained after the confirmatory factor analysis for modified model helped to provide the basis for effective leadership. In this analysis were tested the importance of indicators that had the loadings of 0.7 or higher on the latent variable and the higher importance were achieved for: AB2—maintains a closely knit organization; demonstrates intermember relations ($r = 0.92$, $\alpha = 0.85$); AB9—uses persuasion and argument effectively; exhibits strong convictions ($r = 0.92$, $\alpha = 0.85$); AB3—displays foresight and ability to predict outcome accurately ($r = 0.88$, $\alpha = 0.85$).

Finally, this study concluded that good leaders, which are appreciated by soldiers in the Lithuanian army forces could be perceived as those who take initiatives under rapidly changing and uncertain conditions and are capable of making the right decisions. Moreover, our present-day army is in need of those leaders who could take risks and would not be afraid to take an initiative; who would be able to collect information and process it; who would be loyal, reliable, self-confident, and capable of raising a degree of confidence in those around.

Acknowledgements The second author was supported within the project for "Development of basic and applied research" developed in the long term by the departments of theoretical and applied bases of the FMT UoD (Project code: "VYZKUMFVT" (DZRO K-217)) supported by the Ministry of Defence of the Czech Republic.

References

Anderson, J. C. and Gerbing D.W. (1988): Structural Equation Modeling in Practice: A Review and Recommended Two - Step Approach. *Psychological Bulletin*, 103, 1988, 411–423.

Halpin, Andrew W., and Winer, B. James. A Factorial Study of Leader Behavior Descriptions. Ohio: Bureau of Business Research, The Ohio State University, 1956.

Hasilová, K. (2014) Iterative Method for Bandwidth Selection in Kernel Discriminant Analysis. In: 32nd International Conference Mathematical Methods in Economics MME2014. Olomouc: Palacký University, Olomouc, 2014, 263–268.

Higgs, M and Rowland, D. (2005). All Changes Great and Small: Exploring Approaches to Change and its Leadership. *Journal of Change Management*, 5(2), 121–151.

Hoskova-Mayerova, S. (2016). Leadership – training of military specialists in particular disciplines focused on mathematical modelling. *New Trends and Issues Proceedings on Humanities and Social Sciences*. [Online]. 05, pp 199–204. Available from: http://www.prosoc.eu.

Hošková-Mayerová, Š. (2017) Education and Training in Crisis Management. In: *The European Proceedings of Social & Behavioural Sciences EpSBS, Volume XVI*. Future Academy, 2017, p. 849–856. ISBN 2357–1330.

IBM SPSS AMOS. Available at: http://www-03.ibm.com/software/products/en/spss-amos.

Lithuanian Military Doctrine, (2016). Available at: https://www.google.lt/#q=Lietuvos+karine +doktrina+2016.

Leader Behavior Description Questionnaire – Form XII Self. Available at: https://cyfar.org/sites/ default/files/LBDQ_1962_Self_Assessment.pdf.

Stogdill, R. (1974). Handbook of leadership: A survey of theory and research. New York: Free Press, p. 411.

Stoner, J., Freeman, R., Gilbert, D. (2000). Vadyba. Kaunas: UAB Poligrafija ir informatika.

The military strategy. Nr. V-1305 (2012). Available at: http://www.kam.lt/download//the_ military_strategy_(3).doc.

Chapter 10
Why We Need Mathematics in Cartography and Geoinformatics?

Václav Talhofer

Abstract Cartography and geoinformatics are technical-based which deal with modelling and visualization of landscape in the form of a map. For the given subjects, mathematics is necessary for understanding of many procedures that are connected to modelling of the Earth as a celestial body, to ways of its projection into a plane, to methods and procedures of modelling of landscape and phenomena in society and visualization of these models in the form of electronic as well as classic paper maps. Not only general mathematics, but also its extension of differential geometry of curves and surfaces, ways of approximation of lines and surfaces of functional surfaces, mathematical statistics and multi-criterial analyses seem to be suitable and needful. Underestimation of the significance of mathematical education in cartography and geoinformatics is inappropriate and lowers competences of cartographers and geoinformaticians to solve problems.

Keywords Cartography · Geoinformatics · Mathematics · Education

10.1 Introduction

Over the past couple of years Europe and the whole world have undergone extensive changes. These changes are based on the development of science, technics and economic growth linked with dynamics of innovation of the past decades. This implies more requirements for training and development of human resources. It also entails continuous refining of one's knowledge, skills and gaining new qualifications. Another key phenomenon that already influences the level of education and is likely to do so more and more is a rapid development of information and communication technologies. We have to pay increased attention to the training of human resources. In this relation, one of frequently discussed questions, and

V. Talhofer (✉)
Faculty of Military Technology, Department of Military Geography and Meteorology,
University of Defence, Kounicova 65, 662 10 Brno, Czech Republic
e-mail: vaclav.talhofer@unob.cz

© Springer International Publishing AG 2017
Š. Hošková-Mayerová et al., *Mathematical-Statistical Models and Qualitative Theories for Economic and Social Sciences*, Studies in Systems, Decision and Control 104,
DOI 10.1007/978-3-319-54819-7_10

surely not only in the Czech Republic, is the question of teaching mathematics on all levels of education. The society is still divided into two implacable groups— opponents and supporters of mathematical education.

The former group considers mathematics to be too difficult and useless for everyday life. They use arguments as unnecessary "cramming" of formulas, troubling students, etc. Unfortunately, they do not see and understand (or do not want to see) the main sense of mathematics, that lies especially in forming logical thinking of every individual.

What is really regrettable is the fact that also a lot of medially famous people belong to this group. Not only they are not ashamed of their bad school results in mathematics, on the contrary, they consider appropriate to "boast" about them in mass media. In various interviews or shows they present inglorious stories of their school years and they contribute to the opinion that to get bad marks of mathematics is "cool". Thus the awareness of the necessity and need of mathematics keeps worsening even more in the general public. This "normal behavior", however, has catastrophic consequences for the society. Few fields of human activity nowadays can live without the basics of mathematics.

That is the reason why supporters of necessity of a classic mathematical education patiently keep explaining the positives of mathematics and negative consequences of its restriction at all levels of education. Let us ask a question: How is it, then? Do we really need mathematics? Why do we need it?

When looking for an answer, let us start with beautiful words by Professor Jaroslav Nešetřil, the director of Institute of Theoretical Informatics of Faculty of Mathematics and Physics of Charles University (Nešetřil 2011).

> Mathematics is like salt. There is a nice fairy tale about it ("The Salt Prince" – author's note). Do you eat salt? You do not. A lot of people will tell you they do not salt. But if there were no salt, we would end up just like in the fairy tale. Salt is not spice, it is essence. If children do not learn mathematics, they will be going in the world where there are numbers flashing everywhere but they will not understand. They will not have the opportunity to understand the scheme of things based on mathematics. Mathematics is the essence. As parents and state taking care of the future, we have to make a decision what we want.

First of all, it is necessary to say that the society needs generally educated people. Without mathematics, a person is not generally educated. Such person is excluded not only from mathematics and physics, but also in other natural sciences and the humanities. Maybe with the exception of sciences of language and philosophy.

The same situation applies also in university education of future cartographers and geoinformaticians. Once again words by Prof. Nešetřil (2011):

> Mathematics at university is different from the one on secondary schools. Mathematics has many forms, sociologists at philosophical faculty are lectured something else than is e.g. mathematics for informatics. The essence of higher mathematics is a logical construction which is close to philosophy.

Geoinformaticians and cartographers surely are not any special kind that needs totally different mathematics from other university students. Yet it is necessary to think which parts of mathematics are important and key for these fields, but in fact, they do not get so much space as in other fields. This contemplation is desirable also in respect to the prepared modifications of Body of Knowledge for Cartography, which is one of tasks of Commission on Education and Training of the International Cartographic Association (ICA CET) or in respect to recent up-dating of Body of Knowledge for geoinformation science and technology (DiBiase et al. 2006).

For the given fields mathematics is necessary for understanding a range of procedures which relate to modelling the Earth as a celestial body, to ways of its projection into a plane, to methods and procedures of modelling landscape and processes in the society and visualization of these models in the form of electronic as well as classic paper maps. The following text depicts only some problems that —in my opinion—illustrate the width of the problem in these fields and which require mathematical foundations to be correctly understood. Without these foundations, solution of special problems may not only be unsuitable, but it can lead to mistakes and faults. It is not my ambition to make an overview of all fields, for which mathematics is the basic fundament, but I would like to contribute to the discussion about the importance of mathematics at university studies with my bit.

10.2 Earth as a Celestial Body and Place Where We Live

All life takes place on Earth and its close surroundings. One of the basic man's skills has always been orientation in the area where they live. Without this orientation, prehistoric man would not have remembered where their cave was, where their hunting grounds were and where it was possible to hide. Our current territory is larger and covers basically the whole surface of Earth and its relatively close surroundings (Lauermann and Rybanský 2002). Man's basic skill, however, remains unchanged. Only the ways how we orient in the country, which tools we use for that and how often and for what purpose we need to apply this skill, change. While a prehistoric man used their detailed knowledge of their surroundings to orient in the country, a nowadays man helps themselves also with a map, navigation using global navigation satellite systems (GNSS). The country may not be discovered only in a direct contact with it, but it can be studied again with the help of maps, web map services or on-line maps accessible through mobile phones. They can very easily find a place where they are, using GNSS. At the same time, they assume that everything they find on the above-mentioned means absolutely correctly corresponds to reality. To make this happen, it is necessary that everybody

V. Talhofer

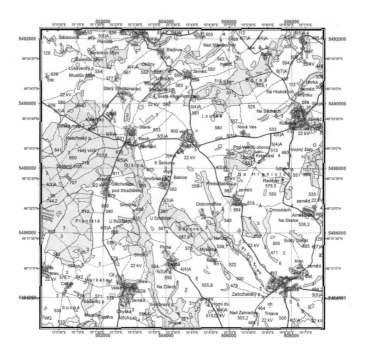

Fig. 10.1 Example of a graphic output of GIS database

who participates on the definition and creation of the given products knows exactly what they do, why they do it and if they use the correct procedures.

The Earth as a celestial body has an irregular shape that is usually substituted by a simpler, mathematically definable body. This shape is a rotational ellipsoid, or a sphere. Positions of objects and landscape phenomena are displayed on the surface of an ellipsoid or sphere. The question is what the position really means and how it is expressed. Objects and phenomena in landscape are modelled as spatial objects that have their own geometry (position and shape), topologic relations between each other (neighborhoods, overlays, connections, etc.) and they have their features that describe characteristics of the given object or phenomenon (kind of woods, construction material of roads, a level of pollution of a watercourse, value of an average temperature at the given meteorological station, etc.). Modern systems of work with this information (geographic information systems, GIS) have at their disposal strong tools for an analysis of relations and features of the saved objects, mostly based on mathematical, statistical and set functions.

The result of work of geoinformaticians and cartographers is usually available to the users in the visual form as a paper map or their electronic version. A simple output may look like in the picture (Fig. 10.1). The following picture (Fig. 10.2) then shows what information is displayed on the graphic output.

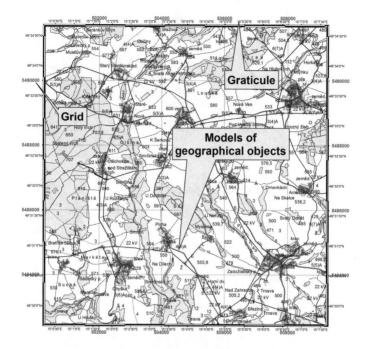

Fig. 10.2 Displayed information on the simple graphic output

10.3 Positional Basics of Maps

10.3.1 Spatial Coordinates

Coordinate systems on referential bodies, usually ellipsoids, are used to express a position. In the past, a lot of ellipsoids that substituted the Earth only in a certain space in a way so that they would maximally be adjacent to its surface, were defined. Only with the development of satellite technology, there occurred the need of definition of so-called global ellipsoids substituting the Earth as a whole. The original ellipsoids, however, are still used because they are still often a part of definitions of state coordinate systems. Ellipsoids differ in the position of their centers against the planetary system, tilting their main and side half-axis and their dimensions. Professionals in our fields, however, must be absolutely sure of the exact definition of the given system, i.e. the used referential body and its features.

To express a position on or above the ellipsoid, geographic coordinates, i.e. geographic longitude λ and latitude φ (Fig. 10.3) are used most frequently. However, also geocentric systems that relate to the given ellipsoid or sphere (Fig. 10.4) are very often used. As real points are not right on the surface of the ellipsoid, but they are in a certain height above it (ellipsoidal height, simply said they lie on a normal to the surface of the ellipsoid in the given point in distance H_{el}, it is often

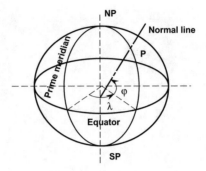

Fig. 10.3 Geographic coordinates on an ellipsoid

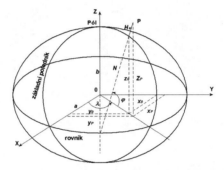

Fig. 10.4 Geocentric coordinates of an ellipsoid

necessary to work also with this height. This height is not negligible; it may reach also tens of thousands of kilometers in case of height of trajectories of satellites of global navigation satellite systems (GNSS).

For successful calculations, it is necessary to know transformations between geographic and geocentric coordinates and height of the given point above the referential ellipsoid H_{el}. For calculation of geocentric coordinates, these relations are used:

$$x = (N + H_{el}) \cos \varphi \cos \lambda,$$
$$y = (N + H_{el}) \cos \varphi \sin \lambda,$$
$$z = \left[N\left(1 - e^2\right) + H_{el} \right] \sin \varphi,$$

and for reverse calculation of geographic coordinates φ and λ equations that are necessary to be solved also with iterations with an appropriate choice of limit increase:

$$\lambda = arctg\left(\frac{y}{x}\right)$$

$$\varphi_0 = arctg\left(\frac{z}{\sqrt{x^2+y^2}}\frac{1}{1-e^2}\right)$$

$$N_0 = \frac{a}{\sqrt{1-e^2\sin^2\varphi_0}}$$

$$H_{el_0} = \frac{x}{\cos\varphi_0\cos\lambda} - N_0 = \frac{y}{\cos\varphi_0\sin\lambda} - N_0$$

$$\varphi_i = arctg\left[\frac{z}{\sqrt{x^2+y^2}}\frac{N_{i-1}+H_{el_{i-1}}}{N_{i-1}(1-e^2)+H_{el_{i-1}}}\right]$$

$$N_i = \frac{a}{\sqrt{1-e^2\sin^2\varphi_i}}$$

$$H_{el_i} = \frac{x}{\cos\varphi_i\cos\lambda} - N_{i-1} = \frac{y}{\cos\varphi_i\sin\lambda} - N_{i-1}$$

Moreover, ellipsoids are also described by radius of curvature in a certain point (meridian M and radius of curvature in prime vertical N) so that it was possible to describe the surface of an ellipsoid in parts. These radiuses may be expressed by equations (Grafarend and Krumm 2006; Talhofer 2007):

$$M = \frac{a(1-e^2)}{(1-e^2\sin^2\varphi)^{3/2}}, N = \frac{a}{(1-e^2\sin^2\varphi)^{1/2}},$$

where a is the size of semi-major axis, e is the first eccentricity and φ is a geographic latitude of the given point.

While roughly till the seventies of the last century, especially geographic coordinates were used for various calculations, in the last decades, geocentric systems have been used more and more, particularly in GNSS technologies or with transformations of coordinates between individual systems. A general transformation between two systems is often solved with a standard 7-parameter similarity transformation (Fig. 10.5).

This transformation is calculated by equation:

$$\begin{bmatrix} x_n \\ y_n \\ z_n \end{bmatrix} = \begin{bmatrix} dx \\ dy \\ dz \end{bmatrix} + \begin{bmatrix} m & 0 & 0 \\ 0 & m & 0 \\ 0 & 0 & m \end{bmatrix} \cdot \begin{bmatrix} 1 & r_z & -r_y \\ -r_z & 1 & r_x \\ r_y & -r_x & 1 \end{bmatrix} \cdot \begin{bmatrix} x \\ y \\ z \end{bmatrix},$$

where (x, y, z) are geocentric coordinates in the original system, (dx, dy, dz) are translations of centers in individual axes, (r_x, r_y, r_z) are rotations of the original system around the axes and m is a change of the size of the ellipsoid.

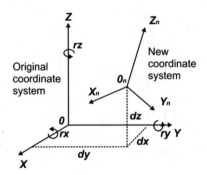

Fig. 10.5 7-parameter similarity transformation

10.3.2   Plane Coordinates

The resulting map image is always projected in a plane, which can be a plane of a map or a computer screen, a navigation system or a mobile phone. With respect to reality, a plane projection is always deformed in lengths, surfaces as well as angles. Already ancient Greeks dealt with a transformation of an ellipsoid or a sphere. To perform such transformation, they used primarily geometric projections, whose procedures were later also expressed in equations. Nowadays, it is based on a geometric image of transformation, but the basis is in using the solution of differential equations with defined initial conditions. According to Talhofer (2007) it is possible to schematically express the basis of transformation of spatial coordinates into plane ones by the following picture (Fig. 10.6):

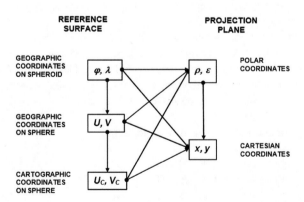

Fig. 10.6 Types of projections

The basis for derivation of the projecting equations is the knowledge of:

- curvatures of referential surface,
- length elements of meridians, parallels and general lines,
- and general laws of distortion of lengths, areas and angles.

With the help of differential geometry of curves and planes, it is possible to define length distortion of an element of a line between differentially close points P and Q on a referential ellipsoid which has geographic azimuth A. If general projection equations are considered (Talhofer 2007):

$$x = f(\varphi, \lambda),$$
$$y = f(\varphi, \lambda).$$

then it is possible to define a generally length distortion as $m = dS/ds$ as a scale (Figs. 10.7 and 10.8).

Differential element of the line on spheroid ds can be expressed as:

$$ds = \sqrt{M^2 d\varphi^2 + N^2 \cos^2 \varphi d\lambda^2}$$

and corresponding element on the plane dS is:

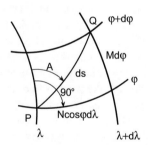

Fig. 10.7 Element of a line on the spheroid

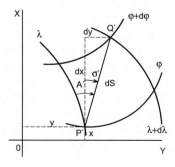

Fig. 10.8 Element of a line on a plane

$$dS = \sqrt{dx^2 + dy^2}.$$

Values x and y are a function of two variables (φ, λ), thus their derivation is necessary to be expressed as a total differential of functions:

$$dx = \frac{\partial x}{\partial \varphi} d\varphi + \frac{\partial x}{\partial \lambda} d\lambda$$

$$dy = \frac{\partial y}{\partial \varphi} d\varphi + \frac{\partial y}{\partial \lambda} d\lambda$$

The resulting equation for calculating of the distortion of lines:

$$m^2 = \frac{E}{M^2} \cos^2 A + \frac{F}{MN \cos \varphi} \sin 2A + \frac{G}{N^2 \cos^2 \varphi} \sin^2 A$$

where E, F, and G are Gauss's symbols expressing differential forms of a plane. Similarly generally, also planar and angle distortions are derived. The projection into a plane itself is then defined as equidistant, equivalent, or conformal. Equidistant projection keeps lengths in a certain direction, equivalent projection keeps the size of planes, and conformal projection keeps angles and shapes of objects. Projection equations are derived based on set conditions for the given distortion which are expressed as differential equations. This equation is solved by a definite or indefinite integral with setting of the initial conditions. For instance, it is possible to define an equidistant conical projection in meridians with two undistorted (standard) parallels. The basic equation is solved by integration of the differential equation expressing length distortion in meridians; initial condition of two undistorted parallels is then solved like a set of two nonlinear equations expressing length distortion in parallels. From the point of mathematics, knowledge of derivations and integrals, as well as a solution of nonlinear equations is necessary. Specific ways of solutions of the individual types of projections can be found e.g. in Grafarend and Krumm (2006) or Talhofer (2007).

Together with knowledge of rules of the individual projections, it is always necessary to choose an optimal projection for the given purpose. Various methods are used to assess the suitability of projections. One of the simple and quite old methods of evaluation is so-called Tissot indicatrix, which is an image of a unit circle on an ellipsoid after its projection to a plane. The following picture shows three projections of the Earth using Tissot indicatrix (Fig. 10.9).

It is necessary to pay attention to the choice of a suitable type of projection, especially nowadays when a large number of users use web map services (WMS) like Microsoft® Bing™ Maps, Google Maps™, ESRI® ArcGIS[SM] Online, etc. A lot of known projections are named after their authors who sometimes derived even more of them. Thus there can occur confusion and subsequent mistakes as pointed out by e.g. Office of Geomatics (2014) when using Mercator and WEB

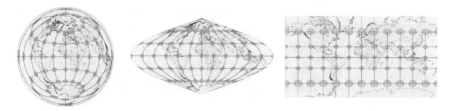

Fig. 10.9 Projection of the Earth, from *left* to *right*, equidistant azimuthal, Sanson's pseudo-cylindrical and Mercator conformal cylindrical projection (using data ESRI 2013)

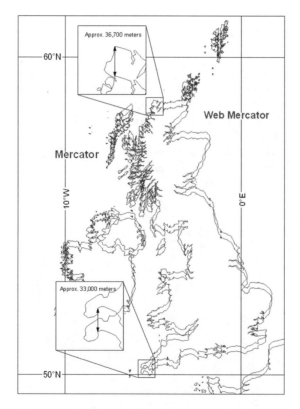

Fig. 10.10 Visual Example of Overlaying Ellipsoid Mercator (*blue*) and Web Mercator (*red*), graticule is WGS 84 (Office of Geomatics 2014)

Mercator projections. The mentioned material also contains a sample of differences in the position of borders of Great Britain in various projections (Fig. 10.10).

10.4 Two-Dimensional and Three-Dimensional Geometry

Positional basics create only conditions for localization of the modelled objects and phenomena in the countryside. The position of the objects themselves is then given with respect to the used data format.

10.4.1 Vector Models

For vector format, the position of objects is set as a point (point objects, points), set of successively adjacent line segments with various initial and ending point (line objects, lines) or a set of successively adjacent line segments with the same initial and ending point (planar objects, polygons). An example of the given models can be seen in the picture (Fig. 10.11).

Thanks to topology, each object usually has a defined relation to its environment. A record of its properties, which are saved in a database as table of properties, are an inseparable part of the object.

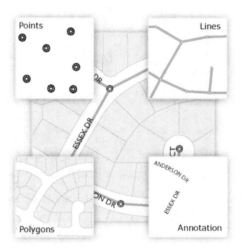

Fig. 10.11 Models of objects in vector format (*source* ESRI 2013)

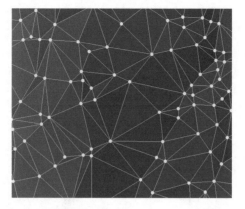

Fig. 10.12 Visualization of a model of surface using TIN method

From the point of spatial dimension, objects may be modelled as two-dimensional (2D), three-dimensional (3D) as well as four-dimensional (4D), where the fourth dimension is usually time. From the point of view of coordinates, it is possible to use Cartesian coordinates, as well as spherical, while it is assumed that localization is maximally accurate from the point of view of the required characteristics of the resulting model. This way, especially discrete objects are modelled.

In the vector format, also continuous objects are modelled, usually as functional surfaces. One of the basic methods used with the mentioned models is so-called method TIN (Triangular Irregular Network, Fig. 10.12). Further information of usage of TIN method can be found in ESRI (2013).

With respect to the characteristics of vector models, it is necessary to understand their projection in various map projections. The picture (Fig. 10.10) shows deformation of the coastline of the British Isles due to different projections. This influence is even more visible in a map image of the Earth used in map portals mentioned in paragraph *Plane coordinates*. The following pictures (Figs. 10.13 and 10.14) show significant differences in projected size of continents, even though the same data are used.

A cartographer and a geo-informatics specialist must then know that it is not only about maximum possible accuracy of setting the position of points, lines and areas, but that visualization of this position is strongly dependent on the used projection.

10.4.2 Raster Models

Qualitatively different way of modelling objects and phenomena is the use of a raster format. Raster format is based on a regular grid, usually square, which

Fig. 10.13 Mercator conformal cylindrical projection

completely fill the given area—raster dataset. The element of the grid is called a pixel. Within a raster dataset each pixel has its unique value which may be for instance existence of an object (field, house, forest) or a value (value in the given classification scale, value of a chosen function, etc.). If discrete objects are modelled by a raster model, or raster dataset models a classification of space, then the value of the given pixel is assigned to the whole pixel and it is usually an integer constant. In places where there are no objects, the pixel values are marked as empty or NO DATA. If functional surfaces (elevation field, atmospheric pressure, etc.) are modelled, the pixel value—which is either measured or calculated using a suitable interpolation function from the measured values in the defined surroundings of the pixel—is adjacent to its center point (see Fig. 10.15).

Raster datasets are localized in same coordinate systems as vector, i.e. in Cartesian as well as spherical. The accuracy of localization of an object or a phenomenon, however, is given by the size of the pixel, within which the accuracy does not increase. The influence of the pixel size on the accuracy of saved data and possibly on calculations of geometric characteristics is shown in the following picture (Fig. 10.16).

Fig. 10.14 Equivalent Lambert cylindrical projection

Value applies to the center point of the cell
For certain types of data, the cell value
represents a measured value at the
center point of the cell. An example is
a raster of elevation

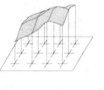

+ 315	+ 319	+ 321	+ 323
+ 317	+ 323	+ 328	+ 326
+ 313	+ 318	+ 325	+ 323

Value applies to the whole area of the cell
For most data, the cell value represents
a sampling of a phenomenon, and the
value is presumed to represent the
whole cell square.

50	45	40	35
35	40	35	25
20	25	30	20

Fig. 10.15 Variants of pixel values in raster models (ESRI 2013)

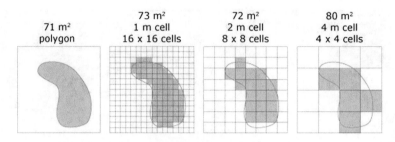

Fig. 10.16 Influence of pixel size on accuracy of saved data in raster models (ESRI 2013)

10.4.3   Work with Geometry

The record of position of the modelled objects and phenomena in the countryside is a basis to be able to answer the question "WHERE *something* is?". *Something* can be substituted by e.g. a municipality, a river, a mountain range, a castle, etc. However, the answer to question WHERE is usually not the end. Most users want to know "HOW big is it?", "HOW far or high is it?", "WHAT is in the surroundings, under or above the object?", etc. As all objects have their localization, answers to the given questions are relatively easy. It is only necessary to know methods of calculation or analysis of spatial relations in the given models.

Most answers to the mentioned and similar questions is based on calculations of distance, on an analysis of the mutual relation of two objects, on the system of work with geometry and on rules which are used in the given system.

10.4.3.1   Geometric Problems in Vector Models

If *vector models* are worked with, calculations are based on plane or spatial trigonometry. Everybody who works in this field should be sure of using basic formulas and theorems. For illustration, it is possible to use the following picture (Fig. 10.17) and generalized equations that solve some problems for plane polygon. The picture shows a triangle *IJK*, whose vertexes have known coordinates in Cartesian system, generally used in mathematics.

If the coordinates of vertexes (x_i, y_i) are known, it is not a problem—using the Pythagoras' theorem—to calculate their distances d_{ij}, directions t_{ij} of their sides and from them also vertex angles. This is taught even on basic schools. What is not normally taught are equations for calculations of side lengths, their directions and vertex angles of polygons with the addition of calculations of circumference, surfaces and centroids. However, in geoinformatics, these problems are very frequent because firstly, objects are usually of an irregular shape (footprint), secondly, it is

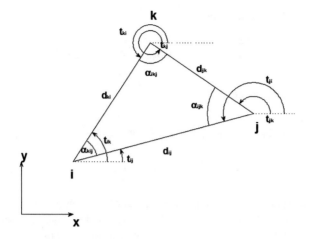

Fig. 10.17 Plane triangle

often necessary to know their above-mentioned spatial characteristics. The following formulas usable for polygons with n vertexes are usually used for calculations:

- calculation of surface of a polygon:

$$P = 0.5 \sum_{i=1}^{n} (x_i + x_{i+1})(y_{i+1} - y_i)$$

or

$$P = 0.5 \sum_{i=1}^{n} x_i(y_{i+1} - y_{i-1});$$

- calculation of side lengths:

$$d_{u,v} = \sqrt{(x_v - x_u)^2 + (y_v - y_u)^2}, \forall u = i, j \wedge \forall v = j, k \wedge u \neq v;$$

- calculation of side directions:

$$t_{u,v} = \arctan \frac{y_v - y_u}{x_v - x_u}, \forall u = i, j \wedge \forall v = j, k \wedge u \neq v;$$

- calculation of internal angles:

$$\alpha_{kij} = t_{ik} - t_{ij}, \alpha_{ijk} = t_{ij} - t_{ik}, \alpha_{ikj} = t_{kj} - t_{ki};$$

- calculation of circumference:

$$O = \sum_{i=1}^{n} \sqrt{(x_i - x_{i+1})^2 + (y_i - y_{i+1})^2};$$

- calculation of centroid:

$$x_c = \frac{1}{6P} \sum_{i=1}^{n} (x_i + x_{i+1})(x_i y_{i+1} - y_i x_{i+1}), y_c = \frac{1}{6P} \sum_{i=1}^{n} (y_i + y_{i+1})(x_i y_{i+1} - y_i x_{i+1}),$$

where $i = 1...n$, $j = 1...n$, $k = 1...n$.

From the point of view of mathematics, the stated calculations are correct, but from the point of view of geoinformatics and cartography the situation might get rather complicated (and it very often does), because of historical and practical reasons classic Cartesian system is not often used, but the position of coordinate axes, their orientation and orientation of directions (here very often called bearing and marked as σ) may vary in different coordinate systems. The following picture (Fig. 10.18) shows the real state in the Czech Republic. As all systems are used in practice, it is necessary to know exactly their characteristics and to be able to deal with modification of the above-mentioned formulas for a specific coordinate system. Also, confusion of coordinates in different systems cannot occur, as it could have fatal consequences.

The stated geometric problems are one of the basis for spatial analyses. Due to the extent of the text, it is not possible to pay attention to all ways of spatial analyses, but only to indicate their methodology. One of the frequent problems is spatial overlay of several objects which from the geometric point of view means calculation of intersection or integration of sets or various combinations of thereof. For documentation it is once again possible to use (ESRI 2013), where various types of intersections, integrations, etc. are solved (see Fig. 10.19).

A real application of the example on the left may be e.g. discovery of decline of land resources—INPUT (classified according to the quality of soil) by construction —ERASE FEATURE. The other example may be e.g. evaluation of influence of a source of radiation—IDENTITY FEATURE on surrounding parcels of land— INPUT. The result are parts of parcels affected by radiation—OUTPUT.

10.4.3.2 Geometric Problems in Raster Models

Similar calculations in *raster format* are relatively easier. As the individual objects or phenomena are basically modelled only by pixels, to whose position they belong, calculations of geometric characteristics are very simple. For instance, surface of an object is expressed as a sum of all pixels of the given object multiplied by the

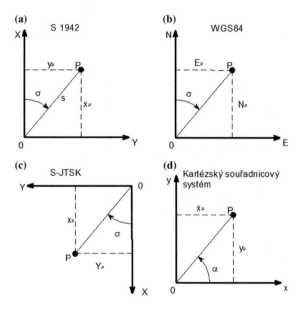

Fig. 10.18 Systems of coordinates used in the Czech Republic: (a) S 1942 (Pulkovo), Gaussian projection; (b) WGS84, UTM projection; (c) S-JTSK, Krovak projection, used mainly for cadaster and civilian state mapping; (d) Cartesian coordinate system

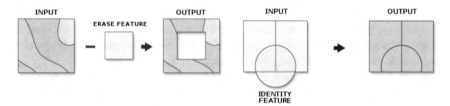

Fig. 10.19 Examples of vector overlay. *Left*—intersection of entry field and erase feature, the result is an entry field without information under erase feature. *Right*—intersection of an entry field and an identity feature, in intersection of both objects the entry field is divided and added with characteristics of identity feature (ESRI 2013)

surface of a pixel. There is no sense in speaking about some internal angles, side lengths, circumference, etc. If it is still necessary to calculate similar characteristics like with vector models, it is possible also in raster models to use the above-mentioned formulas. Coordinates (x_i, y_i) are derived as pixel centers, in which a change of geometry of the modelled object often takes place. However, it is necessary to carefully consider pixel sizes which the objects are modelled with, as there is deformation of the original objects (see Fig. 10.16).

Spatial analyses in raster models, when relations between objects saved in various raster datasets are searched for, are quite complicated with regards to possibilities with the help of which it is possible to model relations. The easiest way is in case that all raster datasets have the same pixels and only the relation between their values in the given position of pixels is evaluated. For instance, if there is a request for an object not to be in a forest, nor meadow, nor pasture, in raster dataset of forests all pixels that are not forest are searched for, and in raster dataset of meadows and pastures again values that are not meadow, nor pasture. The search result is then a raster dataset, in which there are marked suitable pixels that do not have feature of forest, meadow nor pasture, in any entry field. Most spatial analyses in raster models, however, is based on zonal or global analyses. Even though they use quite a lot of mathematics and especially statistics, again with respect to the extent of the text, refer to ESRI (2013), Kresse and Danko (2012) or Udvorka (2006). Spatial analyses are very frequently used also in the field of remote sensing data processing. Many examples of mathematical principles usage is possible to find in Kovarik (2011).

10.4.4 Accuracy of Geometric Problems

Objects and phenomena in the countryside are modelled in the environment of geoinformatics systems only with defined, required or really achievable accuracy. For instance, a footprint of a building can be measured with the accuracy of several centimeters; borderline of forest can be set with the accuracy of meters, but borders between two soil types with the accuracy of several hundred meters. Similarly, also accuracies of setting or deriving thematic features of objects are classified. That is why while solving all problems in GIS environment, it is necessary to take into account characteristics of accuracy and a way how the accuracy in position as well as in thematic features affects the results of various analyses, research, etc. Mathematical basics are once again necessary for understanding of the whole problem.

For example, for evaluation of the accuracy of localization of discrete objects (geodetic points, buildings, poles, etc.) it is possible to use calculus of probability and function of normal distribution. With the use of the mentioned characteristics it is possible to set e.g. probable position of a geodetic point with a defined level of reliability, usually 95%. Another characteristic is mean positioning error, or standard deviation. These characteristics are normally worked with.

Most objects and phenomena in the countryside, however, are not utterly discrete, but they are connected to its surroundings. A house stands on a parcel; it means that where the footprint of the building ends, the object of parcel begins. A forest and a meadow also have a common border. While working with vagueness of such borders, calculus of probability is not an optimal tool, because it is possible to work only with one type of objects. Using fuzzy sets here seems to be very useful. Fuzzy sets enable to mathematically solve a change of membership to the given object given by vagueness of borders of the neighboring objects (Zadeh 1965; Talhofer et al. 2012; Hoskova-Mayerova and Hofmann 2016). Principles of using fuzzy sets in geoinformatics are dealt with e.g. by Kainz (2007) and they are also applied in program systems GIS, e.g. ESRI (2013). According to Kainz (2007), membership to given objects may be modelled by various types of functions. The following picture (Fig. 10.20) shows the linear membership function. This function has four parameters that determine the shape of the function. By choosing proper values for a, b, c, and d, where $a \leq b \leq c \leq d$, we can create S-shaped, trapezoidal, triangular, and L-shaped membership functions.

Memberships may be mathematically expressed as:

$$\mu_A(X) \begin{cases} 0 & x < a \\ \frac{x-a}{b-a} & a \leq x \leq b \\ 1 & b < x < c \\ \frac{d-x}{d-c} & c \leq x \leq d \\ 0 & x > d \end{cases}$$

Using fuzzy sets for a calculation of membership of terrain sections to roads is shown in the picture (Fig. 10.20). In GIS database certain types of roads (field tracks and gravel tracks) are defined only as a point set in the axis of the road. As data about width of roads are missing in thematic features, but in reality this width is usually 5 m, linear membership function to roads in the surroundings of the axis of roads of 5 m to both sides was chosen here (Hofmann et al. 2013) (Fig. 10.21).

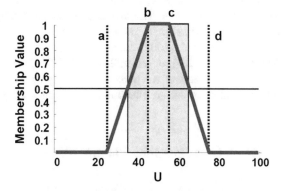

Fig. 10.20 Membership function (Kainz 2007)

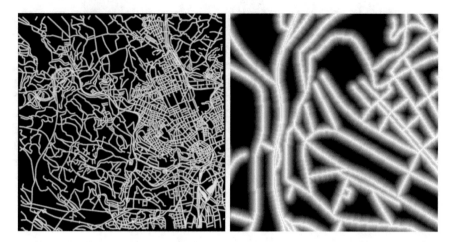

Fig. 10.21 Fuzzyfied roads; on the *left* all roads in the given area, detail on the *right*

10.5 Cartographic Visualization of Objects

The saved objects are presented to users in a graphic form using symbols and texts. In principle, it does not matter what media is used for projection of objects. It may be a classic paper map, computer screen, mobile phone, GNSS navigation or a projection on a projection screen using data projector. However, it will never be possible to project all objects saved in the database together with their features as there are always limitations of the given size of the area to which the content of the database is projected, resolution level of the medium with the help of which the content is projected and finally there are also perception and cognitive possibilities of a person who watches the image. The whole problem of cartographic visualization is described in complex in a range of sources, let us mention e.g. MacEachren (2004), Slocum et al. (2005), Kraak and Ormeling (2010), Lauermann (1974, 1978), Rystedt et al. (2014), or Voženílek et al. (2011). The following text thus gives only chosen examples of methods of cartographic visualization which relate to mathematics and therefore are illustrative from this point of view.

10.6 Principles of Cartographic Visualization

Each map symbol expresses the position and geometry of an object, its qualitative and quantitative features. According to Bertin (1967), it is possible to describe every map symbol with graphic variables, which are position, shape, size, value, pattern, color, and orientation. These symbol variables fully comply with visualization of objects for classic maps and virtual maps that are not of a dynamic character. With the development of dynamic systems, especially with web map services and in GNSS navigations, new symbol variables have been gradually added, such as mainly animations, sounds, interactions and relations to other objects.

The complex of a graphic design of the given map type is called a map style (Voženílek et al. 2011). Figures 10.22 and 10.23 show various map styles for detailed projection of landscape. Figure 10.22 represents the classic map style of topographic maps. The contents of these maps were in the past created mainly by manual or semiautomatic drawing. Nowadays, the contents is created by a choice of objects, modification of their geometry and their symbolization in GIS environment with relativelly significant share of manual finishing.

Virtual maps presented in web map services (Fig. 10.23) have modified map styles which comply with the response speed of web server on user's request. Even though visualization of database contents is usually ready beforehand, it is necessary to quickly choose and modify the contents from the given database according to the required space and its size.

The user's request on the given space is usually placed by the object that is searched for (town, area, etc.). From the spatial point of view, the request is modified in a way so that the searched object was in the middle of the search "window" whose size corresponds to the current size of the area on the computer

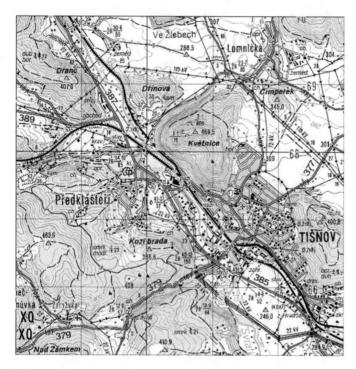

Fig. 10.22 Topographic map style of paper maps (military topographic map of the Czech Army 1:25000—reduced)

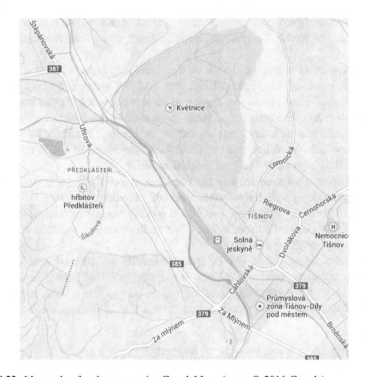

Fig. 10.23 Map style of web map service GoogleMaps (maps © 2016 Google)

screen that is available for the given web service. The size of this area in coordinates of points of the screen is recalculated into spatial coordinates that are then the search key in the spatial database. From the point of view of mathematical principles, these are mainly geometric transformations in the given coordinate systems and people who program and provide these services must be once again absolutely sure of the basics of positioning as well as of the choice of suitable methods of transformations and their usage (Kovarik and Marsa 2014).

The procedure of cartographic visualization itself is shown on an example of network of roads in the following diagram (Fig. 10.24).

10.6.1 Cartographic Generalization

From the picture (Fig. 10.24) it is obvious that for quality and readable visualization it is necessary to modify a lot of presented information. According to Voženílek et al. (2011), with respect to the visualization itself, the size of the area for visualization, character of the projected landscape and used graphic tools, it is necessary to choose objects that shall be visualized, choose and potentially modify their thematic features according to which they will be visualized, and the whole

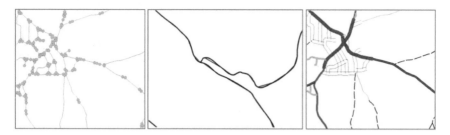

Fig. 10.24 On the *left*—simple visualization of saved linear objects of roads as edges (*black*) and junctions (*green*). In the middle—example of smoothing of the original edges (*black*) using Bézier curves (*red*). On the *right*—example of assignment of symbols to smoothed edges according to traffic use of roads which is stored for each object of the road. Passage through settlements are marked in thick *red*, 1st class roads are *red*, 2nd class roads are *orange*, access roads ochre, streets in settlements are two parallel *black lines*, gravel tracks are grey and fields tracks are *dashed*

visualized content to put in harmony so that there were no overlays of map symbols. This complete procedure is called cartographic generalization. Cartographic generalization is a traditional field of cartography which was developed by many authors, e.g. Srnka (1968), Töpfer (1974), Lauermann (1974). Contemporary approach to cartographic generalizations comes from classic work from the times when maps were created predominantly manually, but it is developed especially under the influence of work with digital data. Most of program systems for work with digital data contain sets of tools for cartographic generalizations. System ArcGIS by ESRI (2013) may be mentioned as an example.

10.6.2 Simplification of Geometry

Tools of cartographic generalization are based on mathematical principles, above all, on applications of analytic geometry, functions of approximation and on tests of mutual position of points, lines or line segments and surfaces. One of the simple tools is e.g. modification of geometry of line objects which may be solved by the reduction of their definition point set. Descriptions of many tools may be found in materials (Bayer 2013). Furthermore, there are given examples of three tools.

Basic mechanical tools are length or angle tests, during which two or three consecutive points are worked with, and with beforehand given minimal (testing) value. In the first case, points of definition point set of line objects are reduced. $P_i, i = 1, \ldots, n-1$, which are closer to each other than given length criterion d_{min}. Vertexes P_{i+1} are omitted. They have smaller Euclidean distance $d(P_i; P_{i+1})$ from the previous vertex P_i than length criterion d_{min}, i.e. $d(P_i; P_{i+1}) < d_{min}$. The work of algorithm and its result can be shown in the following picture (Fig. 10.25).

Another simple algorithm which enables to keep characteristic breaks was suggested by Jenks. During this algorithm, mutual positions of three consecutive

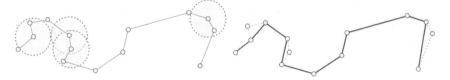

Fig. 10.25 Reduction of a point set using the length test (Bayer 2013)

Fig. 10.26 Jenks algorithm with an angle test (Bayer 2013)

points on line L are tested between which Euclidean distances $d(P_{i-1}; P_i)$ and $d(P_i; P_{i+1})$ and angle $\varpi \angle P_{i-1} P_i P_{i+1}$ are calculated. These values are compared to limit values $d_{1min}, d_{2min}, \omega_{min}$. The test itself is given by statement:

If $d(P_{i-1}; P_i) < d_{1min} \wedge d(P_i; P_{i+1}) < d_{2min} \wedge |\pi - \varpi| < \omega_{min}$, vertex P_i is omitted. Algorithm and its result is again documented in the picture (Fig. 10.26).

In literature (Bayer 2013) it is possible to find several other mechanical algorithms of reduction of point sets.

More complex algorithms based on gradual recursive approximation of the whole line are called global algorithms, Douglas-Peucker algorithm is the most frequently used. Its description and explanation can be found in Bayer (2013), concrete implementation then in tools of ArcGIS (ESRI 2013).

According to Bayer (2013), the algorithm for line L defined by definition points $P_i, i = 1, \ldots, n$ uses a corridor of given width h as a testing area. As opposed to other algorithms, it does not remove vertexes from L that do not meet a geometric condition but gradually adds into it vertexes that meet the geometric condition. The principle of the algorithm may be described in the following steps:

(a) Creation of an edge $e = P_1 P_n$; broken line L substituted by a line segment;
(b) Point P is looked for, it lies outside the corridor in a way that $d(P; e) = max(d(Pi; e))$;
(c) If such point P does not exist, algorithm ends;
(d) If there is at least one such point, edge e is divided into two edges $e_1 = P_1 P$, $e_2 = PP_n$;
(e) Above e_1, e_2 point (a) and (b) are repeated.

Gradual approximation of line L is obvious from the picture (Fig. 10.27).

With similar procedures it is possible to simplify also the geometry of peripheral edges of surface objects.

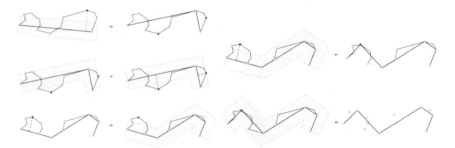

Fig. 10.27 Douglas-Peucker algorithm (Bayer 2013)

10.6.3   Smoothing Algorithms

After reduction of point sets, sets usually approximate to smooth curves which correspond to the reality in the countryside. Water courses, roads, borders of surface objects of natural character and so on do not usually have sharp edges but their course is smooth. That is why also their cartographic visualization should correspond to this fact. This request may sometimes be withdrawn, especially in visualization of landscape in navigation devices. The reason is reduction of data so that their amount was appropriate to performance of the software of the navigation device and the speed of response of the system during drive was high.

Smoothing algorithms are required to remove unsuitable breaks and sudden changes in the shape of line. Algorithms are derived from mathematical principles and their inspiration cartographers often adopted from CAD systems. Basically there are three types of algorithms used: smoothing by averaging of points, smoothing by filtering, smoothing using approximation curves.

As smoothing concerns visualization of objects in the countryside, it is necessary to know the fundaments and functionality of algorithms so that inappropriate reduction of significant shape characteristics of objects was prevented, as well as adding of shapes that the objects do not possess (especially with approximation of polynomials of higher degrees). It is obvious that e.g. a river on the middle course which is marked with frequent meanders has absolutely different geometric character than a highway or railway. That is why it is necessary to choose smoothing algorithms with respect to geographic nature of the modelled object.

As an example of frequent smoothing algorithm using approximation curves we can mention smoothing using Bézier curves.

According to ESRI (2013) Bézier curves are smooth linear transitions between two vertexes. The shape of the curve is defined by the locations of the vertexes and additional control points at the end of blue Bézier handles that radiate from each vertex (Fig. 10.28).

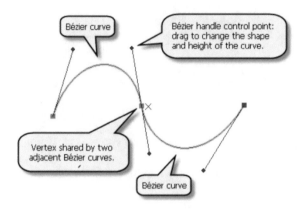

Fig. 10.28 Example of a tool for smoothing a line using Bézier curve (ESRI 2013)

Usually Bézier curve C with two control points is used. Approximation can be expressed mathematically for line L defined by n definition points. From these definition points, there are sequentially chosen 4 consecutive points $P_i, P_{i+1}, P_{i+2}, P_{i+3}, i = 1, \ldots, n-3$, among which the approximation is applied. If we mark $P_i = P_0, P_{i+1} = P_1, P_{i+2} = P_2, P_{i+3} = P_3$, then it is possible to express parametric equation of an approximation curve in the form:

$$C(t) = \sum_{i=0}^{3} \binom{3}{i} t^i (1-t)^{3-i} P_i, t \in \langle 0, 1 \rangle$$

10.6.4 Cartographic Harmonization

Modification of geometry is done progressively with individual thematic layers (bodies of water, roads, built-up area, etc.). After these modifications it is necessary to perform so-called cartographic harmonization whose purpose is to search for all places where the projected objects after merging of all layers into one image cross, overlay, are unfinished, etc. If this was the case, the resulting cartographic visualization would be unreadable and would not correspond to the reality.

To detect the mentioned facts (conflicts), it is possible to use spatial overlays with a calculation of conflict places.

The picture (Fig. 10.29) shows two generalized layers—roads and built-up area. After the symbol for layer of roads has been assigned, it is necessary to "free the space" in the surroundings of buildings so that the symbol does not overlay the buildings. That is why with the use of spatial operations there were places detected where the overlay of symbols of buildings and roads will occur. On the detected

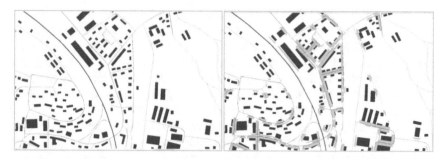

Fig. 10.29 Detection of conflict places—layer of roads and built-up area (data source DMÚ25, ©MoD CR, 2016)

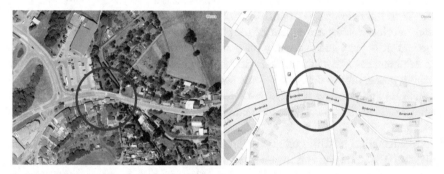

Fig. 10.30 Example of not solving conflicts between buildings and a road—see detail in the circle (https://mapy.cz/zakladni?x=16.0779898&y=49.5613568&z=18 as of 12/8/2016)

places it is then necessary to adjust the geometry of objects, e.g. by transition of less significant objects to a suitable free space in the closest surroundings or by exclusion of the detected objects.

If the conflicts are not solved, the result may be loss of information which demonstrates in a smaller information value of the whole map work (see Fig. 10.30).

10.6.5 Analysis of Thematic Data and Information and Their Visualization

In the previous paragraphs there are examples especially of work with geometry of objects. The power of geoinformatics, however, lies also in keeping thematic features of objects. Basically, it is possible to connect unlimited number of thematic information relating to the object or a group of objects. For instance, for a part of watercourse as a defined object it is possible to generally observe the rate of water flow, instantaneous flow rate, average annual flow rate, width and depth of the bed (Svatonova and Rybansky 2014; Svatonova 2017). Workers responsible for the quality of water will be interested also in the immediate state of water pollution, fishermen will want to know what kind of fish are present in the given segment, and possibly when the closed season for fish begins and ends, canoeists will ask about the passability and demandingness of the course, etc.

Thematic information, however, is not only saved but it is often analyzed, we look for its connections, relations and dependencies. An analysis of spatial information and their thematic features is again a very wide field which rapidly develops either methodically as well as mathematically. Mathematical models of spatial analyses may be found in many sources, e.g. Cressie (1993), Rybansky and Vala (2010), Wilson and Gallant (2000) and others. A lot of principles come from theorems of mathematical analysis, mathematical statistics, work with bulk data, etc. With the help of spatial analysis, for instance, suitable locations for constructions are looked for, movement of people and goods is watched, trends of development of objects and especially phenomena are determined.

Mathematics and mathematical statistics are absolutely essential prerequisites for quality processing of data and information. Without knowledge of their theoretical principles and practical applications it is not possible to complete a quality spatial analysis. Nevertheless, there are many pitfalls that are necessary to take into account.

Mathematical statistics often works with statistical files which have a definable division (Chvalina and Hoskova-Mayerova 2014). Thematic data in geoinformatics, however, are a model of reality in the countryside which is not possible to easily describe in mathematic terms. One of the indicators of distribution of the number of inhabitants in the country is population density calculated as a number of inhabitants for a surface of the area. In every state, there are regions with a higher density and regions with lower, up to small density. Population density is then determined especially by conditions for life of people that may be verbally defined by the amount of roads, level of climate conditions, amount of water and green areas, elevation differences, amount of job opportunities, level of network of schools, shops, cultural and sport facilities, etc. For an analysis of conditions for life is then necessary to consider a lot of factors, describe them by specific criteria, define dependencies between the criteria and only then define the resulting mathematical model for this analysis. A multi-criterial analysis is quite frequently used procedure (Rybansky and Vala 2010).

From this short and simple enumeration, it is possible to imagine the complexity of solution of similar problems. Complex program systems for GIS, e.g. often mentioned ArcGIS, contain strong tools for solution of spatial analyses including the mentioned example ESRI (2013). An analyst, who is able to use services of such program systems, is able to solve such and similar analyses quite easily and to pass the result to the ordering party who may use the result e.g. to make a decision where to invest money to support lagging regions which are becoming depopulated. However, this is very often the pitfall of these solutions. Without understanding geographical relations, it is possible even with correct data to "produce" a misleading analysis which nobody will examine any more but they will only adopt its results. The following simple example documents the possibility of implementation of gross errors and mistakes when using the same data.

In database (ARCDATA 2012) there is information relating to the individual districts in the Czech Republic. There is also information about the number of inhabitants. Let us assign a task to create a thematic map where there will be classification of districts according to the number of inhabitants. First, it is possible to find a basic structure of the statistical file and its characteristics. System ArcGIS is used for this (Fig. 10.31).

From the statistics it is obvious that the mean value of the number of inhabitants in districts is 136,343 but one region (capital city of Prague) has 1,241,664 and two districts have over 300,000 inhabitants. The distribution of frequencies is then irregular and it is necessary to respect it with a suggestion of the number of classification degrees (division into intervals) for the projection of the number of inhabitants in districts. The right method for a division into intervals is the acceptance of local lows and highs in distribution which in cartography and geoinformatics is called "a method of natural borders" or also Jenks method. The picture (Fig. 10.32) shows an example of division of the file into 7 classification classes, which enables to express the differences among the individual districts sufficiently accurately. The following map shows visualization of the number of inhabitants, in which parts of the country inhabitants concentrate (Fig. 10.33).

If we decide to reduce the number of classification classes to 4 while keeping natural borders (Fig. 10.34), the result of the division and the resulting map is then to be seen in the following picture (Fig. 10.35).

Even though the same data were used, the result is totally different and for users it is misleading because 4 classification classes are not enough to express differences in the individual districts.

The last presented variant is a case when the variation span is divided into the same intervals, but the original number of 7 classification classes is kept. The result is obvious in the pictures (Figs. 10.36 and 10.37).

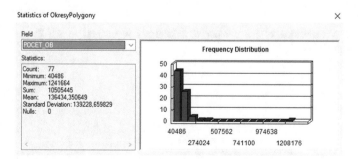

Fig. 10.31 Basic statistics of the number of inhabitants in districts in the Czech Rep.

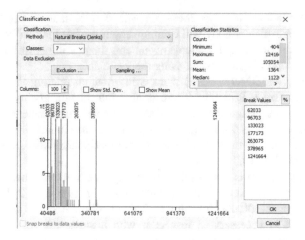

Fig. 10.32 Distribution of frequency of the number of inhabitants in districts of the Czech Rep. into 7 classification classes

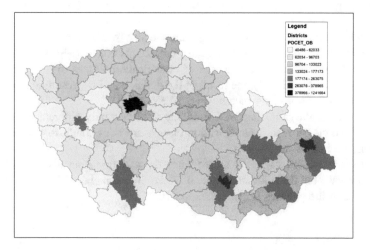

Fig. 10.33 Visualization of the number of inhabitants in districts in the Czech Rep. in 7 classification classes

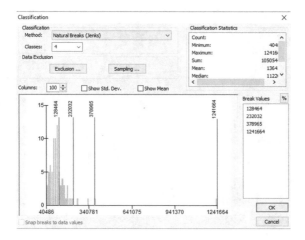

Fig. 10.34 Distribution of frequency of the number of inhabitants in district in the Czech Rep. in 4 classification classes

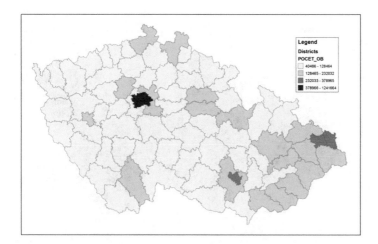

Fig. 10.35 Visualization of the number of inhabitants in districts in the Czech Rep. in 4 classification classes

By this it was "managed" to equalize almost all area of the country and only the capital city of Prague is so big that totally stands out from the statewide average. Such visualization is totally worthless and useless.

The given simple example illustrates trickiness of a mechanical approach to analyses of spatial data. If the worker creating spatial analyses has not acquired sufficient fundamentals of general statistics, extended by geo-statistics, or for some reason is too inactive to merge into the problem, he/she may create an absolutely misleading analysis even from quality data.

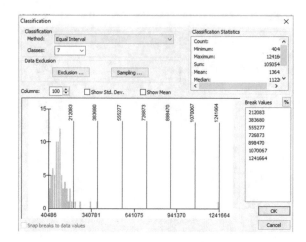

Fig. 10.36 Distribution of frequency of the number of inhabitants in districts of the Czech Rep. in 7 classification classes with the same interval

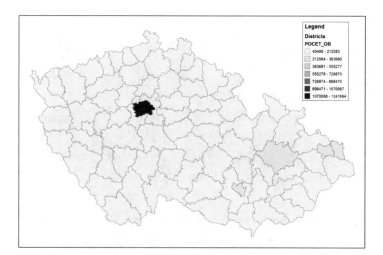

Fig. 10.37 Visualization of the number of inhabitants in districts in the Czech Rep. in 7 classification classes with the same interval

10.7   Conclusion

As I mentioned in the introduction of this text, it was not my ambition to define absolutely all requirements on knowledge of mathematics for students and graduates of cartography and geoinformatics. The extent of the text would not allow me this, anyway. Nevertheless, I hope that I have at least indicated the width of the problem in the given fields for which basics in mathematics and statistics are necessary. Just like at all universities oriented to technical and natural fields, it is obvious that without solid fundamentals in mathematics it is not possible to successfully master professional problematics. Of course, the required level will depend on the level of study, it will be different in bachelor study, other in master study and totally different in PhD programs, from the point of studied fields of mathematics, as well as their depth (Hoskova-Mayerova and Rosicka 2012). However, it will not depend on the way of preparation or language in which mathematics is taught (Hoskova-Mayerova 2011; Rosicka and Hoskova-Mayerova 2014).

From the point of view of fields of mathematics, knowledge and competences of general university mathematics are needed, such as arithmetic, algebra, matrix calculation, goniometric and inverse goniometric functions, differential and integral calculation, etc.

As I tried to point out, extension of mathematics by plane and spherical trigonometry and its solution using especially analytical procedures in various types of coordinate systems seems to be right. Furthermore, usage of differential geometry of curves and surfaces for the solution of procedures of transformation of spatial bodies into a plane is also a good idea. Knowledge of theory and analytical solutions of affinitive and projective transformations mainly for work with visual data from aerial and land cameras and data of remote sensing is also very useful. For work with geographic data and for their analyses, it is useful to know interpolation methods in a plane, as well as in space, and ways of approximation of lines and surfaces by functional surfaces. It seems essential to study the methods and procedures of mathematical statistics and calculus of probability.

Mathematics is fundamental for understanding of many expert problems that may be solved by program tools, but without the knowledge of theory the researcher may make serious mistakes. Mathematics pushes to precision of the solution and precision of its description in documentation of the procedure.

Mathematics itself is a complex subject, it has its theory, its rules and its procedures. It often happens that for a non-mathematician pure mathematics seems incomprehensible which might lead them to turn away from it. To prevent this from happening, also mathematicians must understand requirements of technicians. For technicians, mathematics must be comprehensible and applicable. A mathematician who teaches a technical subject should—at least in basics—understand this subject and indicate what is important for the subject and what theoretical and application ways can be used to solve problems (Hoskova-Mayerova 2012, 2016). Attitude of some mathematicians is actually to the detriment of the cause as they are not interested in applications and do not let the users find applications themselves. In my opinion, this attitude is inappropriate as mathematics must help us solve our problems. A suitable procedure seems to be to include mathematicians in a team of researchers who try to move the theory and practice of cartography and geoinformatics forward.

And finally, I would like to cite Professor Nešetřil (2011):

> On one hand, mathematics is very complex, but on the other hand it is a very simple science in sense there is a simple criterion of truth. Most things a man in an ordinary life does may be answered yes as well as no, everything depends on the extent. In mathematics, it is not the case, the answer there is more or less definite. The basic criterion according to which the quality of work is evaluated is the truth of theorems devised by a mathematician. In mathematics, it is important what a scientist created, it is not evaluated whether he/she was nice or bad. Nice mathematics was produced also by people who were bad in human terms, but it is also true in other subjects.
>
> A significant part of mathematician's work is also an esthetic criterion, elegance, beauty, because mathematics is as close to esthetics as it is to music. Things that are famous in mathematics are also usually very elegant. Mathematics is not a subject that would on purpose create complex, hard-to-understand things. Mathematicians are happy when things are presented simply, the esthetic side of things does not manifest in a formal way, but more in the elegance of connection.

References

ARCDATA. (2012). ArcCR500.gbd. Praha.

Bayer, T. (2013). Educational materials for study of geoinformatics and cartography. Praha: Charles university. In. Czech.

Bertin, J. (1967). *Sémiologie graphique: Les diagrammes - Les Réseaux - Les Cartes.* Paris: Mounton.

Cressie, N. A. (1993). *Statistics for Spatial Data.* New York, USA: John Wiley & Sons, INC.

Chvalina, J., & Hoskova-Mayerova, S. (2014). On certain proximities and preorderings on the transposition hypergroups of linear first-order partial differential operators. *An. Stiint. Univ. "Ovidius" Constanta Ser. Mat., 22*(1), pp. 85–103.

DiBiase, D., DeMers, M., Luck, A. T., Johnson, A., Plewe, B., Kemp, K., & Wentz, E. (2006). Geographic Information Science and Technology: Body of Knowledge 2006. 127. Abington: University Consortium for Geographic Information Science.

ESRI. (2013). User documentation. Copyright © 1995–2013 Esri.

Grafarend, E. W., & Krumm, F. W. (2006). *Map Projections: Cartographic Information System.* Heidelberg, Germany: Springer-Verlag.

Hofmann, A., Hošková-Mayerová, Š. (2016) Development of Applications in the Analysis of the Natural Environment. In: *Aplimat - 15th Conference on Applied Mathematics 2016 Proceedings.* Bratislava, Slovensko: Vydavatelstvo STU Bratislava, 2016, p. 467–481.

Hofmann, A., Hoskova-Mayerova, S., & Talhofer, V. (2013). Usage of fuzzy spatial theory for modelling of terrain passability. *Advances in Fuzzy Systems, 2013*, p. 13.

Hoskova-Mayerova, S. (2011). "Operational program, education for competitive advantage", preparation of study materials for teaching in English. *Procedia Social and Behavioral Sciences, 15*.

Hoskova-Mayerova, S. (2012). Topological hypergroupoids. *Comput. Math. Appl., 64*(9), pp. 2845–2849.

Hoskova-Mayerova, S., & Rosicka, Z. (2012). Programmed learning. *Procedia Social and Behavioral Sciences, 31*, pp. 782–787.

Hoskova-Mayerova, S. (2016). Leadership – training of military specialists in particular disciplines focused on mathematical modelling. *New Trends and Issues Proceedings on Humanities and Social Sciences.* [Online]. 05, pp 199–204. Available from: www.prosoc.eu.

Kainz, W. (2007). *Fuzzy Logic and GIS.* Vienna, Austria: University of Vienna.

Kovarik, V. (2011). Possibilities of Geospatial Data Analysis using Spatial Modeling in ERDAS IMAGINE. *Proceedings of the International Conference on Military Technologies - ICMT'11* (pp. 1307–1313). Brno: University of Defence.

Kovarik, V., & Marsa, J. (2014). Specifics of thematic map production within geospatial support at a politico-strategic level. *Geographia Technica, 9*(1), pp. 52–65.

Kraak, M.-J., & Ormeling, F. (2010). *Cartography: Visualization of Geospatial Data* (Third ed.). Harlow, Essex, England: Pearson Education Limited.

Kresse, W., & Danko, D. M. (2012). *Hanbook of Geographic Information.* Berlin Heidelberg: Springer-Verlag. doi:10.1007/978-3-540-72680-7.

Lauermann, L. (1974). *Technical cartography I.* Brno: Military Academy Brno, in Czech.

Lauermann, L. (1978). *Technical cartography II.* Brno: Military Academy Brno, in Czech.

Lauermann, L., & Rybanský, M. (2002). *Vojenská geografie* (První. vyd.). Praha: Ministrstvo obrany ČR.

MacEachren, A. M. (2004). *How Maps Works: Representation, Visualization and Design.* New York, USA: The Guilford Press.

Nešetřil, J. (2011). Jsem agresivní optimista, říká o sobě profesor Jaroslav Nešetřil. *iFOFUM, časopis Univerzity Karlovy.* Retrieved 26. 07 2016, z https://iforum.cuni.cz/IFORUM-10238-version1.pdf.

Office of Geomatics. (2014). *Implementation Practice Web Mercator Map Projection* (Version 1.0.0 ed.). Washington D.C., USA: National Geospatial-Intelligence Agency (NGA). Retrieved 02 09, 2015, from http://earth-info.nga.mil/GandG/wgs84/web_mercator/%28U%29%20NGA_SIG_0011_1.0.0_WEBMERC.pdf.

Rosicka, Z., & Hoskova-Mayerova, S. (2014). Motivation to study and work with talented students. *Procedia Social and Behavioral Sciences, 114*, pp. 234–238.

Rybansky, M., & Vala, M. (2010). Relief impact on transport. *ICMT'09: International Conference on Military Technologies* (pp. 551–559). Brno: University of Defence.

Rystedt, B., Ormeling, F., et al. (2014). *The World of Maps.* ICA: International Cartographic Association. Retrieved 2014, from http://mapyear.org/the-world-of-maps-book/.

Slocum, T., McMaster, R., Kessler, F., & Howard, H. (2005). *Thematic Cartography and Geographic Visualization* (2 ed.). Upper Saddle River, NJ 07458: Pearson Education, Inc.

Srnka, E. (1968). *Analytical solution in the cartographic generalization,* in Czech) (Habilitation theses ed.). Brno: Military Academy in Brno.

Svatonova, H., & Rybansky, M. (2014). Visualization of landscape changes and threatening environmental processes using a digital landscape model. *IOP Conf. Ser.: Earth Environ. Sci. 18. 18*, pp. 12–18. IOP science.

Svatonova, H. (2017) New Trends in Obtaining Geographical Information: Interpretation of Satellite Data, Recent Trends in Social Systems: Quantitative Theories and Quantitative Models, Decision and Control, Vol. 66, Maturo (Eds.), 173–182.

Talhofer, V. (2007). Basics of maps projections. Brno: Univerzity of Defence, In Czech.

Talhofer, V., Hoskova-Mayerova, S., & Hofmann, A. (2012). Improvement of digital geographic data quality. *International Journal of Production Research, 50*(17), pp. 4846–4859.

Töpfer, F. (1974). *Kartographische Generalisierung.* Leipzig: VEB Hermann Haack.

Udvorka, P. (2006). *Mapová algebra a její využití v geografických analýzách* (Doctoral Thesis). Brno, Czech Republic: University of Defence.

Voženílek, V., Kaňok, J., Bláha, J. D., Dobešová, Z., Hudeček, T., Kozáková, M., & Němcová, Z. (2011). *Metody tematické kartografie, vizualizace prostorových jevů.* Olomouc, Česká republika: Univerzita Palackého.

Wilson, J. P., & Gallant, J. C. (2000). *Terrain Analysis: Principles and Applications.* New York: John Wiley & Sons. Inc.

Zadeh, I. (1965). Fuzy Sets. *Information and Control, 8*, pp. 338 – 353.

Chapter 11
Cognitive Aspects of Interpretation of Image Data

Hana Svatoňová and Radovan Šikl

Abstract Interpretation of image data (one of the basic geographic skills—Řezníčková et al., Standards and research in geography education. Current trends and international issues, pp 37–49, 2014) is a complex of complicated intellectual operations, which is based on visual perception (for example, when working with a map, then we can talk about mapping skills—Hanus and Marada, Geografie 119 (4):406–422, 2014). The theoretical part of the study summarizes the scientific knowledge of processes of visual perception applied in the process of visual interpretation of satellite, aircraft and map image data. The author presents partial phases of image data interpreting process: from the initial recording of the image to detection, identification and objects classification. The complexity of the cognitive process with regard to biological and psychological characteristics of the individual are highlighted. The research section presents the results of image data interpretation research according to gender of individuals/research respondents. The research results show (1) a consistent success rate and (2) a consistent speed of problem solving when dealing with image data of aerial and satellite images. The results were slightly surprising with respect to research results concerning map interpretation where respondents attain different degrees of success rate depending on gender.

Keywords Visual interpretation · Visual perception · Image interpretation · Image data · Satellite data · Airborne data · Aerial photo · Map

H. Svatoňová (✉) · R. Šikl
Department of Geography, Faculty of Education, Masaryk University, Brno,
Czech Republic
e-mail: svatonova@ped.muni.cz

R. Šikl
e-mail: sikl@psu.cas.cz

© Springer International Publishing AG 2017
Š. Hošková-Mayerová et al., *Mathematical-Statistical Models and Qualitative Theories for Economic and Social Sciences*, Studies in Systems, Decision and Control 104,
DOI 10.1007/978-3-319-54819-7_11

11.1 Visual Perception with Regard to Map and Image Data Interpretation

The interpretation of images and maps—image in general *involves a complex of cognitive operations* characterised by *general features*. They are influenced by *particular differences* among individuals. In professional literature a set of mental processes associated with identifying and recognizing what we see is referred to as *visual perception*. The following text is dedicated to the *basics of visual perception, emphasizing the relations and connections with the image and map interpretation.*

Visual perception is a relatively complex mental process involving a number of analytic and synthetic intellectual operations. Although the perception is often carried out naturally and without any conscious effort, it is in fact an extremely complex process employing a significant part of the mind (cognitive, executive and emotional processes), and of the capacity of brain activity (Šikl 2012; Hoskova and Mokra 2010). The main purpose of visual perception is, according to Šikl (2012, p. 10) "speed, efficiency, ability to chart the observed scene in the shortest possible time, and to obtain meaningful and relevant data." Gregory (1966) considers the visual perception a dynamic quest to find the best interpretation of the available data. Crick (1997) defines the visual perception as a creative process during which the brain simultaneously responds to many different "features" of a visual scene and tries to merge them into meaningful wholes. Eysenck and Keane (2008) define the visual perception as a complex process of transformation and interpretation of incoming sensory information. According to Sternberg (2002) the visual perception is in its entirety understood as processes by which the information obtained from the visible range of the electromagnetic radiation falling onto visual sense organs is identified, sorted and is given a meaning. According to Broadbent (1958) the stimulus processing is performed by elementary perceptual processes. During this first phase, which means the input and the entire process opening, a prior experience and human interest in the object are already involved (see the text about individual differences in visual perception). A distinction between bottom-up processing and top down processing of a stimulus is usually drawn (Neisser 1976).

When *applied to map and satellite and airborne image data reading*, we could say that the first approach (bottom-up processing) assumes the parameters of the source— an image (its colour, resolution, display area etc.) to be determining for information processing. It should be noted that although the spatial resolution and other physical parametres of imagery are very important, the interpretability of an image can be severely affected by other influences (Kovarik 2012; Hoskova-Mayerova et al. 2013; Talhofer et al. 2016; Talhofer and Hoskova-Mayerova 2016). The later type— top-down processing assumes an influence of a particular person on the whole process—i.e. their biological determinateness, and mental conditions, and knowledge and skills (images will be read differently by a child, an experienced professional, etc.). The important fact related to the process of information processing dealt with in *this research on visual interpretation of image data and maps is especially* that the symbols are processed using the processes of transformation into other symbols related to objects of the outside world.

To use some simplification concerning the interpretation of cognitive aspects of image interpretation the processes involved in the visual perception can be grouped according to the level of their complexity:

- *Low complexity level* of visual perception: perceptual organization (its goal is to organize the chaos of stimuli based on retinal cells responses to the impact of electromagnetic radiation), the whole (figure) dividing, which means the principal information and subjective perception. It results in the creation of sub-units (entities) made from the original tangle of stimuli. Sub-processes: edge detection, image segmentation, separating an object from its background, partial elements organizing and structuring, a modal perception, parsing (image "compartmentalization").
- *Intermediate complexity level* of visual perception: entities operations, processing of information related to the colour entity, shape, brightness transformation, texture, shadow and size over time.
- *High complexity level* of visual perception: the identification of the object entity. Sub-processes: recognition, classification, categorization, vision-driven mindset.

The image projected on the retina is segmented perceptually completed, then integrated, identified and classified as a representative of a general category. When applying to image interpretation we can imagine that during the process we transform the tangle of forms and colour parts (likewise pixels), create clusters (entities, clusters of pixels), further we attribute them with other characteristics

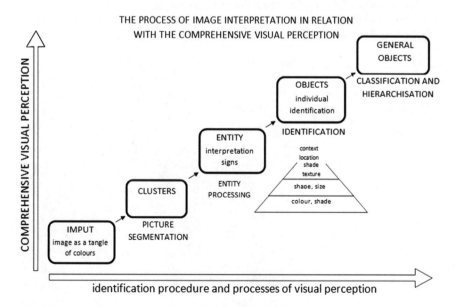

Fig. 11.1 Visual perception and the process of identifying objects in an image

(colours), which help us to recognize them individually (this is an object, e.g. a lake) and subsequently we also categorize them in general (this lake belongs to the class of the lake and this one to a water area).

In relation to the object identification *a scheme* has been processed see Fig. 11.1 *showing the sequential steps—Phases of visual perception during image inter- pretation—object identification.*

11.1.1   The Cues for the Perception of Space in the Picture

The perception of space captured in the image involves the use of so called monocular cues (sometimes referred to as pictorial cues), see Fig. 11.2:

- the shading means the observer correctly concludes that the object casting the shadow must be three-dimensional. This cue is also applied when reading the images.
- texture perspective—a typical spatial arrangement (texture gradient), (Gibson 1979), an example of a textural cue used for image interpretation can be a texture created upon a canopy closure of treetops of deciduous forest, a square or rectangular arrangement of streets, a paving, etc.)
- covering—in pictures some objects may be covered (a brook in the woods, two buildings right behind each other in an oblique image etc.)
- linear perspective—the observer may perceive the lines as if they converged— which generates the impression of depth; this cue applies when reading an oblique picture bearing the perception of perspective,
- the perception of oblique images involves other picture cues, especially the height within the visual field and the atmospheric perspective.

Fig. 11.2 A sample of image cues for perception of a spatial depth—relative size, location in an image plane, perspective, interposition, and shading. (*Source* Atkinson 2003)

11.1.2 Interpretative Signs of Object Recognition in Images

During the image perception the above mentioned cues valid for visual perception in general are applied. The range of cues described above can be involved in the process of *visual interpretation of images*; to recognize the objects within the pictures we use their so called *interpretative signs*. They are mainly the shape and size of the object (or the ratio of width to length), the colour, cast shadow, position or relative position—the context of the position in relation to other objects. We recognize most of the objects in the images by their shapes (contours) and typical details. The objects displayed from the top view are for readers unfamiliar. On that account, it often depends on detail resolution, especially if two different objects have similar plan views. Similarly, the objects with the same layout but different heights and different purposes may be displayed in the same way. The object dimension depends directly on the scale of the image. It is possible to calculate the measuring scale using the real size of the object and its dimension in the picture. The comparing of the respective object dimensions with other well known objects is also used to interpret the images. The length and width ratio helps identify individual buildings, railway lines, highways, roads, streets, etc. (Svatoňová and Lauermann 2010). The colour of the object is an important interpretative sign (Kubicek et al. 2016). For the purpose of interpretation, the specific colour syntheses of satellite data are therefore in true or false colours. The surveys (Svatonova and Rybanský 2014; Svatonova 2016a, b; Hofmann and Hoskova-Mayerova 2016) showed that satellite images in false colours are as successfully interpreted by the readers as the images in true colours. Aerial images are currently acquired mostly in their true colours resembling reality. The object shadow highlights the plasticity of the object body, more precisely its shape. The drop shadow allows determining the depth, width and height of the object. An important fact is that the **drop shadows** of objects and terrain shapes displayed in the image must always point **toward the evaluator** or from **left to right**. In this manner the proper plasticity of the image is required. If these conditions are met, a false perception appears—e.g. a valley seems like ridges and ridges like a valley. We encounter a problem when interpreting the drop shadows of clouds because they partly cover the part of the territory, and the drop shadow of clouds can be easily mistaken for an object (a pond, wet soil…). The object position expresses its spatial relationship to other objects. There exists a causality between the interpreted objects and terrain features in the landscape (a water stream flows through the valley, the dam is built on a watercourse, bridges lead across roads, excavations and embankments are along the roads, railway stations are on the railway lines, etc.). To interpret the image a context position of the object is therefore used: the object is recognized thanks to its usual position in relation with the other organization (i.e. the bridge is always across the watercourse). The context as a parameter of recognition appears to be very important in a series of studies. Many studies have demonstrated that it is precisely the context in everyday life that is an important part of object recognition and not only the individual object which is presented during set laboratory conditions. The studies by Biederman et al. (1982) showed that the inconsistency of the relation "object vs scene context" leads to a decrease of identification accuracy and to prolongation of reaction time. Similar results have been reached by Hock et al. (1974) and Potter and Davenport (2004).

11.1.3   Individual Differences in Relation to the Images' Interpretation

In the process of image interpretation, thus identification of sub-objects in the picture, the properties of the image itself (including interpretive signs of objects that are displayed in the image) are applied, as well as the personal characteristics of the person observer (interpreter), particularly their *biological status* (age, gender, vision—acuity, colour vision, sensitivity to contrast and *mental and social condition* (e.g. priming, motivation and interest).

11.1.3.1   Biological Characteristics of the Individual in Relation with the Image Interpretation

Age: It is the most important for visual perception among all biological characteristics. Increasing age is connected to a series of changes of retinal images, including a decrease of the number of photoreceptors (Scialfa 2002). The measurable characteristics of an older person prove poorer perceptual performance in comparison to a younger one (Šikl 2012). Three year-old children and older ones begin to understand aerial images as a representation of the real world, even if the full understanding is developing up to their adulthood (Blaut 1997; Downs and Liben 1997; Liben and Downs 1997). *In relation with the image interpretation there exist differences in the perceptual performance*; e.g. the age deteriorates detection and identification of objects and stimuli (Madden and Allen 1991; Davis et al. 2002); the reaction time is extended with age (Salthouse and Somberg 1982; Salthouse 2000; Madden 2001), the age-deteriorates the perception of space, orientation and navigation (Baker and Graf 2008). Aspect of age then goes hand in hand with the cognitive aspect of maturity, which in young individuals often develops in accordance with the procedure of the educational process. Age is a crucial factor for working with graphical visualization, in this case with the maps, then this research also identified Hanus and Marada (2016).

Gender: It is another wildly discussed biological characteristic relative to visual perception. According to Šikl (2012, p. 31) "visual perception is one of the few activities of the mind, for which there are no widely accepted stereotypes regarding the difference between men and women. The research findings indicate some differences, but they should not be overestimated." A significant number of studies showed that the differences of average results between groups were smaller than the variance of the results within one group by gender (Mather 2006). Men showed somewhat better results in tests aimed at solving spatial tasks (Linn and Petersen 1985; Voyeur et al. 1995; Marmor and Zaback 1976). Women recognized colours faster and better (Nowaczyk 1982) and recalled the colours better (Pérez-Carpinelli et al. 1998). Hund and Gill (2014) researched the effect of stimuli and memory related to orientation and route search. They found that women are faster if they have more incentives when it comes to spatial tasks; men did not show any difference in time to find the way. A unique study by Campbell et al. (2014) examines the influence of age and gender on a mental topographical map of a user's town. The research involved 63 adult respondents aged between 20 and 79 years in Sydney and its surroundings. The subtest included tasks related to the identification of important elements in the landscape orientation (landmarks), map describing, route plotting and the evaluation of orientation. The best results concerning the orientation were reached by middle-aged men, while in women the orientation deteriorated with age; i.e. the gender and age influenced the orientation in a familiar environment; testing of other tasks concerning the topographical memory of well known sites did not show any differences between women and men, and age. Lloyd et al. (2002) however pointed out that many studies (e.g. by Schaefer and Thomas 1998; Stumpf 1998; Dunn and Eliot 1999) focusing on spatial abilities related to gender rarely used standardized tests. A number of studies evaluate spatial abilities in relation to short-term or long-term memory, possibly with visual memory (Silverman and Eals 1992; Barnfield 1999). The research by Linna and Petersen (1985), Halpern and Crothers (1997) indicate that men are more successful in solving spatial tasks where it is necessary to use short-term memory, whereas women can better recall spatial information from long-term memory (Galea and Kimura 1993; Birenbaum et al. 1994).

Visual impairment: Visual perception of reality can be influenced by various visual impairments: reduced sharpness, sensitivity to contrast and colour vision deficiency. An observer with worsening visual acuity may not have sufficient distinctive ability to find and correctly evaluate all the image features important to understand the map or image. A lower-level ability to distinguish different levels of brightness may result in homogeneous perception of various sections of the map or image. Finally, another important cue for recognizing objects in maps is *colour*, especially if it is a distinctive feature of the object. The inability to detect the colours affects the object identification, its accuracy and speed. Colour-blindness affects about 8% of men and 0.4% of women (Jenny and Kelso 2007). There are different cases of colour blindness. The most common one is the inability to distinguish the colours of long wavelengths. The affected person perceives them as yellowish-brown. Research focused on the relationship between the colour-blindness and map reading and processing were conducted by e.g. Brewer (1997), Olson and Brewer (1997), Harrower and Brewer (2003).

11.1.4   Biological Characteristics and States of the Individual in Relation with the Image Interpretation

Experience: The interpretation of the visual scene is influenced by the experience with an impulse (with map and image). Different knowledge and experience is manifested in various degrees of attention to certain details. Both the experience and the expertise help create an effective structuring of stimulus information (Šimeček and Šikl 2011). Image interpreting may be influenced by a certain form of *anticipation of the object features.* The research by Palmer (1975) has shown that the speed and success of recognition is higher, if the stimulus is semantically consistent with the scene. Maruff et al. (1999) show that the search strategy is being developed with the experience; we search the required information more efficiently in case we have previous experience. The interpretation also positively reflects that *the experience in terms of knowledge of the type of territory,* and anticipation of possible objects set within the territory, i.e. the objects expectation and spatial context are very important in the process of visual interpretation (Hollingsworth and Henderson 1999; Chun 2000). Also interpretation using the typical features of the objects is connected to prior experience—what features might be expected when dealing with a common object (Rosch 1973; Rosch and Mervis 1975; Lloyd et al. 1996). The research test included a task focusing the identification of a nuclear power plant. Without any prior knowledge and visual imagination of the object (the typical sign of a nuclear power plants are cooling towers) the object identification would not be possible. Experienced image readers proceed faster when data interpreting, they have more numerous but shorter eye movements within the image (Rayner 1978).

Experience and skills also relate to a broader concept of *cartographic literacy and literacy applied in image reading.* An experienced analyst of these images uses strong and weak features of each format to get an overview of the area and a variety of valuable information (Small 1999). The geographers are good experts in image reading thanks to their knowledge of object relation in the landscape (Dyce 2013). The study by Davies et al. (2006) focuses on monitoring the models of the visual attention depending on the type of image, image exposure time, the reader's expertise and the type of displayed landscape. Despite researchers' expectations experts were those who tended to have unevenly distributed attention concerning the image interpretation. Muir and Blaut (1969) tested five and six-year-old children (USA) before and after working with black and white vertical images. Children having worked with images, showed a better ability to interpret maps compared to children who did not participate in that activity. Van Coillie et al. (2014) analysed the accuracy of images digitizing by adult respondents with various degrees of experience with image working, with various degrees of their motivation. The accuracy of digitization of monitored respondents was generally very different and even less accurate.

Lloyd et al. (2002) prepared three cognitive experiments conducted with human subjects viewing a series of aerial photographs and categorizing the land use for target locations. Reaction time, accuracy, and confidence were considered as dependent variables related to the success of the categorization process. Subjects had significantly more success with photographs they viewed more than one time. Male subjects were significantly faster, more accurate, and more confident than female subjects at doing the categorization task. By the seventh learning round the male advantage in reaction time and accuracy was no longer significant, but the male advantage in confidence continued through seven learning rounds.

Šikl and a Svatoňová (being prepared) investigated how the fact of being an expert and the type of depicted landscapes in the picture influenced the ability to remember these images. 120 people with various degrees of expertise were tested. The individual's expert experience proved to be significant in the ability to remember the image. The experts were able to correctly remember on average roughly more than 10% of the previously viewed images than laymen. Different results were also achieved in relation to the type of landscape. Concerning the sub-types (urban areas, housing estate areas, parks and gardens, industrial landscapes, transport landscapes), as the most difficult ones to memorize were settlement with a regular structure of built area without any significant elements. However, it is interesting that these differences were not as significant among the experts as they were among the laymen.

11.1.5 The Influence of Stress on Cognitive Processes and Image Interpretation

Stress weakens cognitive functions, concentration, logic and abstract thinking. As the cause of the weakening high levels of emotional activity disturbing the information processing is reported. Mental processes are disrupted by thoughts connected with a possible failure. The weakening of cognitive functions may also result in rigidities or alternative actions. Some people tend to resort to their pre-matured modes of behaviour (Atkinson et al. 1995). Taking into account the psychological description of the effect of stress on thinking in relation with crisis management and image and map interpretation, it is necessary to note especially the first phase of a catastrophic syndrome associated with a loss of orientation concerning the rescued effect of stress on the mind, and weakening of abstract thinking. Therefore, it seems more appropriate to use a simpler basis to transmit spatial information, which is not as demanding as abstract operations.

11.2 The Survey—Comparing the Success of Aerial and Satellite Image Interpretation by Gender

The survey presents the results of the successful visual interpretation of aerial and satellite images assessed by gender and expert experience of individuals. The main aim of the research was to test hypotheses concerning the evaluation of differences in problem solving tasks according to selected biological features and expert knowledge of the respondents. Hypotheses to be verified were: H1—Men and women are equally successful in the interpretation of aerial and satellite data, H2—Men and women are equally quick in interpreting aerial and satellite data. A successful interpretation was assessed with two parameters—the accuracy of solution to the problem and the speed of problem solving by the respondents with regard to their gender and expert experience. The following documents were used to prepare the problem solving tasks concerning the aerial and satellite data and map interpretation: a basic topographic map, scale 1:10 000 and 1:25 000 and 1:50 000; a coloured aerial vertical image, resolution of 0.5 m; orthophoto scale of 1:10 000 resolution 0.5 m; oblique colour image, resolution of 0.5 m and without any resolution; satellite image in natural colours (Landsat 7, RGB 321), scale 1: 100,000, resolution 30 m; satellite image in an unnatural colours (Landsat 7, RGB 742), scale 1: 100,000, resolution 30 m; Aqua satellite image, passive multispectral spectral radiometer MODIS, natural colours, resolution of 250 m. The sources of images and maps were: Map Server "National INSPIRE geo-portal" map server "mapy.cz" NASA website, the website Project Copernicus Emergency Management Service.

11.2.1 Research Respondents—The Structure by Gender, Age and Expert Knowledge

The research included 151 respondents, 67 men and 84 women. 67% of respondents were aged 20–25, 10% of respondents 26 to 30, 23% of respondents were over 31 years old. 59 respondents worked in their job with maps and aerial or satellite data.

11.2.2 Results of Testing

Pairwise testing and the corresponding data processing were used to verify the set hypotheses. The hypotheses were tested at a significance level of 0.05. The hypotheses were tested using non-parametric methods (Wilcoxon test) and parametric methods (t-test, parameter binomial distribution, chi2 test, McNemar test). The testing and data processing was executed using the computing program MATLAB 8.1, results image processing was executed using the program Statistics and MS Excel. The average success rate (accuracy) and the speed of problem solving of an aerial photograph and map were different: it was 82% when reading images, 74% when reading maps. The averages concerning correct answers differed: the image value was 8.2 and map value was 7.4, the median value was equal. The statistical dispersion of responses was higher when reading maps. The respondents were able to solve the tasks quicker when reading aerial images. Both the hypotheses H1 and H2 concerning the comparison of problem solving by gender were confirmed.

11.3   Conclusion

We found that the results of successful solutions of respondent groups are the equal, i.e. women and men are equally successful when identifying objects on aerial photographs, maps, images using true and false colours. The results which prove that men and women are equally successful in solving problems related to map reading are not consistent with some of the other research supporting the greater success of men (Chang and Antes 1987). Šikl (2012) however believes that the results of the comparison by gender should not be unambiguously interpreted; a significant number of studies showed that the differences of the average results between male and female groups were smaller than the variance of the results within one group by gender (Mather 2006). Linn and Petersen (1985), Voyeur et al. (1995), Marmor and Zaback (1976) prove that men showed slightly better results when testing solutions to spatial tasks. Some studies by Kempf et al. (1997), Johnson and McCoy (2000) approach the assessment of greater success from a different point of view; Men are more confident because they are faster and more decisive in problem solving than women who reassure and try to fulfil the task very carefully.

The satellite and aircraft image data represent a realistic view of the landscape, which is what differs fundamentally both the sources from the maps: the map is a model of reality using the signs to represent the objects. The map is therefore more demanding on abstract thinking, and it can be assumed that they are the reasons that men usually succeed more in the map interpretation (Hanus and Marada 2016 demonstrated it too). The results of the interpretation of the satellite and aircraft data images research showed that on this basis both the genders are equally successful.

Acknowledgements This research has been supported by funding from Masaryk University under the grant agreement No. MUNI/M/0846/2015, which is called 'Influence of cartographic visualization methods on the success of solving practical and educational spatial tasks'.

References

Atkinson, R. L., Atkinson, R. C., Smith, E. E., Bemd, D. J. and Nolen-Hoeksema, S. (1995). Psychologie. Praha: Victoria Publishing.
Baker, D. H. and Graf, E. W. (2008). Equivalence of physical and perceived speed in binocular rivalry. Journal of Vision, 8, pp. 1–12.
Barnfield, A. (1999). Development of sex differences in spatial memory. Perceptual and Motor Skills, 89, pp. 339–350.

Biederman, I., Mezzanotte, R. J. and Rabinowitz, J. C. (1982). Scene perception: Detecting and judging objects undergoing relational violations. Cognitive Psychology, 14, pp. 143–177.

Birenbaum, M., Kelly, A. and Levi-Keren, M. (1994). Stimulus features and sex differences in mental rotation test performance. Intelligence, 19, pp. 51–64.

Blaut, J. M. (1997). The mapping abilities of young children: Children can. Annals of American Geographers, 87, pp. 152–158.

Brewer, C. A. (1997). Spectral schemes: Controversial color use on maps. Cartography and Geographic Information Systems, 24, pp. 203–220.

Broadbent, D. E. (1958). Perception and communication. Oxford: Pergamon.

Campbell. J. I., Hepner, I. J. and Miller, L. A., (2014). The influence of age and sex on memory for familiar environment. Journal of Environmental Psychology, 40, pp. 1–8.

Crick, F. (1997). Věda hledá duši. Praha: Mladá fronta.

Chang, K. T. and Antes, J. R. (1987). Sex and cultural differences in map reading. The American Cartographer, 14, pp. 29–42.

Chun, M., 2000. Contextual cueing of visual attention. Trends in Cognitive Science, 4, pp. 170–178.

Davenport, J. L. and Potter, M. C. (2004). Scene consistency in object and background perception. Psychological Science, 15, pp. 559–564.

Davies, C., Tompkinson, W., Donnelly, N., Gordon, L. and Cave, K. (2006). Visual saliency as an aid to updating digital maps. Computers in Human Behavior, 22, pp. 672–684.

Davis, E. T., Fujawa, G. and Shikano, T. (2002). Perceptual processing and search efficiency of young and older adults in a simple-feature search task: A staircase approach. Journal of Gerontology: Psychological science, 57, pp. 324–337.

Downs, R. M. and Liben, L. S. (1997). The final summation: The defence rests. Annals of American Geographers, 87, pp. 178–180.

Dunn, A. and Eliot, J., (1999). An exploratory study of undergraduates' attributions of success or failure on spatial tests. Perception and Motor Skills, 89, pp. 695–702.

Dyce, M. (2013). Canada between the photograph and the map: Aerial photography, geographical vision and the state. Journal of Historical Geography, 39, pp. 69–84.

Eysenck, M. W.and Keane, M. T. (2008). Kognitivní psychologie. Praha: Academia.

Galea, L. and Kimura, D. (1993). Less skilled readers have less efficient suppression mechanism. Psychological Science, 4, pp. 294–298.

Gibson, J. J. (1979). The ecological approach to visual perception. Boston: Houghton-Mifflin.

Gregory, R. L. (1966). Eye and brain: The psychology of seeing. New York: McGraw-Hill.

Halpern, D. and Crothers, M. (1997). The sex of cognition. In L. Ellis (Ed.) Sexual orientation: Toward biological understanding. Westport, CT: Praeger, 187–197.

Hanus, M., Marada, M. (2014): Map skills: definition and research. Geografie, 119, No. 4, pp. 406–422.

Hanus, M., Marada, M. (2016): What does a map-skills-test tell us about Czech pupils? Geografie, 121, 2, 279–299.

Harrower, M. and Brewer, C. A. (2003). ColorBrewer.org: An online tool for selecting colour schemes for maps. The Cartographic Journal, 40, pp. 27–37.

Hock, H. S., Gordon, G. P. and Whitehurst, R. (1974). Contextual relations: The influence of familiarity, physical plausibility, and belongingness. Perception and Psychophysics, 16, pp. 4–8.

Hollingworth, A. and Henderson, J., (1999). Object identification is isolated from scene semantic constrain: Evidence from object type and token discrimination. Acta Psychological Sciences, 10, pp. 319–343.

Hošková, Š., Mokrá T. (2010) Alexithymia among students of different disciplines, Procedia–Social and Behavioral Sciences, 9(2010), p. 33–37.

Hoskova-Mayerova, S., Talhofer, V., Hofmann, A., Kubíček, P, (2013). Mathematical model used in decision-making process with respect to the reliability of geodatabase. Studies in Computational Intelligence, 448, pp. 127–142.

Hoskova-Mayerova, S., Hofmann, A. (2016). Development of Applications in the Analysis of the Natural Environment. In: *Aplimat - 15th Conference on Applied Mathematics 2016 Proceedings*. Bratislava, Slovensko: Vydavatelstvo STU Bratislava, 2016, p. 467–481.

Hund, A. M. and Gill, D. M., (2014). What constitutes effective wayfinding directions: The role of descriptive cues and task demands? *Journal of Environmental Psychology*, 38, pp. 217–224.

Jenny, B. and Kelso, N. V., 2007. Color design for the color vision impaired. *Cartographic perspectives*, 57, pp. 61–67.

Johnson, W. and McCoy, N. (2000). Self-confidence, self-esteem, and assumptions of sex role in young men and women. Perceptual and Motor Skills, 90, pp. 751–756.

Kempf, D. S., Palan, K. M. and Laczniak, R. N. (1997). Gender differences in information processing confidence in an advertising context: A preliminary study. Advances in Consumer Research, 24, pp. 443–449.

Kovarik V. (2012). Effects and limitations of spatial resolution of imagery for imagery intelligence. In: *Proceedings of the International Conference on Military Technologies and Special Technologies - ICMT'-2012*. Trenčín: Alexander Dubček University of Trenčín, pp. 363–368.

Kubicek, P., Sasinka, C, Stachon, Z., Sterba, Z., Apeltauer, J and Urbánek, T. (2016). Cartographic Design and Usability of Visual Variables for Linear Features. The Cartographic Journal, 1/2016, pp. 1–11.

Liben, L. S.and Downs, R. M. (1997). Canism and can'tianism: A straw child. Annals of the Association of American Geographers, 87, pp. 159–167.

Linn, M. C. and Petersen, A. C. (1985). Emergence and characterization of sex differences in spatial ability: A meta-analysis. *Child Development*, 56, pp. 1479–1498.

Lloyd, R., Patton, D. and Cammack, R., (1996). Basic-level geographic categories. *The Professional Geographer*, 48, pp. 181–194.

Lloyd, R., Hodgson, M. E. and Stokes, A. (2002). Visual categorization with aerial photographs. *Annals of the Association of American Geographers*, 92, pp. 241–266.

Madden, D. J. (2001). Speed end timing of behavioral processes. In J. E. Birrenand K. W. Schaie (Eds.) Handbook of the Psychology of aging. 5th edn. San Diego, CA: Academic Press, pp. 288–312.

Madden, D. J. and Allen, P. A. (1991). Adult age differences in the rate of information extraction during visual search. *Journal of Gerontology: Psychological Sciences*, 46, pp. 124–126.

Marmor, G. and Zaback, L., (1976). Mental rotation by the blind: Does mental station depend on visual imagery? *Journal of Experimental Psychology: Human Perception and Performance*, 2, pp. 515–521.

Maruff, P., Danckert, J., Camplin, G. and Currie, J. (1999). Behavioral goals constrain the selection of visual information. *Psychological Science*, 10, pp. 522–525.

Mather, G. (2006). *Foundations of Perception*. Psychology Press, Hove, UK.

Muir, M. and Blaut, J. (1969). The use of aerial photographs in teaching mapping to children in first grade: An experimental study. *The Minnesota Geographer*, 22, pp. 4–19.

Neisser, U. (1976). Cognition and reality. San Francisco, CA: W. H. Freeman.

Nowaczyk, R. H. (1982). Sex-related differences in the color lexicon. *Language and Speech*, 25, pp. 257–265.

Olson, J. M. and Brewer, C. A. (1997). An evaluation of color selections to accommodate map users with color-vision impairments. *Annals of the Association of American Geographers*, 87, pp. 103–134.

Palmer, S. E. (1975). The effects of contextual scenes on the identification of objects. Memory and Cognition, 3, pp. 519–526.

Pérez-Carpinell, J., Baldoví, R., de Fez, M. D. and Castro, J. (1998). Color memory matching: Time effect and other factors. *Color Research and Applications*, 23, pp. 234–247.

Rayner, K., (1978). Eye movements in reading and information processing. *Psychological Bulletin*, 85, pp. 618–660.

Rosch, E., (1973). Natural categories. *Cognitive Psychology*, 4, pp. 328–350.

Rosch, E. and Mervis, C. (1975). Family resemblances: Studies in the internal structure of categories. *Cognitive psychology*, 7, pp. 573–605.

Řezníčková, D., Marada, M., and Hanus, M. (2014). Geographic skills in Czech curricula: analysis of teachers´opinions. In. D., Schmeinck, J., Lidstone (Eds.): *Standards and Research in Geography Education. Current Trends and International Issues*. Mensch und Buch, Berlin, pp. 37–49.

Schaefer, P. and Thomas, J. (1998). Difficulty of spatial task and sex differences in gains from practice. *Perceptual and Motor Skills*, 87, pp. 56–58.

Scialfa, C. T. (2002). The role of sensory factors in cognitive aging research. Canadian Journal of Experimental Psychology, 56, pp. 153–163.

Salthouse, T. A. (2000). Aging and measures of processing speed. Biological Psychology, 54, pp. 35–54.

Salthouse, T. A. and Somberg, B. L. (1982). Isolating the age deficit in speeded performance. Journal of Gerontology, 37, pp. 59–63.

Silverman, I. and Eals, M. (1992). Sex differences in spatial abilities: Evolutionary theory and data. In J. Barkow, L. Kosmidesand J. Tooby (Eds.) The Adapted Mind. New York: Oxford University Press.

Sternberg, R. J. (2002). Images of Mindfulness. Journal of Social Issues. 56, pp. 11–26.

Stumpf, H. (1998). Gender-related differences in academically talented student's test scores and use of time on tests of spatial ability. *Gifted Child Quarterly*, 42, pp. 157–171.

Svatoňová, H. and Lauermann, L. (2010). *Dálkový průzkum Země – aktuální zdroj geografických informací*. Masarykova univerzita, Brno.

Svatonova, H. and Rybansky, M. (2014). Children observe the Digital Earth from above: How they read aerial and satellite images. In IOP Conference Series: *Earth and Environmental Science* 18 012071. Malaysia: ISDE, p. 7.

Svatonova, H. (2016a). Analysis of visual interpretation of satellite data. *International Archives of the Photogrammetry, Remote Sensing and Spatial Information Sciences - ISPRS Archives,* 41, pp. 675–681.

Svatonova, H. (2016b). New Trends in Obtaining Geographical Information: Interpretation of Satellite Data, *Recent Trends in Social Systems: Quantitative Theories and Quantitative Models*. Studies in System, Decision and Control 66. Antonio Maturo et al. (Eds), Springer International Publishing, pp. 173–182.

Šimeček, M. and Šikl, R. (2011). How we estimate the relative size of human figures when seen on a photography. Perception. 40, pp 148–148.

Šikl, R. (2012). *Zrakové vnímání*. Grada Publishing, Praha.

Talhofer, V., Hoskova-Mayerova, S., Hofmann, A. (2016) Verification Tests of Mathematical Models of Passage throug Terrain. In: *Aplimat - 15th Conference on Applied Mathematics 2016 Proceedings*. Bratislava, Slovensko: Nakladatelstvo STU Bratislava, 2016, p. 1010–1025.

Talhofer, V., Hoskova-Mayerova, S. (2016). Towards efficient use of resources in military: methods for evaluation routes in open terrain. *Journal of Security Sustainability Issues*, 2016, 6 (1), p. 53–70.

Van Coillie, F., Gardin, S., Anseel, F., Duyck, W., Verbeke, L. and De Wulf, R. (2014). Variability of operator performance in remote-sensing image interpretation: The importance of human and external factors. *International Journal of Remote Sensing*, 35, pp. 754–778.

Voyer, D., Voyer, S. and Bryden, M. P. (1995). Magnitude of sex differences in spatial abilities: A meta-analysis and consideration of critical variables. *Psychological Bulletin*, 117, pp. 250–270.

Electronical Sources

Copernicus emergency management service: A service in support of European emergency response [online]. [cit. 5. 1. 2014]. Available from: <http://emergency.copernicus.eu/>.

Mapový portál Mapy.cz [online]. © Seznam.cz, a.s. [cit. 2. 11. 2013]. Available from: <http://www.mapy.cz>.

Národní geoportál INSPIRE [online]. [cit. 10. 10. 2013]. Available from: <http://geoportal.gov.cz>.

NASA Education: For educators [online]. [cit. 4. 1. 2014]. Available from: <http://www.nasa.gov/audience/foreducators>.

Peace, A. and Peace, B. (2001). *Why men don't listen and women can't read maps: How we're different and what to do about it.* Netley: Pease International Pty Ltd [online]. [cit. 29.8.2014]. Available from: <http://nguyenthanhmy.com/courses/2013/WhyMen.pdf>.

Chapter 12
Misconceptions Regarding Providing Citations: To Neglect Means to Take Risk for Future Scientific Research

Engin Baysen, Šárka Hošková-Mayerová, Nermin Çakmak and Fatma Baysen

Abstract Academic integrity is one of the fundamental values of being honest. The present study is aimed at finding out citation understandings of Czech (n = 283) and Turkish (n = 182) secondary and high school students (13–20-year old). Except for a few students, secondary and high school students have misconceptions concerning providing citations. Students are unintentionally vulnerable to plagiarize while reporting. The study showed that only secondary and high school education is not enough for implementing honesty regarding citation and we will face a significant risk for future scientific research, unless we treat, educate and inform our secondary and high school students properly and continuously about honesty in research and plagiarism.

Keywords Citations · Academic honesty · Plagiarism · Misconceptions · Secondary · High school students · Risk

E. Baysen · F. Baysen
Department of Primary School Education, Near East University, Nicosia, Cyprus
e-mail: bengin71@gmail.com

F. Baysen
e-mail: fatma.baysen@neu.edu.tr

Š. Hošková-Mayerová (✉)
Department of Mathematics and Physics, University of Defence, Brno,
Czech Republic
e-mail: sarka.mayerova@unob.cz

N. Çakmak
Library Department of Chamber of Architects of Turkey, Ankara, Turkey
e-mail: ncakmak73@gmail.com

© Springer International Publishing AG 2017
Š. Hošková-Mayerová et al., *Mathematical-Statistical Models and Qualitative Theories for Economic and Social Sciences*, Studies in Systems, Decision and Control 104,
DOI 10.1007/978-3-319-54819-7_12

12.1 Introduction

One of the prominent aims of education is to bring up honest people. Academic integrity is one of the fundamental values (Schmelkin et al. 2010) of being honest. Although it is thought that increased politicy, guides, teaching methods and plagiarism detection tools should turn in increased awareness regarding plagiarism (Bacha et al. 2012), plagiarism, cheating, academic dishonesty and the similar acts have widened (Schmelkin et al. 2010) and became an important problem for education institutes (Bacha et al. 2012; Ho 2015; Sobhy 2015; Rosicka 2008a, b).

Jensen et al. (2002) defined academic dishonesty as students' attempt to present someone else's work as their own and include cheating in exams, copying other students' work and plagiarism. Those research regarding secondary and high school students is found to be focused mostly on cheating, categorised as academic dishonesty (e.g. Kessler 2003; Ma et al. 2007; Strom and Strom 2007; Sobhy 2015; Barnhardt 2016; Hasilová 2014; Rosicka 2006). This study, on the contrary, is focused on the act of plagiarism. Plagiarism (2015) in Merriam-Webster Online Dictionary is defined as: *"the act of using another person's words or idea without giving credit to that person; the act of plagiarizing"*. Walker (2010) categorised the act of plagiarism in three classes: (1) Presenting the information when directly citing, utilizing other words or changing the words found in the original source without using quotation marks (Sham paraphrasing); (2) Presenting the information as yours copied verbatim from the source without citing the source (Verbatim) and (3) To present a part or whole work of other student's as his/her own within or without other students' knowledge (Purloining).

According to Sureda and Comas (2010), plagiarism, copying, deceiving and cheating, have been acted for a very long time in classes. These issues and phenomena have reached an alarming rate parallel to proliferation of internet use all around the World. Bibliometric indicators has shown clearly that plagiarism has increased excessively (Sureda et al. 2015). Research dealing with plagiarism has increased in return in the act of increased plagiarism. While research found in the literature is largely related with academic dishonesty of university students and academicians (Bennet 2005; Gullifier and Tyson 2010; Estow, Lawrence and Adams 2011; Löfström 2011; Evering and Moorman 2012; Chen and Chou 2015; Horová et al. 2013), there is a scarcity in those regarding secondary and high school students (Dant 1986; Kessler 2003; Ma et al. 2007; Lai and Weeks 2009; Sureda et al. 2015; Vieyra and Weaver 2016). Vieyra and Weaver (2016) revealed in their research that secondary and high school students do have confusions concerning plagiarism, which resist at higher schools. Vieyra and Weaver explained the reasoning behind plagiarism confusions at secondary and high school students as lack in education and inconsistent instruction. They stated that students include reference page, however, less than 35% only include in-text citations in their research projects. Analysing 6th and 12th grade students' scientific papers concerning attribution, prevalence and quality, Vieyra and Weaver also found that 12th grade students are not more successful in providing proper citations than 6th grade students. According to Vieyra and Weaver, teachers and institutes being in charge of supervising student research should guarantee teaching of providing proper citations. Thus, those problems depending on such habits generated long time ago before higher school regarding citations can be prevented. Bacha et al. (2012) stressed that, if high school teacher awareness is increased concerning students' way of cheating and plagiarism, then students are well prepared for higher education where the sanctions are more strictly applied. Although Jensen et al. (2002) stated that in twenty years' time of academic dishonesty among higher schools it was increasing in a fast rate; they emphasized that the act of academic dishonesty is more prevailing among secondary school students. Jensen et al. (2002) stated types of academic dishonesty, which are most widely spread such as: cheating in tests and homework, and plagiarism in books and articles. In addition, they stated that lower grades do plagiarise more than higher ones.

Sureda et al. (2015) also found out that plagiarism is prevalent among secondary and high school students. Comas and Sureda (2010) classified the reasoning for plagiarism such as: individual factors ("academic performance, procrastination, gender, motivation, etc."), institutional factors ("the existence of academic regulations that address the issue of plagiarism, the ethical culture of education centre, the existence and use of detection programmes, etc."), instructional factors ("types of assignments that are given, number of assignment given, follow-up on assignments by the teacher, etc.") and, finally, out-of-school factors ("levels of political corruption, crisis in the system of values, etc.").

In parallel with improvements in information and communication, students' easy access to limitless internet and similar technologies caused increase in copy-paste type plagiarism among secondary and high school students (Ma et al. 2007; Sisti 2007; Lai and Weeks 2009; Lehman 2010; Thomas and Sassi 2011; Williamson and McGregor 2011). Consistently, Lai and Weeks (2009) in 2006 found out that there were more than 250 'paper mills' and claimed that these websites increased online plagiarism among students.

Literature analysis regarding secondary and high school students showed that researchers deal with intentional plagiarism. The present study, on the other hand, is focused on misconceptions regarding plagiarism, classified as unintentional plagiarism (Baysen et al. 2016). In general, research concerning plagiarism dealing with misconceptions is bound to higher education, not to secondary and high school students (Belter and Pre 2009; Graveline 2010; Mahmood et al. 2010; Ahmad et al. 2012; Cheak et al. 2013; Çakmak 2015; Baysen et al. 2016). Thus, the present study is focused on citation misconceptions of secondary and high school students. The study aims to reveal whether secondary and high school students do have misconceptions regarding citing and, if so, what kind of misconceptions they have.

Finally, the present study selected secondary and high school students to reveal their understanding concerning plagiarism; thus, it becomes urgent to enhance students' knowledge of plagiarism, academic honesty, citation process as well as to improve their scientific writing skills and emphasize and encourage the academic ethics. The present study aimed at revealing precious information for teachers, education institutes, policy-makers and researchers.

12.2 Method

Descriptive quantitative research approach is carried out to reveal misconceptions among students concerning citation. The present study has adopted the approach by Baysen et al. (2016). The used questionnaire was as follows:

Questionnaire

1. You can copy the information on the internet without providing citation because it is anonymous.

2. If you citate/quote an author than you have to get permission from him/her.

3. There is no need to provide citation whenever a figure or a table is copied.

4. There is a need to cite if only most of the research paper is used.

5. There is a need to cite if only original words used in the source are copied.

6. If you write the author's thinking in your own words, than there is no need to cite.

7. If you change the words of the author than there is no need to cite.

8. If the information is known by everyone and is widely used or common, there is no need to provide citation.

9. If the sentences copied are short, there is no need to provide citation.

10. If you translate the information (from other language), there is no need to provide citation.

11. If you summarize part of author's sentences, you can provide citation at the end of your text.

12. If you summarize more than one paragraph of an author and write your own paragraph, there is no need to provide citation.

13. If you use a friend's written work, there is no need to provide citation.

14. If the data is from interviews and conversation conducted by you, there is no need to provide citation.

12.2.1   Participants

Total of 465 Czech and Turkish secondary and high school students participated in the conducted study voluntarily, aged between 13–20 years; of these 283 are Czech, and 182 are Turkish students. The Czech group included 143 girls and 140 boys while Turkish group included 111 girls and 71 boys. Both Czech and Turkish students are from different secondary and high schools in Brno and Nicosia, respectively. These students did not pass any special program or seminar concerning plagiarism.

12.2.2   Data Collection

A questionnaire had been applied consisting of 14 Likert type (Yes- No- I am not sure) questions implemented by Baysen et al. (2016). The items were comprehensible by secondary and high school students. Comprehensiveness had been tested by a pilot study before the research. Students answered the questions in their classes. Researchers were communicable, both in the classes and during the applications. Students were given information about the researchers, when they needed. Students finished answering in approx. 10 min.

12.2.3   Data Analysis

Number of conceptions and misconceptions were count and Chi square ($p = 0.05$ or $p = 0.001$) is calculated for each item to interpret the differences between the two and between Czech and Turkish students. The numbers of students being not sure regarding each item were also summed for further interpretation.

12.2.4   Results

In Turkish group, only two students were found out answering all the questions correctly. One of them answered all the items such as: "I am not sure". Thus, this fact can be interpreted that about all the Turkish participants in the present study may show (or will show) the act of plagiarism (Table 12.1). In addition, significantly more Turkish students answered wrong items 2, 4 and 5. Significantly more Turkish students answered correctly the items 8, 9, 10 and 11. Finally, the number of Turkish students answering correctly is not significantly different from the number of students answering wrong the items 1, 3, 6, 7, 12, 13 and 14. For Turkish group students, items number 2 and 4 were the most challenging, while items

Table 12.1 Turkish and Czech students (mis)conceptions and significance calculations in providing citations

Question no.	Turkish students						Sig.	Czech students						Sig.
	Conception (C)		Misconception (M)		Not sure			Conception (C)		Misconception (M)		Not sure		
	f	%	f	%	f	%		f	%	f	%	f	%	
1	54	30	56	31	72	40	No Diff.; p > 0.05	173	61	61	22	49	17	C > M; p < 0.001
2	45	25	89	49	48	26	M > C; p < 0.001	218	77	38	13	27	10	C > M; p < 0.001
3	50	28	70	39	63	34	No Diff.; p > 0.05	104	37	124	44	55	19	No Diff.; p > 0.05
4	37	20	89	49	56	31	M > C; p < 0.001	124	44	96	34	63	22	No Diff.; p > 0.05
5	45	25	84	46	53	29	M > C; p < 0.05	71	25	171	60	41	15	M > C; p < 0.001
6	57	31	76	42	49	27	No Diff.; p > 0.05	76	27	175	62	32	11	M > C; p < 0.001
7	69	38	61	34	52	29	No Diff.; p > 0.05	96	34	142	50	45	16	M > C; p < 0.05
8	96	53	40	22	46	25	C > M; p < 0.001	134	47	89	31	60	21	C > M; p < 0.05
9	76	42	41	23	65	36	C > M; p < 0.05	176	62	53	19	54	19	C > M; p < 0.001
10	78	43	48	26	56	31	C > M; p < 0.05	159	56	64	23	60	21	C > M; p < 0.001
11	97	53	40	22	45	25	C > M; p < 0.001	169	60	41	15	73	26	C > M; p < 0.001
12	64	35	59	32	59	32	No Diff.; p > 0.05	126	45	81	29	76	27	C > M; p < 0.05
13	70	39	49	27	63	35	No Diff.; p > 0.05	169	60	70	25	44	16	C > M; p < 0.001
14	53	29	64	35	65	36	No Diff.; p > 0.05	97	34	126	45	60	21	No Diff.; p > 0.05

Table 12.2 Comparison between the number of conceptions between Czech and Turkish students regarding each item

Quest. no.	Sig.	In favour of	
		Czech students	Turkish students
1	Yes; p < 0.001	√	
2	Yes; p < 0.001	√	
3	**No Diff.**		
4	Yes; p < 0.001	√	
5	**No Diff.**		
6	Yes; p < 0.05		√
7	Yes; p < 0.05		√
8	Yes; p < 0.05	√	
9	Yes; p < 0.05	√	
10	Yes; p < 0.05	√	
11	Yes; p < 0.05	√	
12	**No Diff.**		
13	Yes; p < 0.05	√	
14	**No Diff.**		

number 8 and 11 were the easiest ones for this group. However, 10 (3%) Czech students were found out answering all the questions correctly. However, this percentage is accepted as low as well. The question number 6 is the most difficult item for this group, while the question number 2 is the easiest one. Significantly, more Czech students answered wrong item 5, 6 and 7 (Table 12.1); more Czech students answered items 1, 2, 8, 9, 10, 11, 12 and 13 correctly. Finally, the number of Czech students answering correctly is not significantly different from the number of students answering wrong the items 3, 4 and 14. Czech students are more successful concerning items 1, 2, 4, 8, 9, 10, 11 and 13 comparing to their counterparts. On the contrary, Turkish students are more successful regarding items 6 and 7. Finally, there are no significant differences between the number of Czech and Turkish students answering correctly regarding the items 3, 5, 12 and 14 Table 12.2.

12.3   Conclusion and Discussion

Except for a few students, both Czech and Turkish secondary and high school students have misconceptions concerning providing citations. Thus, the study revealed an important risk for future scientific research, unless we treat, educate and inform our secondary and high school students properly and continuously about honesty in research and plagiarism. This study showed that only secondary and high school education is not enough for implementing honesty in terms of citation. Those future higher education students and future scientists will plagiarise in their research, which is consistent with the findings of Baysen et al. (2016). This is surely thought to preclude improvement both in science and technology. There are differences between Czech and Turkish students' conceptions concerning providing citations, reminding the cultural differences. Czech students seemed more knowledgeable concerning providing citations and to refrain from plagiarism than their counterparts, Turkish students. The results of the present study are consistent, indeed, with the results of Baysen et al. (2016). Hereby, the present study has widened the age range concerning the relation between providing citations (plagiarism) and misconceptions. Thus, theory asserting that the misconceptions are the causes of plagiarism, in other words "Students plagiarism act is unintentional" has been strengthened.

Acknowledgements The second author was supported within the project for "Development of basic and applied research" developed in the long term by the departments of theoretical and applied bases of the FMT UoD (Project code: "VYZKUMFVT" (DZRO K-217)) supported by the Ministry of Defence of the Czech Republic.

References

Ahmad, U. K, Mansourizadeh, K. and Ai, G. K. M. (2012). Non-native university students' perception of plagiarism. Advances in Language and Literacy Studies, 3 (1), 39–48. Retrieved from http://journals.aiac.org.au/index.php/alls/article/view/42/39.

Bacha, N. N., Bahous, R., & Nabhani, M. (2012). High schoolers' views on academic integrity. Research Papers in Education, 27(3), 365–381. Retrieved from http://www.tandfonline.com/doi/pdf/10.1080/02671522.2010.550010?needAccess=true.

Barnhardt, B. (2016). The "epidemic" of cheating depends on its definition: A critique of inferring the moral quality of "cheating in any form". Ethics & Behavior, 26(4), 330–343. Retrieved from http://www.tandfonline.com/doi/pdf/10.1080/10508422.2015.1026595?needAccess=true.

Baysen, E., Hoskova-Mayerova, S., Çakmak, N. and Baysen, F. (2016). Misconceptions of Czech and Turkish university students in providing citations. In Maturo, A., Hoskova-Mayerova, S., Soitu, D.-T., Kacprzyk, J. (Series Eds.), Studies in Systems, Decision and Control Series: Vol. 66 Recent trends in social systems: Quantitative theories and quantitative models (pp. 183–190). Switzerland: Springer. doi:10.1007/978-3-319-40585-8_16.

Belter, R. W. and Pré, A. (2009). A strategy to reduce plagiarism in an undergraduate course. Teaching of Psychology, 36 (4), 257–261. doi:10.1080/00986280903173165.

Bennet, R. (2005). Factors associated with student plagiarism in a post-1992 university. Assessment & Evaluation in Higher Education, 30 (2), 137–162. Retrieved from http://www.tandfonline.com/doi/pdf/10.1080/0260293042000264244?needAccess=true.

Cheak, A. P. C., Sze, C. C., Ai, Y. J., Min, C. M. and Ming, S. J. (2013, 26–27 September). Internet plagiarism: University students' perspective. In: International Research Conference 2013 Retrieved from http://scholar.google.com.tr/scholar?q=Internet+plagiarism%3A+university+students%27+perspective&btnG=&hl=tr&as_sdt=0%2C5.

Chen, Y. & Chou, C. (2015). Are we on the same page? College students' and faculty's perception of student plagiarism in Taiwan. Ethics & Behaviour, 00(00), 1–21. Retrieved from http://www.tandfonline.com/doi/pdf/10.1080/10508422.2015.1123630?needAccess=true.

Comas, R. & Sureda, J. (2010). Academic plagiarism: Explanatory factors from students' perspective. Journal of Academic Ethics, 8(3), 217–232. doi:10.1007/s10805-010-9121-0.

Çakmak, N. (2015). Undergraduates' misconceptions concerning plagiarism. Turkish Librarianship 29(2), 212–240. Retrieved from http://www.tk.org.tr/index.php/TK/article/view/2543.

Dant, D. R. (1986). Plagiarism in high school: A survey. The English Journal, 75(2), 81–84. Retrieved from http://www.jstor.org/stable/pdf/817898.pdf?_=1471307035702.

Estow, S., Lawrence, E. K., & Adams, K. A. (2011). Practice makes perfect: Improving students' skills in understanding and avoiding plagiarism with a themed methods course. Teaching of Psychology, 38(4), 255–258. Retrieved from http://top.sagepub.com/content/38/4/255.full.pdf +html.

Evering, L. C. & Moorman, G. (2012). Rethinking plagiarism in the digital age. Journal of Adolescent & Adult Literacy, 56(1), 35–44. doi:10.1002/JAAL.00100.

Gullifer, J. & Tyson, G. A. (2010). Exploring university students' perceptions of plagiarism: a focus group study. Studies in Higher Education, 35 (4), 463–481. Retrieved from http://www.tandfonline.com/doi/pdf/10.1080/03075070903096508?needAccess=true.

Graveline, J. D. (2010). Debunking common misconceptions and myths. College & Undergraduate Libraries, 17(1), 100–105. doi:10.1080/10691310903584650.

Hasilová, K. (2014) Iterative Method for Bandwidth Selection in Kernel Discriminant Analysis. In: 32nd International Conference Mathematical Methods in Economics MME2014. Olomouc: Palacký University, Olomouc, 2014, 263–268.

Ho, J. K. K. (2015). An exploration of the problem of plagiarism with the cognitive mapping technique. Systems Research and Behavioral Science, 32(6), 735–742. doi:10.1002/sres.2296.

Horová, I., Koláček, J. & Hasilová, K. (2013) Full bandwidth matrix selectors for gradient kernel density estimate. Computational Statistics & Data Analysis, 57(1), 364–376.

Jensen, L. A., Arnett, J. J., Feldman, S. S., & Cauffman, E. (2002). It's wrong, but everybody does it: Academic dishonesty among high school and college students. Contemporary Educational Psychology, 27(2), 209–228. doi:10.1006/ceps.2001.1088.

Kessler, K. (2003). Helping high school students understand academic integrity. The English Journal, 92(6), 57–63. Retrieved from http://www.jstor.org/stable/pdf/3650536.pdf.

Lai, K. W., & Weeks, J. J. (2009). High school students' understanding of e-plagiarism: Some New Zealand observations. CINZS: LTT, 21(1), 1–15. Retrieved from http://www.otago.ac.nz/cdelt/otago067253.pdf.

Lehman, K. (2010). Stemming the tide of plagiarism: One educator's view. Library Media Connection. Oct/Nov, 29(2), 44–46. Retrieved from http://eds.a.ebscohost.com/eds/pdfviewer/pdfviewer?vid=11&sid=90610ee6-f3a9-4ddf-b871-55ba439c5a28%40sessionmgr4010&hid=4213.

Löfström, E. (2011). "Does plagiarism mean anything? LOL." Students' conceptions of writing and citing. Journal of Academic Ethics, 9(4), 257–275. doi:10.1007/s10805-011-9145-0.

Ma, H., Lu, E. Y., Turner, S., & Wan, G. (2007). An empirical investigation of digital cheating and plagiarism among middle school students. American Secondary Education, 35(2), 69–82. Retrieved from http://www.jstor.org/stable/pdf/41406290.pdf.

Mahmood, S. T., Mahmood, A., Khan, M. N. & Malik, A. B. (2010). Intellectual property rights: Conceptual awareness of research students about plagiarism. International Journal of Academic Research, 2(6), November, 193–198.

Plagiarism. (2015). In Merriam-Webster.com. Retrieved June 5, 2016, from http://www.merriam-webster.com/dictionary/plagiarism.

Rosicka, Z. (2008a). Confrontation and overcoming cultural barriers. Journal STOP, 10 (3), 59–62, ISSN 1212-4168.

Rosicka, Z. (2008b). Risk Assessment Related to Information Uncertainty Components. Reliability & Risk Analysis: Theory & Applications, 1(2), 110–116, San Diego, USA. ISSN 1932-2321.

Rosicka, Z. (2006). Safety, Crisis Awareness and Information Management. Proceedings of the fourth International Scientific Conference "Challenges in Transport and Communication", 221–226, University of Pardubice, Jan Perner Transport Faculty, ISBN 80-7194-880-2.

Schmelkin, L. P., Gilbert, K. A., & Silva, R. (2010). Multidimensional scaling of high school students' perceptions of academic dishonesty. The High School Journal, 93(4), 156–165. Retrieved from http://eds.b.ebscohost.com/eds/pdfviewer/pdfviewer?vid=2&sid=7be0c184-85bb-4773-8a0d-aff48b6b35d9%40sessionmgr106&hid=119.

Sisti, D. A. (2007). How do high school students justify internet plagiarism? Ethics & Behavior, 17 (3), 215–231. doi:10.1080/10508420701519163.

Sobhy, M. M. (2015). Academic dishonesty prevalence and perception in the high school setting. Conatus: The Academic Journal of Dowling College, 5, 12–18. Retrieved from http://wwwx. dowling.edu/wikis/uploads/LibraryArchives/conatus2015c.pdf#page=12.

Strom, P. S., & Strom, R. D. (2007). Cheating in middle school and high school. The Educational Forum, 71(2), 104–116. doi:10.1080/00131720708984924.

Sureda, J., Comas, R. & Oliver, M. F. (2015). Academic plagiarism among secondary and high School students: Differences in gender and procrastination. Comunicar, English ed., 22(44), 103–110. doi:10.3916/C44-2015-11.

Thomas, E. E., & Sassi, K. (2011). An ethical dilemma: Talking about plagiarism and academic integrity in the digital Age. The English Journal, 100(6), 47–53. Retrieved from http://www. jstor.org/stable/pdf/23047881.pdf?_=1471336994175.

Vieyra, M., & Weaver, K. (2016). The prevalence and quality of source Attribution in middle and high school science papers. Issues in Science and Technology Librarianship. doi:10.5062/ F4FB50Z1.

Walker, J. (2010). Measuring plagiarism: Researching what students do, not what they say they do. Studies in Higher Education, 35(1), 41–59. Retrieved from http://www.tandfonline.com/doi/ pdf/10.1080/03075070902912994?needAccess=true.

Williamson, K., & McGregor, J. (2011). Generating knowledge and avoiding plagiarism: Smart information use by high school students. School Library Media Research, 14. Retrieved from http://files.eric.ed.gov/fulltext/EJ954598.pdf.

Chapter 13
Social Aspects of Teaching: Subjective Preconditions and Objective Evaluation of Interpretation of Image Data

Hana Svatoňová and Šárka Hošková-Mayerová

Abstract Between both learning and teaching processes, there is an ongoing relationship showing specific teaching and bidirectional relationships between teachers and their students. The teachers can influence their communication and interactions with students in a very favourable or, on the contrary, unfavourable way. Speaking about the area of attitudes and values shaping the relationship between teachers and their pupils, the article focuses on the part of subjective assumptions concerning pupils' progress when working with specific materials—aerial and satellite image and map interpretation. The survey respondents were primary school teachers and primary school students aged 11 and 15. Teachers and pupils' subjective assumptions were mutually compared; subjective assumptions were compared with objective data—the test results concerning pupils' work with image data. The survey has proved that (1) there is a strong correlation between the teachers' assumptions concerning the difficulty to interpret an image by themselves and the assumption concerning the difficulties encountered by their pupils; (2) the pupils' assumption concerning the difficulty when working with some material (satellite images in false colours) significantly differs from the objective success rate: students expect very difficult tasks whereas they are actually very successful; (3) the teachers and students differ about their opinion concerning the use of satellite and aerial images during the teaching process; the teachers declare a higher level of significance than students; (4) the teachers expect the satellite and aerial image data to be far more attractive than the pupils expect.

Keywords Teacher's attitude · Subjective assumptions concerning success · Aerial and satellite photograph interpretation

H. Svatoňová (✉)
Faculty of Education, Department of Geography, Masaryk University,
Brno, Czech Republic
e-mail: svatonova@ped.muni.cz

Š. Hošková-Mayerová
Faculty of Military technology, Department of Mathematics and Physics,
University of Defence, Brno, Czech Republic
e-mail: sarka.mayerova@unob.cz

© Springer International Publishing AG 2017   187
Š. Hošková-Mayerová et al., *Mathematical-Statistical Models and Qualitative Theories for Economic and Social Sciences*, Studies in Systems, Decision and Control 104,
DOI 10.1007/978-3-319-54819-7_13

13.1 Introduction

Learning is a lifelong process, during which people change their knowledge of the natural and human environment, their behaviour, activities and their personal features and their self-image (Mareš 1998). The above mentioned changes are primarily based on experience, i.e. the results of previous activities, which are transformed into systems of knowledge. The experience is either individual or it is based on the individual's identification with social experience. The teaching process is a system of planned activities aimed at reaching a set of educational objectives. The system includes sub-components—subsystems: a control system (teacher), a controlled system (pupil), curriculum, teaching resources and material conditions. The sub-components are linked with bonds; within them a transmission of various types of information is implemented. Between both learning and teaching processes, there is an ongoing relationship showing specific teaching and bidirectional relationships between teachers and their students (Dunn and Dunn 1979). The teacher can act in a very favourable or unfavourable way, choosing deliberately educational activities. The teacher's communication and interaction with students seems to be very significant (Čáp and Mareš 2001). The interaction between a teacher and a student is connected not only with their relationships and the curriculum, but also with the teacher's interpretation of a pupil's behaviour and the pupils' interpretations of teacher's behaviour. Another important aspect that is reflected in the interaction and communication between a teacher and a pupil is the subjective perception of the other person. The teacher's attitude often affects the way the students perceive their position from the teacher's point of view. Subjective factors in the teaching profession are important due to the number and variety of tasks and the large number of factors that affect the teaching process (Janik et al. 2014).

The teachers' quality is generally considered a key factor that significantly affects the quality of school education (Crone and Teddlie 1995; Kratochvílová 2007; Hoskova-Mayerova 2017). When speaking about the attitudes, values and personal features of high quality teachers, most authors focus on the following characteristics: a devotion to the profession, enthusiasm, a high level of commitment and the amount of energy dedicated to their work, a strong commitment to work with pupils, empathy, a positive attitude to pupils, conviction, faith (belief) that all pupils can be successful under certain conditions and that the teachers are able to help them fulfil their potential and succeed (e.g. Minor 2002, Kramáreková et al. 2016a). The presented study focuses teachers' attitudes, especially their expectations concerning the pupils' success rate. It deals with a teaching process involving the use of aerial and satellite images and maps. It uses information from similar researches that were conducted in the Czech Republic and in the Slovak Republic, too (e.g. Kramáreková et al. 2016b). These resources have become a subject of research and other projects. For example (Král and Řezníčková 2013) examined the current situation and the main barriers to the expansion of geographic information systems, including aerial and satellite photographs in teaching in secondary schools in the country. The second example is research to verify the disparity of opinion geography teacher elementary, middle and high schools on the importance of learning various skills, including use of aerial and satellite images (Řezníčková et al. 2011, 2014; Baysen et al. 2016).

The study evaluates identical or different assumptions of both the teachers and the pupils during a learning process, and the teaching process involving the support of material concerning the remote exploration of the Earth. It compares teachers' and pupils' subjective assumption concerning the difficulty of working with aerial and satellite images, the relation between the teachers' anticipated difficulty for teachers and their opinion about the difficulty for their pupils; it notices the similarity or differences of opinion about the attractiveness of aerial and satellite images and its use in the teaching process.

13.2 Objectives and Research Methodology

Both, the teachers' and the pupils', subjective assumptions are involved in the teaching process. The presented study continues the extensive research concerning the image data interpreting by pupils and adolescents (11–19 years) see Svatonova and Rybansky (2014). The research involved research concerning a group of pupils and students and their aerial and satellite image data interpretation. The aim of the research was to determine the way the users who have been confronted with a satellite image on the Internet or on television since their early childhood are able to interpret aerial photographs and satellite images in true or false colours. The research participants—378 pupils and students aged 11, 15 and 19—dealt with spatial tasks in various types of images and maps. The differences in the success rate of problem solving based on the type of source image (a map—an aerial photograph, a satellite image in true colours, a satellite image in false colours) were measured. The success rate was compared by age and gender of respondents. The students and pupils were asked about their subjective evaluation of difficulty concerning the reading of various types of image data, about their personal preferences that influenced their selection of a map or image as the source of information. The study found evidence which subsequently opened up some ideas for follow-up research: (a) pupils work successfully with satellite and airplane map and image data; their success rate varies over the years (from 11 to 19 years of age) according to both—the documents (map, image) and the gender; (b) the pupil's subjective evaluation of the difficulty of working with image data does not correspond to objective success: the students subjectively consider working with satellite images in natural colours much simpler, yet they are objectively slightly more successful at solving problems in the pictures with unnatural colours (Potůček and Hošková-Mayerová 2017).

The main objective of the research was to compare the subjective assumptions of teachers with the subjective assumptions of pupils; then to compare the subjective assumptions of teachers with the objective results of pupils.

The research questions were compiled as follows:

1. How do the teachers and pupils evaluate the use of aerial and satellite photographs in the teaching process?
2. How interesting the work with aerial and satellite data is from the teachers and pupils' point of view?
3. Do the teachers' and pupils' subjective evaluations of the difficulty of working with aerial and satellite data match?
4. Do the subjective assumptions correspond to the objective results?

13.3 Survey Respondents

Two groups of respondents took part in the survey: teachers and their pupils. The teachers were represented by 24 primary school teachers certified to teach geography, including 13 males and 11 females. The structure according to the age of teachers and the length of their teaching experience in percentages is showed in Figs. 13.1 and 13.2. The teachers were asked about their college training concerning the work with aerial and satellite image data. 8 teachers said they had worked with aerial and satellite image data during their university studies, 2 teachers claimed they had never worked with satellite image data during their university studies. 11 teachers reported that they had worked with geographic information systems (GIS) during their university studies. School equipment can be considered very good: all the teachers can use the projector while teaching, 15 teachers use specialized geography classrooms. The distribution of schools by size

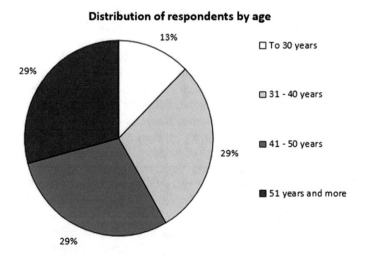

Fig. 13.1 Structure of teachers by age

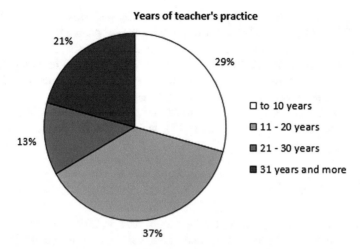

Years of teacher's practice

Fig. 13.2   Number of years of teachers' teaching experience

of municipalities was even: 6 schools in villages up to 5000 inhabitants; 8 schools in the towns from 5 to 50 thousand inhabitants; 3 schools in the towns from 50 to 100 thousand inhabitants; 7 schools in cities with over 100,000 inhabitants.

The second group of respondents were pupils and students aged 11 and 15. Overall 242 students were tested, of which 122 were boys and 120 girls. 108 children were aged 11 and 134 children 15. Pupils aged 11 and 15 were taught in eight classes in total (always two classes for each age group in the school).

13.4   Test Preparation

With regard to the questions of the research, a questionnaire for teachers, and a test and questionnaire for pupils were prepared. The questionnaire survey included questions related to the subjective evaluation of the difficulty of interpretation of aerial and satellite images and maps for both the teacher and their pupils. An example of the pupils' test was included. The pupils' test consisted of three main parts—the first part of the test contained *the tasks to evaluate the success of interpretation of aerial photograph and map data*, the second part dealt with *the evaluation of the successful interpretation of satellite images*, the third part of the test contained questions about the respondent's *subjective evaluation*. Thanks to the availability of images for both the respondents and the researchers, the images and maps available on the map server the Czech National Geoportal INSPIRE were used as the main source of image data. A cut from LANDSAT images provided by TopGIS Company was used to test satellite image data. From a range of options concerning unnatural colours of LANDSAT images, the combination 742 was chosen, which indicates water surfaces blue to black, watercourses blue, trees and forests green, the fields in pink-green surface mosaic, built-up areas in purple-pink shades. The combination RGB 742 (Red, blue, green of channels 7, 4, 2) allows a very good resolution of water areas, watercourses and forests. The questionnaire survey and testing materials included the following material: a basic topographic map, scale 1:10 000, the image data: a vertical aerial image of resolution of 0.5 m; an oblique aerial image of resolution of 0.5 m, a satellite image in natural colours (LANDSAT 7, RGB 321), scale 1: 100,000, resolution 30 m, a satellite image in unnatural colours (LANDSAT 7, RGB 742), scale 1: 100,000, resolution of 30 m.

13.4.1 Course of Testing

The questionnaire survey among teachers took place in 2015; an interview and a questionnaire where the teacher assumed both positions of an interviewee and a researcher. School testing took place in 2013. The pupils and students were given written individual tests following the initial presentation of the test purpose and organizational instructions. The pupils were not limited in time; they dealt with the test for about 20 min on average. Unclear information was explained during the testing phase. The evaluation of the answers was with regard to the "paper" form of the test made manually. The test evaluation provided objective data concerning the pupils' success rate. Figure 13.3 is a sample of a pupils' test. The pupils worked with a vertical coloured aerial image showing the school surroundings: they were expected to identify selected objects and draw the required tasks in the image.

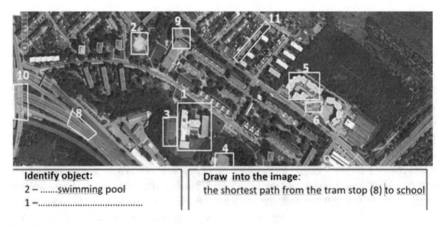

Fig. 13.3 Sample of a pupils test: The original test used a coloured aerial photograph

13.5   Results

In the first part of the test, in addition to the questions about the respondent's characteristics were the questions about the use of aerial and satellite image during the teaching process. The responses of teachers show that 88% of them use the teaching aids during the teaching process. A more detailed analysis of responses is shown in Table 13.2.

The pupils were asked whether they worked with aerial and satellite data during the classes. Both groups of pupils (aged 11 and 15) reported in about half the cases the answer "NO", i.e. they do not use the material concerning the remote exploration of the Earth. A more detailed analysis of responses is shown in Table 13.1.

A partial conclusion: the teachers claim they use the aerial and satellite image data. About 50% of pupils claim they do not work with the image data. However, the results may be distorted by the fact that the surveyed pupils were not the interviewed teachers' pupils—the survey was conducted with pupils in other schools.

Other questions for students and pupils pointed at their evaluation of working with aerial and satellite image data. A sample of the teachers' question: Do you think working with aerial and satellite image data is interesting for pupils? A sample of the pupils' question: Do you consider working with aerial and satellite images interesting? Specific results are given in Table 13.2. Most teachers think that work with aerial and satellite data is interesting for students, because they probably consider map and satellite servers attractive. The pupils reported more or less positive response in approximately 70% of cases, see Table 13.2. The pupils also said they viewed the aerial and satellite images in their free time (a) regularly (27% children aged 11 and 42% children aged 15) or (b) occasionally (42% children aged 11 and 37% children aged 15). Overall, around 80% of children work with such images in their free time, which demonstrates the attraction of interest of these documents and supports the teachers' evaluation results.

Table 13.1 The use of aerial and satellite photographs in the classroom—pupils

Responses:	11 year-old pupils/%	15 year-old pupils/%	Teachers/%
Yes, in detail	6	10	88
Yes, marginally	34	34	8
No	50	56	4

Table 13.2 How interesting the work with aerial and satellite data is from the teachers' and pupils' point of views?

Responses:	Teachers/%	11 years-old pupils/%	15 year-old pupils/%
Yes	71	33	29
Yes, but a little	17	38	44
No	12	22	19
I do not know	0	7	8

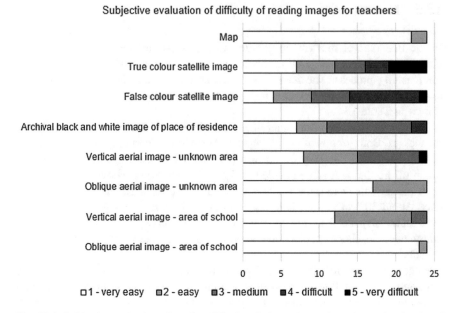

Fig. 13.4 Subjective evaluation of reading difficulty of viewed image from the teachers' point of view

Both the pupils and teachers evaluated aerial and satellite image using a scale of 1–5, where 1 represented the value of the simplest image type and 5 the most difficult type of image. For comparison, a map was also evaluated. The diagram in Fig. 13.4 shows that the most difficult satellite image was evaluated in its unnatural colours.

The teachers evaluated how difficult the reading of different types of aerial and satellite photographs for their pupils and for themselves might be. The teachers expected a higher level of difficulty when this image is used by younger children, the least difficulty when used by themselves. The easiest ones seem to be oblique colour aerial images, not maps, showing the school surroundings that, according to teachers, are easier to be interpreted by children.

Table 13.3 shows the teachers' evaluation of particular types of aerial and satellite images read by themselves and their pupils. For the gained data a correlation was calculated; the correlation coefficient of 0.90 shows a strong link between the assumptions of image difficulty for themselves (teachers) and for pupils.

The pupils evaluated the difficulty level of selected aerial and satellite image data for themselves. Also the value specified by teachers is inserted in order to make the results clearer. The correlation coefficient between the pupils' and teachers' evaluations is 0.75. The teachers and pupils coincidently suggest that an oblique aerial image is the easiest one to be interpreted. The evaluation of results concerning the subjective pupils' and teachers' assumptions is presented in Table 13.4.

Table 13.3 The teacher's subjective evaluation of the difficulty level of reading aerial and satellite image data

The teacher's subjective evaluation of the difficulty			
Image	For 11-year-old pupils	For 15-year-old pupils	For teachers
An oblique coloured aerial image—school surroundings	1.63	1.33	1.04
A vertical coloured aerial image—school surroundings	2.25	1.71	1.58
An oblique coloured aerial image of an unknown area	2.46	1.67	1.29
A vertical coloured aerial image of an unknown area	2.88	2.38	2.13
An archival black and white aerial image of the residence place	3.54	2.67	2.33
A satellite image in unnatural colours	3.54	3.04	2.92
A satellite image in natural colours	3.21	2.54	2.75
Map	2.00	1.50	1.08
Average evaluation	2.69	2.10	1.89

Table 13.4 Subjective evaluation of the difficulty level of reading aerial and satellite image data

The pupil's subjective evaluation of the difficulty	11 year-old pupils	Teacher: for 11-year-old pupils	15 year-old pupils	Teacher: for 15-year-old pupils
An oblique satellite image in natural colours	1.75	1.63	2.14	1.33
A vertical satellite image in natural colours	2.37	2.88	2.14	2.38
A satellite image in natural colours	2.88	3.54	2.38	3.04
A satellite image in unnatural colours	3.54	3.21	3.04	2.54
Map	2.28	2.00	2.04	1.50
Average	2.56	2.65	2.34	2.16

The pupils were also tested on how successful they are in problem solving tasks when using map and aerial and satellite image data. Although the expected difficulty level of reading satellite image in unnatural colours was very high, they were objectively successful—see Table 13.5. Contrarily, working with maps was more difficult for pupils than reading aerial and satellite image data. The objective success rate demonstrates a shift in the success rate of children with regard to age—pupils aged 15 are more successful than younger classmates.

Table 13.5 Objective success rate of pupils

Objective success rate of pupils	11 year-old pupils (%)	15 year-old pupils (%)
Map	53	70
A vertical coloured image	71	80
A satellite image in true colours	69	83
A satellite image in false colours	74	89

13.6 Conclusion

The study focused on teachers' and students' attitudes concerning the teaching process, teachers and students' assumptions concerning the work with selected resources. The aerial and satellite image data that have lately appeared in the teaching process was used in the survey evaluation process. The pupils and teachers evaluated how interesting these resources are, how difficult it is to work with them, and their actual use in the teaching process. The pupils were given a test concerning the work with aerial and satellite image data, from which the objective data on the pupils' progress rate was obtained. The following conclusions are based on the respective research:

(1) There is a difference between the teachers' and pupils' evaluations of the use of aerial and satellite image data during the teaching process. About 50% of pupils say they do not use the image during the teaching process. Only 4% of teachers said they do not use the image data.

(2) The teachers and pupils agree that the use of aerial and satellite image data is interesting. Pupils also use them in their free time. The teachers expect the respective teaching resources to be more attractive than the pupils, while the pupils are more reserved in the relation to these resources.

(3) The teachers assume that working with aerial and satellite image data is less difficult for themselves than for their pupils. The teachers also take into account pupils' ages—they assume the image data understanding is more difficult for younger children. The assumption concerning the difficulty for teachers themselves strongly correlates with their assumption concerning the difficulty for their pupils. The research shows that it is necessary to influence the teachers' training—if we want the teachers to use certain teaching aids, they should not consider them of a great difficulty for their pupils.

(4) The pupils and teachers agree on assumptions concerning the difficulty of working with some teaching aids. This is particularly evident in the images in unnatural colours. The data to be further researched is whether the evaluation concerning the terms of "natural" and "unnatural" colours would achieve different results. The authors' experience in teaching students—the future teachers —shows that the term "unnatural" commands students' respect—a concern that the work will be of great difficulty.

(5) The subjective assumptions concerning the difficulty of image work is fundamentally different from the objective results (As proved research

Řezníčková et al. 2013). Although the pupils and the teachers had expected considerable difficulty in dealing with satellite image data in unnatural colours, the pupils reached the highest success rate. The colour combination of unnatural colours was chosen deliberately, so that e.g. water (rivers and ponds in the image) was blue. In the Czech Republic, the natural colour of water in rivers is mainly brown-green and in this colour it is displayed in images with natural colours. The pupils, however, encountered a problem in finding rivers and ponds with such colours in the images. They were much more successful in reading images in unnatural colours, where they used not only the anticipated— and learned skill concerning such colours in maps (bright blue) but also contrasts in relation to the surroundings. The testing showed that the use of unnatural colours when reading satellite image is reasonable when being used by laymen.

The teachers' faith and belief in the success of their pupils is in general, according to many studies (e.g. Minor 2002, Řezníčková a kol. 2013), an important part of teachers' attitudes and values that apply in teaching. The teachers' approach influences the development of reading skills concerning aerial and satellite image. Active teachers' approach, their corrections, and interpretation clarifications of a particular image including the pupils' work evaluations is positively reflected in the results interpretation. The development of cartographic skills varies not only with age, but also with pupils' practice and teachers' support (Hanus and Marada 2014, 2016).

The presented study is a contribution to the research of attitudes and values applied by teachers in the teaching process, and to a comparison of the assumptions and their objective results.

Acknowledgements This research has been supported by funding from Masaryk University under the grant agreement No. MUNI/M/0846/2015, which is called 'Influence of cartographic visualization methods on the success of solving practical and educational spatial tasks'.

The work presented in this chapter was also supported within the project "Development of basic and applied research in the long term developed by the departments of theoretical and applied foundation FMT" (Project code VÝZKUMFVT) supported by the Ministry of Defence the Czech Republic.

References

Baysen, E., Hoskova-Mayerova, S., Cakmak, N., Baysen, F. (2016) Misconceptions of Czech and Turkısh Unıversity Students in providing Citations. Switzerland: Springer International Publishing, 2016, p. 183–191. *Antonio Maturo et al. (Eds): Recent Trends in Social Systems: Quantitative Theories and Quantitative Models. Studies in System, Decision and Control 66.*
Crone, L. J., Teddlie, Ch. (1995). Further Examination of Teacher Behavior in Differentially Effective Schools: Selection and Socialization Process. *Journal of Classroom Interaction, 30 (1)*, 1–12.
Čáp, J., Mareš, J. (2001). Psychologie pro učitele. Praha: Portál, p. 656.

Dunn, R., Dunn, K. (1979). Learning Styles/Teaching Styles: Should They... Can They ... Be Matched? *Educational Leadership* [*online*]. 36(4), [cit. 2016-10-01], 238–244.

Hanus, M., Marada, M. (2014): Map skills: definition and research. Geografie, 119(4), 406–422.

Hanus, M., Marada, M. (2016): What does a map-skills-test tell us about Czech pupils? Geografie, 121(2), 279–299.

Hoskova-Mayerova, S. (2017) Education and Training in Crisis Management. In: *The European Proceedings of Social & Behavioural Sciences EpSBS, Volume XVI.* Future Academy, 2017, 849–856. ISBN 2357-1330.

Janík, T., Píšová, M. and Spilková, V. (2014). *Standards in the teaching profession: Foreign approaches and analysis of their effects.* Orbis Scholae, 8(3), 133–158.

Kramáreková, H., Nemčíková, M, Rampašeková, Z., Svorad, A., Dubcová, A. and Vojtek, M., (2016). *Cartographic Competence of a Geography Teacher - Current State and Perspective.* In: 6th International Conference on Cartography & GIS : 13–17 June 2016, Albena, Bulgaria. - Sofia : Bulgarian Cartographic Association, 200–209.

Kramáreková, H., Nemčíková, M., Vojtek, M., Dubcová, A., Gajdošíková, B. and Konečný, M. (2016). *Comparison of Cartographic Language of Pupils in the 4th Grade of Primary School (Case study of the Slovak Republic and Czech Republic).* In: 6th International Conference on Cartography & GIS : 13–17 June 2016, Albena, Bulgaria. Sofia : Bulgarian Cartographic Association, 2016, 176–187.

Král L., Řezníčková, D. (2013). Rozšíření a implementace GIS ve výuce na gymnáziích v Česku. *Geografie,* 2013, 118(3), 265–283.

Kratochvílová, J. (2007). Učitelé škol v nové roli "tvůrců" školního kurikula. *Orbis scholae,* 1(1), 101–110.

Mareš, J. (1998). *Styly učení žáků a studentů.* Praha: Portál, p. 240.

Minor, L.C. (2002). Preservice Teachers´Beliefs and Their Perceptions of Characteristics of Effective Teachers. *Journal of Educational Research,* 96(2), 116–127.

Potůček, R. Hošková-Mayerová, Š. (2017) Qualitative and quantitative evaluation of the entrance draft tests from mathematics. Switzerland: Springer International Publishing, 2017, p. 53–64. *Antonio Maturo et al. (Eds): Recent Trends in Social Systems: Quantitative Theories and Quantitative Models. Studies in Systems, Decision and Control, Vol. 66.*

Řezníčková, D. a kol. (2013). *Dovednosti žáků v biologii, geografii a chemii,* P3 K, Praha.

Řezníčková, D., Marada, M., Hanus, M. (2011). Porovnání představ a názorů pedagogů různých stupňů škol na standardy geografických dovedností. In Janík, T., Knecht, P., Šebestová, S. (Eds.), *Smíšený design v pedagogickém výzkumu: Sborník příspěvků z 19. výroční konference České asociace pedagogického výzkumu.* (Brno: Masarykova univerzita. pp. 304–309.

Řezníčková, D., Marada, M., & Hanus, M. (2014). Geographic skills in Czech curricula: analysis of teachers´opinions. In. D., Schmeinck, J., Lidstone (Eds.): *Standards and Research in Geography Education. Current Trends and International Issues,* Mensch und Buch, Berlin, pp. 37–49.

Svatonova, H., Rybansky, M. (2014). Children observe the Digital Earth from above: How they read aerial and satellite images. In IOP Conference Series: *Earth and Environmental Science* 18 012071. Malaysia: ISDE, p. 7.

Chapter 14
Do Institutional or Foreign Shareholders Influence National Board Diversity? Assessing Board Diversity Through Functional Data Analysis

Fabrizio Maturo, Stefania Migliori and Francesco Paolone

Abstract This study analyses the external antecedents of board diversity. We use the resource-based view and the agency theory to investigate the impact of institutional and foreign shareholders on national board diversity. Because we highlight that the most used diversity indices neglect the multidimensional aspect of diversity, we refer to diversity profiles and suggest the functional data analysis approach for diversity assessment in corporate governance studies. Specifically, we use the parametric functional analysis of variance in analysing the influence exerted by institutional and foreign shareholders on national board diversity. Focusing on a sample of 1,230 Italian medium-large firms, our results show that institutional shareholders do not influence national board diversity, while foreign shareholders strongly affect it, especially when they hold more than 50% of shares. Thus, we address the research gap on the determinants of national board diversity and enrich comparative European research on this topic.

Individual Contributions of the Authors
The authors have contributed to the chapter as follows:
Fabrizio Maturo: Sections 14.3.3, 14.4.1, 14.4.2.
Stefania Migliori: Introduction, Sections 14.1; 14.2.1, 14.2.2, 14.2.3.
Francesco Paolone: Sections 14.3.1, 14.3.2, Abstract, Conclusions.

F. Maturo (✉) · S. Migliori
Department of Management and Business Administration, University of Chieti-Pescara, Pescara, Italy
e-mail: f.maturo@unich.it

S. Migliori
e-mail: s.migliori@unich.it

F. Paolone
Law Department, University of Naples, Parthenope, Napoli, Italy
e-mail: francesco.paolone@uniparthenope.it

© Springer International Publishing AG 2017
Š. Hošková-Mayerová et al., *Mathematical-Statistical Models and Qualitative Theories for Economic and Social Sciences*, Studies in Systems, Decision and Control 104,
DOI 10.1007/978-3-319-54819-7_14

Keywords Corporate governance · Diversity profiles · Foreign shareholders · Functional data analysis · Institutional shareholders · National board diversity

14.1 Introduction

In the corporate governance field, board diversity and the "value-in-diversity" have been widely debated (Hillman 2015; Milliken and Martins 1996). Recently, European institutional guidelines also stressed the relevance of board diversity for the effectiveness of firms' corporate governance and the selection process of board members: "Diversity in the members' profiles and backgrounds gives the board a range of values, views, and sets of competencies. It can lead to a wider pool of resources and expertise. Different leadership experiences, national or regional backgrounds or gender can provide effective means to tackle 'group-think' and generate new ideas. More diversity leads to more discussion, more monitoring and more challenges in the boardroom" (EU Commission Green Paper 2011, p. 5, 6). This perspective highlights the influence of boards' human capital diversity on firms' strategic behaviour (Johnson et al. 1996; Stiles 2001) and suggests that board diversity can increase board independence, thus improving the monitoring on managers (Bysinger and Hoskisson 1990; Lane et al. 1998; Pearce and Zahara 1992).

In the literature, several empirical studies identify board diversity as an important indicator of success for international corporate practices. However, most of these studies analyze the effects of board diversity on firm performance (Ararat et al. 2015; Bell et al. 2011; Carter et al. 2003; Dimovski and Brooks 2006; Erhardt et al. 2003; Ujunwa et al. 2012), on board functioning and its outcome (Van Ees et al. 2009; Veltrop et al. 2015), or on mediating and moderating variables in the relationship between board diversity and firm performance (Ararat et al. 2015; Miller and Triana 2009). Nevertheless, many other aspects remain poorly investigated. First, there are few studies on the determinants or external antecedents of board diversity (Anderson et al. 2011; Arnegger et al. 2014; Bianco et al. 2015; Hillman et al. 2007). Instead, a more refined knowledge of board diversity determinants can contribute to a better understanding of the role of the board within a firm's corporate governance system. Specifically, many studies in corporate governance field are based on the assumption that shareholders can influence firms' performances through board (composition) diversity, but we little know on whether and how different types of shareholders choose diverse directors' profiles. Based on a fiduciary responsibility, the board should reflect the interest of their shareholders. Consequently, heterogeneity in shareholders' composition should be reflected in high board diversity. Following Sur et al. (2013), each type of ownership has different imperatives and may prefer different types of board directors to fulfill their goals and governance needs. Second, studies on board diversity focus on different aspects, distinguishing task-related (educational or functional background) and non-task-related diversity (gender, nationality, age, ethnic, race) (Adams et al. 2015). However, even though most research investigates gender board diversity because of the spread of the "quota law" in several countries, many other aspects can bring diversity into the boardroom (Hillman 2015). In particular, in this study we focus on national board diversity (NBD-expressed by the presence of members of different nationalities in the board). It represents a relevant dimension of board diversity in Europe (Oxelheim and Randøy 2003; Ruigrok et al. 2007) and in firms' corporate governance systems (European Commission 2011; Heidrick and Struggles 2009). However, NBD has received little attention in the empirical literature (Choi et al. 2012).

Based on such considerations, the aim is to discover whether and how shareholders' composition is a determinant of NBD; in particular, we consider separately the effect of foreign and institutional shareholders. We focus on institutional and foreign shareholders for two reasons. First, the increase in institutional investors' ownership in companies across the world has influenced shareholder composition over the last decades (Ferreira and Matos 2008; Mallin 2012). Second, the process of globalizing corporate governance systems has increased the number of foreigners in shareholders' composition (Oxelheim and Randøy 2003). Thus, there is a growing interest in the role that these types of shareholders can play in corporate governance systems, so an analysis of their influence on NBD can help understand whether and how they use board diversity to protect their interests.

Using the agency and the resource dependence theories, we investigate the effect exerted by institutional and foreign shareholders on NBD using a sample of 1,230 Italian medium-large firms. Our results show that institutional shareholders do not influence NBD. On the contrary, foreign shareholders strongly affect NBD; furthermore, this effect increases when foreign shareholders hold more than 50% of shares.

This chapter offers several contributions to the diversity and corporate governance literature. First, it responds to the call for research on the antecedents of diversity in organizations (Harrison and Kelin 2007). Second, it extends the investigation on NBD to a context (country and corporate governance system) not yet investigated in previous studies, thus enriching the European comparative research on this topic. Third, we provide more refined knowledge about the effect of shareholder composition on corporate governance mechanisms. Specifically, we distinguish between different types of shareholders to understand which of them use NBD as a mechanism to improve boards' monitoring function or their decision processes. Moreover, our research supports the idea that NBD is contingent because it is also related to shareholders and ownership compositions in a single time instant (Desender et al. 2010). Fourth, our study makes an important methodological contribution to the call regarding the relationship between the diversity types investigated and their appropriate operationalization (Harrison and Klein 2007). Specifically, different board diversity indices exist in the literature, but none is able to capture the multi-dimensional aspect of diversity; moreover, they often lead to mixed results (Adams et al. 2015) and different diversity rankings (Di Battista et al. 2016a, b, c).[1] To overcome these limits, we propose extending the ecological

---

[1]The main issue in measuring diversity is that richness and evenness are confounded when a single index is considered. Indeed, a board with few species and high evenness could have the same

concept of biodiversity to the field of management studies, that is able to consider simultaneously the richness (i.e., the number of different species in a given area) and the evenness (i.e., a measure of the relative abundance of each species in an area) using a single index. Specifically, we focus on the beta diversity profile to inspect the multidimensional aspects of diversity and refer to the functional data analysis (FDA) (Ramsay and Silverman 2005; Di Battista et al. 2016a, b).

Section 14.2 presents a brief literature review on NBD and discusses the link between shareholders and NBD, which helps us to formulate our hypotheses. Section 14.3 shows the data and the variables, and it proposes a new model for the board diversity assessment. Section 14.4 presents the application of our model and tests our hypotheses. Finally, we present our results and conclusions. .

14.2   Literature Review and Hypotheses

14.2.1   National Board Diversity

Frequently, the term "diversity" is used as a synonym of heterogeneity, dispersion, variety, or disparity. In the organizational context, diversity is "a unit-level, compositional construct used to describe the distribution of differences among the members of a unit with respect to the common attribute X, such as tenure, ethnicity, conscientiousness, task attitude, or pay" (Harrison and Klein 2007, p. 1200).

In the field of corporate governance, several studies have focused on the concept of diversity applied to the board directors' profiles; board diversity can be analysed considering different aspects, and many studies have identified board diversity as an important indicator of success for international corporate practice (Kiel and Nicholson 2003; Rose 2007; Dahya and McConnel 2007). Some scholars have highlighted that, to date, the majority of studies have investigated gender diversity and its effect (Hillman 2015). Indeed, there are few empirical studies have addressed other types of diversity, such as national, educational, functional, or age diversity. However, corporate governance practices highlight a growing emphasis on NBD. Specifically, the Heidrick and Struggles report on corporate governance (2009) shows that the percentage of foreign board members doubled during the global financial crisis (2007–2009). Moreover, the EU Commission Green Paper (2011) shows that in a sample of large European listed companies 29% of board members, on average, are non-national. This shows that companies recognize the increasing importance in making a board diverse as a key variable for efficiency and effectiveness of firms' corporate governance. In recent decades, corporate governance systems were characterized by a process of globalizing (Aguilera and

---

(Footnote 1 continued)

diversity as a board with many species and low evenness. Thus, different indices could lead to different firms ranking according to the nationality of their boards.

Cuervo-Cazurra 2000; OECD 1998); thus, the board composition reflects the ongoing process of ownership internationalization that is generated by the reduction of barriers to trade and capital equity flows (Oxelheim and Randøy 2003). Furthermore, as suggested by the recent literature, NBD is also a way to import foreign corporate governance systems, which are considered better for transparency and effectiveness (Oxelheim and Randøy 2003). Therefore, the process of globalization that is sweeping firms' corporate governance systems across the world may strongly affect NBD; it is then expected that NDB will gain a growing relevance in the field of corporate governance research.

To date, there have been few studies on NBD, and those that do exist often focus on its effects on firm performance (Ararat et al. 2015; Oxelheim and Randøy 2003; Rose 2007; Ujunwa et al. 2012), board monitoring (Ararat et al. 2015) and audit quality (Lee et al. 2012), corporate expropriation (Hamzah and Zulkafli 2014), or share ownership of executive directors (Filatotchev and Bishop 2002). Most studies refer to the northern European context, while others refer to Nigeria, Korea, Malaysia, Swiss and Turkey. All of these studies highlight the importance of NBD in the system of corporate governance, but we still know little about NBD determinants.

In the literature, another main topic is the potential cost, or benefit, related to board diversity (Anderson et al. 2011); in particular, there are two opposing perspectives. The first argues that foreign members can bring different expertise, skills and heterogeneity in the companies through their different opinions, perspectives, backgrounds, languages, and religions; furthermore, from the agency theory perspective, foreign members can reduce the power of existing board members and assure foreign minority investors that their interests are protected (Oxelheim and Randøy 2003). In addition, if they belong to countries with strong shareholder rights, they may also bring strong notions on the roles of board control (Lee et al. 2012; Ararat et al. 2015). The second perspective highlights the disadvantages related to different languages and the lack of knowledge that foreign board members may have on the domestic market. Empirical research on Scandinavian countries shows that NBD exerts a positive effect on firm performance (Oxelheim and Randøy 2003; Rose 2007); however, these results are not easily comparable because these researchers have used different indices in assessing diversity. Specifically, the most frequently used methods to measure NBD are as follows: (a) a dummy variable (equal to "one" if the firm has a foreign directors, and "zero" otherwise) (Choi et al. 2012; Oxelheim and Randøy 2003); (b) a ratio scale (foreign members to total board size) (Anderson et al. 2011; Ujunwa et al. 2012); (c) the number of different nationalities within the board (Eulerich et al. 2014); and (d) the Simpson-Blau's index (Blau 1977; Ararat et al. 2015). A more detailed description of the common issues of these indices is shown in Di Battista et al. (2016a, b, c), where we also explain the proposed solution to overcome these limitations.

14.2.2 Institutional Investors and National Board Diversity

Starting from the seminal work of Berle and Means (1932), the phenomenon of separation between ownership and control that is associated with ownership dispersion has changed. In recent decades, institutional investors have acquired increasing portions of companies' share capital; furthermore, they have shareholding sizes and monitoring expertise that differ from those of atomistic investors (Elyasiani and Jia 2010). Recently, Ferreira and Matos (2008) used a sample of equity holdings of 27 different countries to show a strong presence of institutional investors during the period 2000–2005. In addition, the OECD's principles for Corporate Governance (2004, 2011) highlighted that the effectiveness of corporate governance systems will depend on the function exerted by institutional investors; moreover, several studies have suggested that institutional owners promote good governance and affect firms' strategic choices (Filatotchev et al. 2001; Filatotchev and Wright 2011). Institutional investors have become key players in the corporate governance of liberal market economies; thus, they are highly engaged in several corporate governance activities (Gillan and Starks 2007; Goyer and Jung 2011). This phenomenon raises the question of whether and how institutional investors influence corporate governance systems. Although the debate on the active, or passive, role of institutional investors is still lively in the literature (Klein and Zur 2009; Whitley 2009), there is a general consensus on their ability to influence governance mechanisms (Feldmann and Schwarzkopf 2003; Mallin 2012; McNulty and Nordberg 2015). Moreover, Mallin (2012) remarked that after the recent big scandals, the role and importance of institutional investors became very important. Specifically, given the fiduciary responsibilities of institutional investors towards the ultimate owner, they took an increasingly active role within the corporate governance systems. To ensure that companies and boards act according to the perspective of firms' long-term survival, institutional investors have increased their interest in being actively involved in companies where they invest (Mallin 2012). Institutional investors have different ways of influencing firms' actions: directly through the voting rights and dialogue on firms' objectives, or indirectly by encouraging the improvement of corporate governance systems (Mallin 2012). Therefore, a possible way is to influence the board composition to ensure that directors are leading the company in the interest of shareholders. Indeed, several studies suggest that the presence of institutional investors may influence board composition (Feldman and Schwarzkopf 2003). In line with the agency theory, many scholars have showed that firms change their board composition in response to institutional shareholder pressure. Specifically, they have focused on the relationship between the presence of institutional investors and board size, turnover of board members (Feldman and Schwarzkopf 2003), or number of outside directors (Feldman and Schwarzkopf 2003). Thus, it is reasonable to expect that institutional investors can also influence the NBD because it may be a way to improve the monitoring role and the effectiveness of board. Thus, foreign directors may be appointed as "outsiders" who can improve board independence and not protect a specific group of shareholders. Moreover, NBD is also a way to import corporate governance practices from foreign systems that may be considered superior in terms of transparency and effectiveness (Oxelheim and Randøy 2003). Since institutional investors can influence board diversity, it is reasonable to expect that a high percentage of shares owned by institutional investors could result in high NBD. As suggested by Sur et al. (2013), we argue that aggregate ownership (i.e. all the type of owners and their respective shareholdings) predict the composition of the board in a given firm. Specifically, a different aggregated ownership configuration have different risk concerns and imperatives with respect to firms where they invest; these aspects are reflected in different needs about directors' profiles and attributes. Institutional owners aim to safeguard their financial investments; thus, their primary imperative is to reduce agency conflicts through the effectiveness of board monitoring (Sur et al. 2013). Thus, we formulate the following hypotheses:

H1a: *The presence of institutional shareholders has a positive effect on NBD.*
H2a: *When the aggregate percentage of shareholdings by institutional owner increases, the NBD increases.*

14.2.3   Foreign Shareholders and National Board Diversity

In recent decades, a general process of globalization has involved companies' corporate governance systems (OECD 1998) and removed barriers to trade, capital flows and cross-border information gaps (Useem 1998; Randøy et al. 2001). Thus, companies have achieved easy access to capital and worldwide shareholding. This trend has generated important insights for the study of intersections between national and global financial markets (Aguilera and Jackson 2003, 2010; Goyer and Jung 2011). In addition, the process of globalization has broken down legal and institutional barriers to foreign investment, resulting in lower costs of equity capital and positive effects on firms' equity value (Oxelheim and Randøy 2003). Therefore, foreign shareholders represent a profitable and easy way for companies to acquire foreign capital. Moreover, this context is a way for foreign shareholders to acquire large stakes in enterprises. Moreover, the increased importance of foreign shareholders is associated with the greater propensity of firms to implement strategies to enhance shareholder value; at the same time, it is subject to a strong mediating role of national corporate governance features (Goyer and Jung 2011). In many studies, the presence of foreign shareholders is associated with corporate governance improvement (Aggarwal et al. 2011) and high efficiency of monitoring systems (Ferreira and Matos 2008). Of course, to protect their investments, foreign shareholders seek guarantees on the effectiveness of firms' corporate governance. Thus, the presence of foreign board members may be observed as a sign of effective representation of foreign shareholders' interests, which would disallow managers from acting only in favour of domestic shareholders or particular groups.

On the other hand, following the resource based view, many studies have shown that diversity generates high knowledge, creativity, innovation, effective problem solving, and then strong firm performance (Carpenter 2002). Therefore, the presence of foreign board members could add more qualified candidates available to a board (Ujunwa et al. 2012) as well as more diverse industry expertise that domestic board members could not possess (Lee and Farh 2004); it could also promote the exchange of information into the international network (Oxelheim and Randøy 2003). Furthermore, foreign board members could be a mechanism to assure foreign investors that the firm possesses the skills and abilities to compete in the global market. Furthermore, we need to consider that the ability to influence the composition of the board is greater for large foreign shareholders (Shleifer and Vishny 1986). Therefore, as argued for institutional investors, if foreign investors possess a high percentage of shares (Sur et al. 2013), they have better chances of transferring their goals and imperatives on board composition; thus, we suppose that they could also influence NBD and formulate the following hypotheses:

H1b: *The presence of foreign shareholders has a positive effect on NBD.*
H2b: *When the aggregate percentage of shareholdings by foreign owners increases, the NBD increases.*

14.3   Data and Methodology

14.3.1   Data

Our data are gathered from the AIDA dataset (Italian Database of Companies), which is the Italian provider of the Bureau Van Dijk European Database; it is the most complete and reliable economic and financial information source about non-public Italian companies. The AIDA database contains detailed information with up to 10 years of history of approximately 1,170,000 Italian listed and non-listed companies. We adopt the EU (2005) definition of small, medium, and large companies, but we refer only to medium-large corporations. Moreover, we select only firms with the following characteristics: (a) currently operating in manufacturing industries (classes 10–30 of the ATECO code are used for the industrial classification) because many studies have shown that firms belonging to a particular industry are more similar and (b) not involved in a bankruptcy process. We focus on the period 2013–2014; in particular, our empirical design relies on shareholders as lagged (t-1) compared to board diversity (t). This choice reflects the perspective that shareholders select directors' profiles that match their goals and imperatives and seek to mitigate potential endogeneity concerns (Sur et al. 2013; Muller-Kahle and Lewellyn 2011). The above criteria provide a sampling of 1,565 firms. Companies without all data available for the functional analysis are excluded; thus, a final sample of 1,230 is obtained.

14.3.2 Variable Description

NBD is the dependent variable that considers the diverse nationalities of the board members. Contrary to previous research, we refer to FDA because diversity is a function rather than a scalar measure (Di Battista et al. 2016a). In this context, the functional variable is measured by the diversity profile (Gattone and Di Battista 2009; Patil and Taillie 1979). With this method, it is possible to overcome the limitations of the traditional indices and take into account both richness and evenness. Thus, diversity is a function defined in the beta fixed domain in which beta is a parameter indicating the importance of richness or evenness. We refer to shareholders using two different variables: institutional or foreign. In hypothesis H1a, the variable "*institutional shareholders*" is measured as a dummy variable; it takes the value "zero" if there are no institutional shareholders, and it takes the value "one" if the firm has at least one institutional shareholder. In hypothesis H2a, the variable institutional shareholders is measured as a categorical variable; it takes the values "zero" if there are no institutional shareholders, "one" if institutional shareholders hold less than 50.01% of shares, and "two" if institutional shareholders hold more than 50% of shares. In hypothesis H1b, the variable "*foreign shareholders*" is measured as a dummy variable; it takes the value "zero" if there are no foreign shareholders and the value "one" if the firm has at least one foreign shareholder. In hypothesis H2b, the variable foreign shareholders is calculated as a categorical variable, it takes the values "zero" if there are no foreign shareholders, "one" if the foreign shareholders hold less than 50.01% of shares, and "two" if the foreign shareholders hold more than 50% of shares.

14.3.3 Method

The problem in defining the concept of "diversity" is reflected on the difficulties in measuring; indeed, the debate about the right method that can capture the different dimensions of diversity is still open. Harrison and Klein (2007) also emphasize the link between the conceptualization of "diversity" and its measurement methodology. In particular, in the economic-management field, previous studies have always used indices that neglect the multidimensional aspect of diversity. For this reason, we introduce a new method for diversity assessment in corporate governance studies that takes advantage of the ecological literature. Indeed, in the field of ecology, diversity assessment has been already widely deepened over the decades, and it is widely held that biodiversity is a multidimensional concept accounting for both species richness (the number of different species present in an ecological community), and species evenness (a measure of the relative abundance of each species in an area). Thus, it is well known that the use of a single index greatly reduces the complexity of the ecological systems (Gattone and Di Battista 2009; Patil and Taillie 1979). An optimal solution is the use of diversity profiles that are parametric families of diversity indices (Hill 1973; Patil and Taillie 1979; Di Battista et al. 2014, 2015). A diversity profile is a curve depicting several diversity indices in a single graph, and it includes the most commonly used indices: the Shannon index (Shannon 1948), Simpson index (Simpson 1949), and species richness. Therefore, the diversity profile is a function dependent on a single continuous parameter that is sensitive to both rare and common species. The plot of diversity profiles plays a fundamental role in comparing different communities because if the diversity profiles do not intersect, the higher curve corresponds to the community with greater diversity (Di Battista et al. 2016a). In particular, the β diversity profile proposed by Patil and Taillie (1979) is applied: $\Delta_\beta = \sum_{j=1}^{s} \frac{(1-p_j^\beta)}{\beta} p_j$ with $\beta > -1$, where, p_j is the relative abundance of the j-th specie (j = 1, 2, .., s), with $p_j = \frac{P_j}{\sum_{j=1}^{s} P_j}$ such that and $\sum_{j=1}^{s} p_j = 1$; Pj is the absolute abundance of the j-th species (the number of individuals belonging to the species j), and the value of β denotes the relative importance of richness and evenness (Patil and Taillie 1979). The most common diversity indices are special cases of the β diversity profile: $\beta = -1$ generates the richness index; $\lim\beta \to 0$ leads to the Shannon diversity index; $\beta = 1$ provides the Simpson index. Further details about the limitations of the classical indices are described in Di Battista et al. (2016a). Because the diversity profile expresses diversity as a function of the relative abundance vector in a fixed domain, it can be analysed in a functional context (Gattone and Di Battista 2009). FDA addresses problems in which the observations are described by functions rather than finite dimensional vectors (Ramsay and Silverman 2005; Ferraty and Vieu 2006; De Sanctis and Di Battista 2012; Maturo et al. 2015). In the economic context, S could be the family of diversity profiles, and for each i-th firm, i = 1, 2, …, N, each relative abundance vector can be assumed as a single parameter (Di Battista et al. 2016a; Maturo et al. 2016). To quantify each of the effects exerted at multiple levels on the functional observation by some factors, the functional analysis of variance (fANOVA) is proposed (Ramsay and Silverman 2005; Di Battista et al. 2014, 2016a, b, c)[2] . This procedure involves testing for possible differences among population mean curves under G different treatments over the whole functional domain. We introduce propose this method and apply it to beta diversity profiles to understand whether two or more groups of functions are statistically distinguishable. For this purpose, we want

---

[2]We assume that there is a single factor with K different groups (k = 1, 2,..., K) and $N = \sum_{i=1}^{K} n_k$ total observations, with n_k observations within each group; then, the model for the i-th observation (i = 1, 2,...., N) in the k-th group can be expressed by $\Delta_{ik}(\beta) = \mu(\beta) + \gamma_k(\beta) + u_{ik}(\beta)$, where $\Delta_{ik}(\beta)$ is the functional response in the k-th group, $\mu(\beta)$ is the grand mean function, $\gamma_k(\beta)$ is the functional effect of being in a specific group, and $u_{ik}(\beta)$ is the residual functions (unexplained variations for the i-th observation within the k-th group).

to test the null hypothesis that the functional groups have the same functional mean against the alternative hypothesis that there is some difference among them.

14.4   Application and Hypotheses Testing

14.4.1   The Functional Effect of Institutional Shareholders on NBD and Results

We aim to assess whether the presence of institutional shareholders influences NBD. Thus, the fANOVA model is applied in the economic context to quantify the effects exerted on functional diversity by the presence of institutional shareholders. In this context, the presence of institutional shareholders is the "factor", and the two "treatments" are "presence" and "not presence". Figure 14.1a shows the diversity profiles of the 1,230 Italian corporations; it is not possible to identify a clear ranking between them in terms of diversity because the profiles often cross one another. In this case, it is possible to establish a unique order only in fixed parts of the domain because different values of β bring different rankings; in this chapter, the sorting problem is not our focus, so we will not dwell on the possible solutions to this issue (Di Battista et al. 2016a).

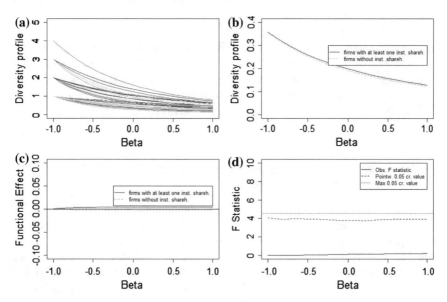

Fig. 14.1 a The functional beta profiles of the 1,230 Italian corporations. **b** The estimated β profiles of the two groups according to the institutional shareholders area. **c** Functional effects of institutional shareholders on board diversity. **d** The functional F-test of the fANOVA model using institutional shareholders as factor

To quantify the amount of board diversity variation explainable by the presence of institutional shareholders, the fANOVA model has been applied. Figure 14.1b displays the beta profiles of the two groups; it is clear that the group of the companies without institutional shareholders is the less diverse group (dashed yellow curve), while the group of companies with at least one institutional shareholder is the more diverse group (solid red curve). This first indication is consistent with what we expected from the analysis. However, this finding does not mean that the presence of the institutional shareholders influences board diversity because it is necessary to check whether this difference between groups is statistically significant before asserting that hypothesis H1a is supported.

Figure 14.1c shows the functional effects of being in a specific group. Of course, the functional effect is strongly linked to Fig. 14.1a; the only difference is that the curves are shifted around the axis of abscissas. The institutional shareholders group exerts a positive effect on board diversity. On the contrary, the effect of being in the group of companies without institutional shareholders is negative.

Figure 14.1d shows the functional statistics[3] for testing the null hypothesis that there are no significant differences between the mean group functions. The solid line represents the observed F statistic; the dashed curve indicates the 0.05 point wise critical value computed with the permutation test, and the dotted line is the 0.05 maximum critical values. This test is based on the null distribution that has been constructed using 2,000 random permutations of the curve labels. It is evident from Fig. 14.1d that the observed F statistic is everywhere under the significance level, so we can conclude that there are no significant differences between the groups in terms of their mean functions.

Relying on the functional F test we can conclude that the presence of the institutional investors does not influence NBD in a statistically meaningful way as we expected; thus, hypothesis H1a is not supported. Because the presence of institutional investors does not significantly influence NBD, it is useless to consider the group of companies in which institutional investors own more or less than the 50% of shares; thus, we can also reject hypothesis H2a.

14.4.2 The Functional Effect of Foreign Shareholders on NBD and Results

The fANOVA model is applied to evaluate the effect exerted on board diversity by the presence of foreign shareholders. In the economic context, the "factor" is the presence of foreigners among the shareholders and the three "treatments" are "companies without foreign shareholders", "companies with foreign shareholders holding less than 50.01% of shares" and "companies with foreign shareholders

---

[3]To assess whether there are significant differences between the groups, a point wise F statistic can be used (Ramsay and Silverman 2005; Di Battista et al. 2016b).

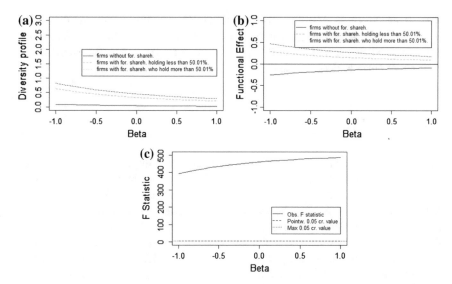

Fig. 14.2 **a** The estimated β profiles of the three groups according to the presence of foreign shareholders. **b** Functional effects of the presence of foreign shareholders on board diversity. **c** The functional F-test of the fANOVA model using the presence of foreign shareholders as factor

holding more than 50% of shares". However, even considering only the "presence" or "absence" of foreign shareholders in the two groups, we obtain the same results, so we only present the fANOVA model with three groups. The fANOVA model is applied to quantify the amount of board diversity variation explainable by the presence of foreigners among the shareholders. Figure 14.2a displays the β profiles of the three groups; it is clear that the group in which the foreign shareholders hold more than 50% of shares is the most diverse (dotted grey curve), while the group of companies without foreign shareholders is the less diverse (solid red curve). The group with foreign shareholders holding less than 50.01% of shares (dashed yellow curve) presents an intermediate level of NBD. These preliminary results are coherent with our hypotheses H1b and H2b. However, it is necessary to check whether these differences between group means are statistically significant. Figure 14.2b shows the functional effects of being in a specific group. The presence of foreign shareholders exerts a positive effect on board diversity; in fact, the dotted and the dashed curves lie over the mean, while the solid curve is under the mean. Thus, the absence of foreign shareholders has a negative impact on NBD. Figure 14.2c show the functional statistics to test the null hypothesis that there are no significant differences among groups. The solid line is the observed F statistics, the dashed curve is the 0.05 point wise critical value computed with the permutation test, and the dotted line is the 0.05 maximum critical values. In addition, this test is based on the null distribution constructed using 2,000 random permutations. Figure 14.2c highlights that the observed F statistic is far above the significance level everywhere; thus, there are significant differences among groups in terms of their mean functions. Therefore, the functional F test confirms our hypotheses H1b and H2b; thus, we can assert that the presence of foreign shareholders significantly influences NBD and that diversity appreciably increases as the percentage owned by foreign shareholders increases.

14.5 Conclusion

Several empirical studies identify board diversity as an important indicator of success for international corporate practices. These studies have focused on board diversity from different perspectives and with different purposes. Most of the studies have analyzed the effects of board diversity on firms' performances, while other studies have considered the mediating variables in the relationship between board diversity and firm performances. On the contrary, few studies have focused on the determinants of board diversity. This chapter aims to address this gap by analysing the variables that could affect NBD; in particular, we focus on shareholders because they have the power to appoint and ultimately elect board members. We analyse whether there is NBD in the boardroom, i.e., whether some types of shareholders choose to appoint foreign members to the board. Our study highlights that different types of shareholders influence NBD in different ways. In particular, our results show that the presence of institutional shareholders does not have significant effects on board diversity. These results suggest that even if they hold more than 50% of shares, institutional shareholders do not consider the appointment of foreign board members as a way to affect corporate governance. Thus, foreign directors are not perceived as "outsiders" who can enhance board independence, nor are they perceived as a way to improve the monitoring role and the effectiveness of the board. Indeed, it is reasonable to suppose that institutional investors could prefer other ways to safeguard board diversity, such as gender diversity (Bianco et al. 2015), or the numbers of outside directors (Feldman and Schwarzkopf 2003). On the other hand, our study highlights that foreign shareholders are strongly related to NBD. These findings demonstrate that, as suggested by the agency and resource based theories, foreign board members may represent the interests of foreign shareholders and assure foreign investors that a firm possesses the skills and ability to compete in the global market. Moreover, this result is strengthened by the fact that NBD is higher when foreign shareholders hold more than 50% of shares. This confirms the arguments proposed by Sur et al. (2013), that a different ownership configuration may affect the board composition in different ways.

This study could have several practical implications concerning the effect of shareholder composition on corporate governance mechanisms focusing on the distinction between different types of shareholders to understand which of them use NBD as a mechanism to improve boards' monitoring function or their decision processes. This study also contributes to the existing body of literature on corporate governance providing a new way of measuring the national board diversity since different board diversity indices are not able to capture the multi-dimensional aspect of diversity. The use of the beta diversity profile to inspect the multidimensional aspects of diversity may constitutes a more accurate measure of diversity for investors. It opens up the firm, and its management, to pressures other than that from shareholders. It also emphasizes the need to look at more accurate measure of board diversity.

For future research, we consider that a longitudinal analysis, or the use of control variables, can better verify the relationship among diversity and foreign, or institutional, shareholders. However, we must note that research on parametric FDA is still currently under development in the statistic field; thus, a panel analysis using vectors of functions, rather than vectors of numbers, is still a goal of advanced search. The same issue arises if we would consider control variables. Another possible future development could lead to consider diversity as a fuzzy concept and thus develop a model of analysis of variance by adopting fuzzy regression models and fuzzy numbers (Maturo and Hošková-Mayerová 2017; Maturo and Maturo 2013, 2014, 2017; Maturo and Fortuna 2016; Maturo 2016).

This chapter focuses on only two typologies of shareholders (institutional and foreign); however, other shareholders and ownership categories can influence NBD (family or non-family shareholders and ownership; individual ownership, etc.). For a more complete analysis, we therefore call for a future investigation on other typology of shareholders and ownership. Finally, our results suggest that foreign shareholders strongly affect NBD. This result is supported both by the agency theory and the resources based view; however, we do not know the real motivations that lead foreign shareholders to appoint foreign directors. To understand better the relationship between foreign shareholders and board diversity, we suggest that future studies combine qualitative and quantitative research by using techniques such as shareholder questionnaires and interviews.

References

Adams, R. B., de Haan, J., Terjesen, S. and van Ees, H. (2015). 'Board diversity: moving the field forward'. *Corporate Governance: An International Review*, **23**, 77–82.

Aggarwal, R., Erel, L., Ferreira M. and Matos, P., (2011). 'Does governance travel around the world? Evidence from institutional investor'. *Journal of Financial Economics*, **100**, 154–81.

Aguilera, R. V. and Jackson, G. (2003). 'The cross-national diversity of corporate governance: Dimensions and determinants'. *The Academy of Management Review*, **28**, 447–65.

Aguilera, R. V. and Jackson, G. (2010). 'Comparative and international corporate governance'. *The academy of management annals*, **4**, 485–556.

Anderson, C. R., Reeb, D. M., Upadhyay A. and Zhao, W. (2011). 'The economics of director heterogeneity'. *Financial Management*, **40**, 5–38.

Ararat, M., Aksu, M. H. and Cetin, A. T. (2015). 'How board diversity affects firm performance in emerging markets. Evidence on channels in controlled firms'. *Corporate Governance: An International Review*, **23**, 83–103.

Arnegger, M., Hoffman, C. and Vetter, C., (2014). 'Firm size and board diversity'. *Journal of Management and Governance*, **18**, 1109–1135.

Bell, S. T., Villado, A. J., Lukasik, M. A., Belayu, L. and Briggs, A. L. (2011). 'Getting specific about demographic diversity variable and team performance relationship: A meta-analysis'. *Journal of Management*, **37**, 709–43.

Berle, A. A. and Means, G. C. (1932). *The modern corporation and private property*. New York, McMillan.

Bianco, M., Ciavarella, A. and Signoretti, R. (2015). 'Women on corporate boards in Italy: The role of family connections'. *Corporate Governance: An International Review*, **23**, 129–44.

Blau, P. M., (1977). *Inequality and heterogeneity*. Glencoe, IL: free press.

Bysinger, B. and Hoskisson, R. E., (1990). 'The composition of board of directors and strategic control: effects on corporate strategy'. *Academy of Management Review*, **15**, 72–87.

Carpenter, M. A. (2002). 'The implication of strategy and social context for the relationship between top management team heterogeneity and firm performance'. *Strategic Management Journal*, **23**, 275–84.

Carter, D. A., Simkins, B.J. and Simpson, W.G. (2003). 'Corporate Governance, Board Diversity, and Firm Value'. *Financial Review*, **38**, 33–53.

Choi, W. Y., Su, W. and Min, S. K., (2012). 'Foreign board membership and firm value in Korea'. *Management Decision*, **50**, 207–33.

Dahya, J. and McConnell, J. J. (2007). 'Board Composition, Corporate Performance and Cadbury Committee Recommendation'. *Journal of Financial and Quantitative Analysis*, **42**, 535–64.

De Sanctis, A. and Di Battista, T. (2012). 'Functional analysis for parametric families of functional data'. *International Journal of Bifurcation and Chaos*, **22**.

Desender, K. T., Aguilera, R. V., Crespi, R. and Garcìa-Cestona, M. (2010). 'When does ownership matters. Board characteristics and behaviour'. *Strategic Management Journal*, **34**, 823–42.

Di Battista, T., Fortuna, F., and Maturo, F. (2014). Parametric functional analysis of variance for fish biodiversity, in: International Conference on Marine and Freshwater Environments, iMFE 2014. URL: http://www.scopus.com.

Di Battista, T., Fortuna, F. and Maturo, F. (2015). 'Diversity in Regional Economics - A Case Study of Abruzzo'. *Global and Local Economic Review*. **19**. 37–54.

Di Battista, T., Fortuna, F. and Maturo, F. (2016a). 'Environmental monitoring through functional biodiversity tools'. *Ecological Indicators*, **60**, 237–47.

Di Battista, T., Fortuna, F., and Maturo, F. (2016b). Parametric functional analysis of variance for fish biodiversity assessment. Journal Of Environmental Informatics. doi:10.3808/jei. 201600348.

Di Battista, T., Fortuna, F., and Maturo, F. (2016c). BioFTF: An R package for biodiversity assessment with the functional data analysis approach. Ecological Indicators. doi:10.1016/j. ecolind.2016.10.032.

Dimovski, W. and Brooks, R. (2006). 'The gender composition of board after IPO'. *Corporate Governance*, **6**, 11–17.

Elyasiani, E. and Jia, J. J. (2010). 'Distribution of institutional ownership and corporate firm performance'. *Journal of Banking and Finance*, **34**, 606–20.

Erhardt, N.L., Werbel, B. C. and Shrader, C.B. (2003). 'Board of Director Diversity and Firm Financial Performance'. *Corporate Governance: An International Review*, **11**, 102–11.

Eulerich, M., Velte, P., and Van Uum, C. (2014). The impact of management board diversity on corporate performance - an empirical analysis for the german two-tier system. Problems and Perspectives in Management, 12(1), 25–39. Retrieved from: http://www.scopus.com.

Europen Commission, Green Paper "The EU Corporate Governance Framework", Bruxelles, COM 2011, 5–6.

Feldman, A. D. and Schwarzkopf, D. L. (2003). 'The effect of institutional ownership on board and audit committee composition'. *Review of Accounting and Finance*, **2**, 87–109.

Ferraty, F. and Vieu, P. (2006). *Nonparametric functional data analysis*. Springer, New York.

Ferreira, M. A. and Matos, P. (2008). 'The colours of investor's money: The role of institutional investors around the world'. *Journal of Finance Economics*, **88**, 499–533.

Filatotchev, I., Buck, T., Demina, N. and Wright, M. (2001). 'Corporate governance and exporting in the former Soviet Union'. *Journal of international business studies*, **32**, 853–71.

Filatotchev, I. and Wright, M., (2011). 'Agency perspectives on corporate governance of multinational enterprises'. *Journal of management studies*, **48**, 471–86.

Filatotchev, I. and Bishop, K. (2002). 'Board composition, share ownership and 'underpricing' of UK IPO firms'. *Strategic Management Journal*, **23**, 941–55.

Gattone, S. A. and Di Battista, T. (2009). 'A functional approach to diversity profiles'. *Journal of the Royal Statistical Society*, **58**, 267–84.

Gillan, S. L. and Starks, L. T. (2007). 'The evolution of shareholder activism in the United States'. *Journal of Applied Corporate Finance*, **19**, 55–73.

Goyer, M. and Jung, D. K. (2011). 'Diversity of institutional investors and foreign blockholdings in France: The evolution of an institutionally hybrid economy'. *Corporate Governance: An International Review*, **19**, 562–84.

Hamzah, A. H. and Zulkafli, A. H. (2014). 'Board diversity and corporate expropriation'. *Procedia-Social and Behavioural Sciences*, **164**, 562–68.

Harrison, D. A. and Klein, K. J. (2007). 'What's the difference? Diversity construct as separation, variety, or disparity in organizations'. *Academy of Management Review*, **32**, 1199–228.

Heidrick, and Struggles (2009). *Corporate Governance Report 2009. Boards in turbulent times.* Available at: http://www.heidrick.co/PublicationReports/PublicationsReports/CorpGovEurope 2009.pdf.

Hill, M. (1973). 'Diversity and evenness: a unifying notation and its consequences'. *Ecology.* **54**, 427–32.

Hillman, A. J. (2015). 'Board Diversity: Beginning to Unpeel the Onion'. *Corporate Governance: An International Review*, **23**, 104–07.

Hillman, A. J., Shropshire, C. and Cannella, A. A. (2007). 'Organizational predictors of women on corporate board'. *Academy of Management Journal*, **50**, 941–52.

Johnson, J. L., Daily, C. M. And Ellstrand, A. E. (1996). 'Board of directors: a review and research agenda'. *Journal of Management*, **22**, 409–38.

Kiel, G.C. and Nicholson, G.J. (2003). 'Board Composition and Corporate Performance: how the Australian experience informs contrasting theories of corporate governance'. *Corporate Governance: An International Review*, **11**, 189–205.

Klein, A. and Zur, E. (2009). 'Entrepreneurial shareholder activism: Hedge funds and other private investors'. *The Journal of Finance*, **64**, 187–229.

Lane, P. J., Cannella A. A. and Lubaktin, M. H. (1998). 'Agency problems as antecedents to unrelated mergers and diversification: Amihud and Lev reconsidered'. *Strategic Management Journal*, **19**, 555–78.

Lee, C. and Farh, J. L. (2004). 'Joint effects of group efficacy and gender diversity on group cohesion and performance'. *Applied psychology: An international Review*, **53**, 136–54.

Lee, S. C., Rhee, M. and Joon, J. (2012). *The effect of foreign monitoring on audit quality: the compositional dynamic of organizational group. Evidence from Korea.* Available at SSRN: http://www.ssrn.com/abstract=2001782.

Mallin, C. A. (2012). 'Institutional investors: the vote as a tool of governance'. *Journal of Management & Governance*, **16**, 177–96.

Maturo F., Migliori S., Consorti A. (2015). Corporate Board Diversity. In: A.P. Haller and M. Galea. Proceedings of the International Conference Humanities and Social Sciences Today. Economics. Pro Universitaria Publishing House. Bucharest. pp. 113–125. ISBN: 978-606-26-0413-4

Maturo, F. (2016). Dealing with randomness and vagueness in business and management sciences: the fuzzy probabilistic approach as a tool for the study of statistical relationships between imprecise variables. Ratio Mathematica 30, 45–58.

Maturo, F., Di Battista, T., Fortuna, F. (2016). BioFTF: Biodiversity assessment using functional tools. https://www.cran.r-project.org/web/packages/bioftf/index.html.

Maturo, A., Maturo, F., (2013). Research in Social Sciences: Fuzzy Regression and Causal Complexity. Springer Berlin Heidelberg, Berlin, Heidelberg. pp. 237–249. URL: http://dx.doi.org/10.1007/978-3-642-35635-3 18, doi:10.1007/978-3-642-35635-3 18.

Maturo A., Maturo F. (2014). Finite Geometric Spaces, Steiner Systems and Cooperative Games. Analele Universitatii "Ovidius" Constanta. Seria Matematica. Vol. 22(1), pp. 189–205. (2014). ISSN: Online 1844-0835. doi:10.2478/auom-2014-0015.

Maturo, A., Maturo, F., (2017). Fuzzy Events, Fuzzy Probability and Applications in Economic and Social Sciences. Springer International Publishing, Cham. pp. 223–233. URL: http://dx.doi.org/10.1007/978-3-319-40585-8 20, doi:10.1007/978-3-319-40585-8 20.

Maturo, F., Fortuna, F. (2016). Bell-Shaped Fuzzy Numbers Associated with the Normal Curve. Springer International Publishing, Cham. pp. 131–144. URL: http://dx.doi.org/10.1007/978-3-319-44093-4 13, doi:10.1007/978-3-319-44093-4 13.

Maturo, F., Hošková-Mayerová, S. (2017). Fuzzy Regression Models and Alternative Opertions for Economic and Social Sciences. Springer International Publishing, Cham. pp. 235–247. URL: http://dx.doi.org/10.1007/978-3-319-40585-8 21, doi:10.1007/978-3-319-40585-8 21.

McNulty, T. and Nordberg, D. (2015). 'Ownership, Activism and Engagement: Institutional Investors as Active Owners'. *Corporate Governance: An International Review*, doi:10.1111/corg.12143.

Miller, T. and Triana, M.C. (2009). 'Demographic Diversity in the Boardroom: mediators of the Board Diversity-Firm Performance Relationship'. *Journal of Management Studies*, **46**, 755–86.

Milliken, F. and Martins, L. (1996). 'Searching for common threads: understanding the multiple effects of diversity in organizational groups'. *Academy of Management Journal*, **21**, 402–33.

Muller-Kahale, M. I. and Lewellyn, K. B. (2011). 'Did board configuration matter? The case of US subprime lenders'. *Corporate Governance: An International Review*, **19**, 405–17.

OECD (1998). Corporate governance: Improving competitiveness and access to capital in global markets. *Organization for Economic Cooperation and Development Document*, Paris.

OECD (2004). Principles of corporate governance, *Organization for Economic Cooperation and Development Document*, Paris.

OECD (2011). The role of institutional investors in promoting good corporate governance. *Organization for Economic Cooperation and Development Document*, Paris.

Oxelheim, L. and Randøy, T. (2003). 'The impact of foreign board membership on firm value'. *Journal of Banking & Finance*, **27**, 2369–392.

Patil, G. and Taillie, C. (1979). 'An overview of diversity'. In: Grassle, J., Patil, G., Smith, W., Taillie, C. (Eds.). *Ecological Diversity in Theory and Practice*. International Co-operative Publishing House, Fairland, MD, 23–48.

Pearce, J. A. and Zahara, S. A. (1992). 'Board composition from a strategic contingency perspective'. *Journal of Management Studies*, **29**, 411–38.

Ramsay, J. and Silverman, B. (2005). *Functional Data Analysis*, 2nd edition, Springer, New York.

Randøy, T., Oxelheim L. and Stonehill, A. (2001). 'Global financial strategies and corporate competitiveness'. *European Management Journal*, **19**, 659–69.

Rose, P. (2007). 'The Corporate Governance Industry'. *Journal of Corporation Law*, **32**, 887–26.

Ruigrok, W., Pec, S. and Tacheva, S. (2007). 'Nationality and Gender Diversity on Swiss Corporate Boards'. *Corporate Governance: An International Review*, **15**, 546–57.

Shannon, C. (1948). 'A mathematical theory of communication'. *Bell system technical journal*, **27**, 379–423.

Shleifer, A. and Vishny R. (1986). 'Large shareholders and corporate control'. *Journal of Political Economy*, **94**. 461–88.

Simpson, E. (1949). 'Measurement of diversity'. *Nature*. **163**, 688.

Stiles, P. (2001). 'The impact of board on strategy: an empirical examination'. *Journal of Management Studies*, **38**, 627–50.

Sur, S., Lvina, E. and Magnan, M. (2013). 'Why do boards differ? Because owners do: assessing ownership impact on board composition'. *Corporate Governance: An International Review*, **21**, 373–89.

Ujunwa, A., Okoyeuzu, C. and Nwakoby, I. (2012). 'Corporate Board Diversity and Firm Performance. Evidence from Nigeria'. *Review of International Comparative Management*, **13**, 605–20.

Useem, M., (1998). 'Corporate leadership in a globalizing equity market'. *Academy of Management Executive*, **12**, 43–59.

van Ees, H., Gabrielsson and J., Huse, M. (2009). 'Toward a behavioural theory of boards and corporate governance'. *Corporate Governance: An International Review*, **17**, 307–19.

Veltrop, D., Hermes, N., Postman, T. J. B. and de Haan, J. (2015). 'A tale of two factions: Why and when factional demographic faultlines hurt board performance'. *Corporate Governance: An International Review*, **23**, 145–60.

Whitley, R. (2009). 'US capitalism: A tarnished model?'. *Academy of Management Perspective*, **23**, 11–22.

Chapter 15
Inequalities in the Provinces of Abruzzo: A Comparative Study Through the Indices of Deprivation and Principal Component Analysis

Domenico Di Spalatro, Fabrizio Maturo and Lorella Sicuro

Abstract The indices of deprivation are a valuable tool to measure the socioeconomic disadvantage in certain geographical areas of interest. This study aims to compare inequalities between the provinces of Abruzzo over the last two decades suggesting some indices of deprivation to capture the key aspects of the great wealth of information relating to population census. Specifically, we propose three indices of deprivation to measure the material and social disadvantage. Moreover, a principal component analysis is performed using the most know indicators of deprivation. Using these methods, we observe an increase in the proportion of disadvantaged areas in the Abruzzo region from 1991 to 2011 in its four provinces.

Keywords Deprivation indicator · Disadvantaged areas · IDM · IDS · IAS

15.1 Introduction

Over the last decades, the debate on "Equitable and Sustainable Well-being" has received many attentions. Indeed, the debate on the ability of the gross domestic product (GDP) to provide a proper picture of reality is still lively. GDP is a quantitative measure of economic system but it does not offer a comprehensive view of the progress of the society. It should be integrated with other indicators reflecting the condition of citizens, such as health, safety, subjective well-being,

D. Di Spalatro (✉) · L. Sicuro
National Institute of Statistics - ISTAT, Pescara, Italy
e-mail: dispalat@istat.it

L. Sicuro
e-mail: sicuro@istat.it

F. Maturo
Department of Business Administration -, "G. d'Annunzio" University of Chieti-Pescara, Pescara, Italy
e-mail: f.maturo@unich.it

© Springer International Publishing AG 2017   219
Š. Hošková-Mayerová et al., *Mathematical-Statistical Models and Qualitative Theories for Economic and Social Sciences*, Studies in Systems, Decision and Control 104,
DOI 10.1007/978-3-319-54819-7_15

working conditions, the economic well-being, inequality, the state of the environment. For this reason, in Italy, CNEL (the National Council for Economy and Labor), ISTAT (the National Statistics Institute), and the scientific community, have selected a set of 134 indicators to represent 12 dimensions of the fair and sustainable welfare. Specifically, these aspects of well-being are the following: health, education and training, work and life schedules, material well-being, social relations, politics and institutions, security, subjective well-being, landscape and cultural heritage, environment, research and innovation, quality of services.

These research questions have stimulated the development of the concept of "deprivation" and its related measures (Blane et al. 1989). An index of deprivation is a "composite aggregate indicator" (composed of several basic indicators) (Carstairs and Morris 1989) capable of synthesizing the possession of social and material resources related to well-defined geographical units.

Within these geographical units, the deprivation index measures the proportion of families who have a combination of features indicating a low standard of living, a high demand for services, and therefore a situation of economic and social disadvantage. The geographically-based deprivation indices have many advantages: first, they constitute one of the few operational available tool to measure briefly the socioeconomic conditions of a given territory; second, they are cheap because it is easy to directly obtain them from census data; finally, they are based on objective and easily available information.

The use of a deprivation index implies at least three choices (Jarman 1983; Testi et al. 2005). The first choice concerns the geographical unit of reference. In Italy, the minimum available reference unit is the census tract, but this breakdown is rarely used (Michelozzi et al. 1999); indeed, it is frequently preferred to use the municipal level (Cadum et al. 1999; Minerba and Vacca 2006) because the town is a homogeneous area, under the physical environment and social profiles, and represents a set of values, traditions and shared norms, which anthropologically make up the overall effect of the place of residence on health (Macintyre et al. 2002).

The second choice is related to the variables, or the partial indicators, for the construction of the index of deprivation. Because there is no single definition of the concept of deprivation or a single method for its measurement in the literature, there are no standard procedures for the selection of the elementary variables. Therefore, we proceed by taking into account all the available variables that in some way can be considered related to the condition of deprivation, and then we choose the most suitable. Those potentially linked to deprivation and most used in the literature are unemployment, low education, overcrowding, lack of home ownership, ethnic minorities, the dependency ratio and the presence of single parent families (Table 15.1).

In specific cases, the partial indicators can be weighted (Forrest and Gordon 1993; Gordon 1995), taking into account exogenous information provided by experts. If the variables are expressed in different units of measurement, before making the sum, to avoid that some have a greater weight than the others, standardization must be carried out (Bartley and Blane 1994). A more refined way to

Table 15.1 Variables, or partial indicators, of deprivation available at the municipal level and most used in the literature

Variable	Definition
Unemployment	Percentage of economically active people are unemployed
Low education	Percentage of people with a lower or equal to the middle school diploma
Overcrowding	Average number of occupants per dwelling
Home ownership	Percentage of the non-owning families of the house where they live
Ethnical minorities	Percentage of foreign residents
Dependency ratio	Percentage of children under 5 years or older and seniors over 65
Single parents	Percentage of families with a single parent

obtain the index of deprivation is to proceed with a factor analysis, which allows identification of the primary variables providing important contributions to explain deprivation.

15.2 Materials and Methods

Following the most used approach for the construction of the index of deprivation (Minerba and Cow 2006), we proceeded to the identification of the index of material deprivation (IDM), which measures the state of the need of a population from a material objective point of view.

Material deprivation leads to a lack of goods, services, resources, comfort normally enjoyed or at least widely accepted as primary goods. The IDM can be refined by introducing an additional indicator that turns it into a real social deprivation index (IDS) that measures the state of the social needs of a population. It is calculated, for each municipality, as the sum IDM and the indicator relating to single-parent families. The refinement procedure can be completed by introducing an additional indicator that turns the IDS in a disadvantaged area index (IAS). This represents an overall measure for the evaluation of geographical differences because it summarizes the dimensions of socioeconomic disadvantage, which is useful for spatial planning of social and health interventions. It is calculated for each municipality as the sum IDS and an indicator that expresses the potential burden of care and the need for social support.

Using the new "8milaCensus" database of ISTAT that collects the census data of the Italian population, we have identified eleven indicators (Table 15.2) improving the information of the most used indices (Table 15.1).

The selection of the elementary indicators has been driven to find indicators with a good degree of validity (i.e. Capable of effectively representing the main dimensions of meaning) among the numerous variables available from the census detection.

Table 15.2 Selected variables, or partial indicators, of deprivation available at the municipal level

Indicator name	Indicator structure	Aggregate index in which it is used
Incidence of adults with middle school	Percentage ratio of the population aged 25–64 with middle school and the resident population aged 25–64	IDM-IDS-IAS
Unemployment rate	Percentage ratio of the resident population aged 15 and over in search of employment and the resident active population aged 15 and more	IDM-IDS-IAS
Incidence of employment on low-skill level occupations	Percentage ratio of the employed occupational activity (professions unqualified) to total employment	IDM-IDS-IAS
Incidence of housing not owned	Percentage ratio of not owned occupied housing and the total of occupied dwellings	IDM-IDS-IAS
Incidence of improper accommodation	Percentage ratio of the other types of accommodation and the total of the houses	IDM-IDS-IAS
Incidence of the population in conditions of overcrowding	Percentage ratio of the population living in dwellings with a surface area less than 40 m^2, and more than 4 occupants, or 40–59 m^2 and more than 5 occupants, or 60–79 m^2 and more than six occupants, and the total of the population living in occupied housing	IDM-IDS-IAS
Incidence of families with potential economic hardship	Percentage ratio between the number of families with children with the reference person aged up to 64 years in which no member is employed, or retired from work, and the total of families	IDM-IDS-IAS
Incidence of families in hardship assistance	Percentage ratio between the number of families with at least two components, without cohabiting, with all components over 65 years and with the presence of at least one member 80 years or more, and the total of families	IDS-IAS
Incidence of young single-parent families	Percentage ratio between the number of families with a single nucleus of young single-parent type (father/mother with less than 35 years), with and without isolated members, and the total number of single parent families, with and without isolated members	IDS-IAS
Incidence of single-parent aged families	Percentage of single-parent families made up of only one core.	IDS-IAS
	Elder (father/mother aged 65 and over), with and without isolated members, and the total number of single parent families, with and without isolated members	
Elderly dependency ratio	Percentage ratio of the population aged 65 and over and the population aged 15–64 years	IAS

Therefore, for each geographical unit, we calculated the three indices of deprivation: IDM, IDS and IAS, to represent in more detail, at the municipal level, the heterogeneity in the health and social burden. For the aggregation of the elementary indices, with appropriate adjustments, we used the methodology adopted by ISTAT for the construction of the index of social and material vulnerability (http://ottomilacensus.istat.it/).

We attributed to each municipality, a value comparable in historical and territorial series for the three census surveys (1991–2001–2011). By construction, the values are all within the range (70–130). For the IDM and IDS we considered an area as "deprived" if it assumes a value greater than, or equal to, 100. Similarly for IAS, we considered a disadvantage if the index assumes values greater than, or equal to, 100. It was also performed the Principal Component Analysis (ACP) on eleven selected indicators in order to identify the variables providing greater contributions to the explanation of the phenomenon.

15.3 Results

15.3.1 IDM, IDS and IAS

By analyzing the IDM (Table 15.3) in 1991, 148 municipalities of Abruzzo on 305 (48.52%) result materially deprived. The province of Pescara has the poorest performance because 60.87% of municipalities have great hardship.

In 2001, the regional result is almost unchanged: there are 149 of the 305 municipalities of Abruzzo (48.85%) in conditions of deprivation. The most deprived province is Teramo, where 31 of 47 municipalities (65.96%) are deprived.

Table 15.3 Abruzzo provinces IDM class. Number of municipalities and percentage within the province (1991, 2001, 2011)

Provinces	l'Aquila		Teramo		Pescara		Chieti		Abruzzo	
	n.	%	n.	%	n.	%	n.	%	n.	%
IDM 1991										
Not deprived	46.0	42.6	25.0	53.2	18.0	39.1	68.0	65.4	157.0	51.5
Deprived	62.0	57.4	22.0	46.8	28.0	60.9	36.0	34.6	148.0	**48.5**
Total	108.0	100.0	47.0	100.0	46.0	100.0	104.0	100.0	305.0	100.0
IDM 2001										
Not deprived	64.0	59.3	16.0	34.0	17.0	37.0	59.0	56.7	156.0	51.1
Deprived	44.0	40.7	31.0	66.0	29.0	63.0	45.0	43.3	149.0	**48.9**
Total	108.0	100.0	47.0	100.0	46.0	100.0	104.0	100.0	305.0	100.0
IDM 2011										
Not deprived	24.0	22.2	3.0	6.4	4.0	8.7	30.0	28.8	61.0	20.0
Deprived	84.0	77.8	44.0	93.6	42.0	91.3	74.0	71.2	244.0	**80.0**
Total	108.0	100.0	47.0	100.0	46.0	100.0	104.0	100.0	305.0	100.0

Table 15.4 Abruzzo provinces IDS class. Number of municipalities and percentage within the province (1991, 2001, 2011)

Provinces	l'Aquila		Teramo		Pescara		Chieti		Abruzzo	
	n.	%	n.	%	n.	%	n.	%	n.	%
IDS 1991										
Not deprived	32.0	29.6	31.0	66.0	23.0	50.0	65.0	62.5	151.0	49.5
Deprived	76.0	70.4	16.0	34.0	23.0	50.0	39.0	37.5	154.0	**50.5**
Total	108.0	100.0	47.0	100.0	46.0	100.0	104.0	100.0	305.0	100.0
IDS 2001										
Not deprived	31.0	28.7	16.0	34.0	14.0	30.4	46.0	44.2	107.0	35.1
Deprived	77.0	71.3	31.0	66.0	32.0	69.6	58.0	55.8	198.0	**64.9**
Total	108.0	100.0	47.0	100.0	46.0	100.0	104.0	100.0	305.0	100.0
IDS 2011										
Not deprived	11.0	10.2	1.0	2.1	2.0	4.3	10.0	9.6	24.0	7.9
Deprived	97.0	89.8	46.0	97.9	44.0	95.7	94.0	90.4	281.0	**92.1**
Total	108.0	100.0	47.0	100.0	46.0	100.0	104.0	100.0	305.0	100.0

In 2011, there is a significant increase in the number of municipalities (80%), which are situated in areas of great hardship. In this context, Teramo reinforces its negative record: 44 of 47 municipalities (93.62%) are located in areas of great material deprivation.

By analyzing the results of the IDS (Table 15.4) of 1991, 154 municipalities of Abruzzo (50.49%) are in condition of social deprivation. The province of L'Aquila records the worst situation with 76 municipalities of 108 (70.37%) situated in areas of great discomfort.

In 2001, the regional datum shows an increased number of municipalities, 198 of 305 (64.92%) in conditions of social deprivation. The negative record still belongs to the province of L'Aquila in which 77 municipalities (71.30%) are located in areas of deprivation.

Also for the IDS, the worst result is in 2011 because there is a significant increase in the number of municipalities, 281 of 305 (92.13%), which are situated in areas of social deprivation. The negative record goes to Teramo because almost all municipalities (46 of 47) are in the areas of high hardship.

Finally, by analyzing the results of the IAS (Table 15.5) in 1991, 159 municipalities of Abruzzo (52.13%) are disadvantaged. The province of L'Aquila has the poorest performance because 83 municipalities of 108 (76.85%) are placed in disadvantaged areas.

In 2001, the regional datum shows a growing number of municipalities of Abruzzo (201, 65.90%), who are disadvantaged. Again, it is the province of L'Aquila to record the highest number of municipalities (81, 75%) placed in disadvantaged areas.

Also for the IAS, the worse result is in 2011 where we observe a significant increase in the number of municipalities, 286 of 305 (the 93.77%), which are

Table 15.5 Abruzzo provinces IAS class. Number of municipalities and percentage within the province (1991, 2001, 2011)

Provinces	l'Aquila		Teramo		Pescara		Chieti		Abruzzo	
	n.	%	n.	%	n.	%	n.	%	n.	%
IAS 1991										
Not deprived	25.0	23.1	34.0	72.3	24.0	52.2	63.0	60.6	146.0	47.9
Deprived	83.0	76.9	13.0	27.7	22.0	47.8	41.0	39.4	159.0	**52.1**
Total	108.0	100.0	47.0	100.0	46.0	100.0	104.0	100.0	305.0	100.0
IAS 2001										
Not deprived	27.0	25.0	19.0	40.4	15.0	32.6	43.0	41.3	104.0	34.1
Deprived	81.0	75.0	28.0	59.6	31.0	67.4	61.0	58.7	201.0	**65.9**
Total	108.0	100.0	47.0	100.0	46.0	100.0	104.0	100.0	305.0	100.0
IAS 2011										
Not deprived	7.0	6.5	1.0	2.1	2.0	4.3	9.0	8.7	19.0	6.2
Deprived	101.0	93.5	46.0	97.9	44.0	95.7	95.0	91.3	286.0	**93.8**
Total	108.0	100.0	47.0	100.0	46.0	100.0	104.0	100.0	305.0	100.0

situated in deprived areas. As for the IDM and IDS, also for IAS, Teramo (with 46 of 47 municipalities) is the most disadvantaged province.

Figure 15.1 displays that the province of l'Aquila, in 1991, recorded the highest values of social deprivation and disadvantage compared to the other provinces of the region. However, in 2001, L'Aquila is the only province to record a decrease in

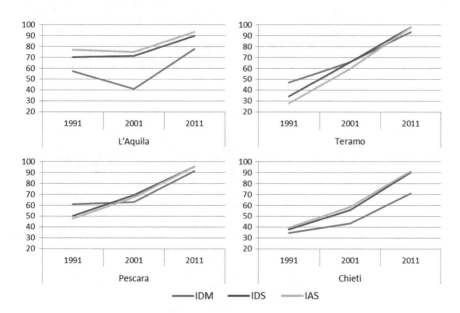

Fig. 15.1 Percentage of municipalities in conditions of deprivation. IDM, IDS, IAS by province for the years 1991, 2001, 2011

IDM and IAS; indeed, material deprivation decreases from 1991 to 2001, making the province, the less materially deprived of Abruzzo. In 2011, we observe a sharp deterioration of the conditions of material deprivation (increase to 91%); thus, it decreases the distance between the IDM and the other two indices.

In the provinces of Pescara and Teramo, we highlight a linear increase of the IDS and IAS over the years. Moreover, in 2011, these provinces are characterized by an overlap of the values of the IDM, IDS and IAS.

Chieti is the province with the lowest percentage of deprived municipalities; however, also for this area, we observe deterioration from 2001 to 2011. Moreover, in 2011, Chieti is the province with the highest gap between the IDM and the other two indexes.

15.3.2 Principal Component Analysis (PCA)

Regarding the PCA, we use as starting variables the eleven indicators of Table 15.2. In 1991 (Table 15.6 and 15.7), the first factor explains 24% of the total variance, while the second explains 17%, thus they capture 41% of the total variability. The first principal component, which might be called the "elderly discomfort", is positively correlated with the elderly dependency ratio (0.876), the incidence of families in hardship assistance (0.775) and the incidence of elderly single-parent families (0.656). The second axis, which explains the "material disadvantage", is highly correlated with the incidence of households with potential economic hardship (0,746), the incidence of the population in crowded conditions (0.648) and the unemployment rate (0.638).

In 2001 (Tables 15.8 and 15.9), the first two factors explain respectively 23% and 15% of variability, thus 38% of the total variance. The indicators highly and

Table 15.6 PCA 1991. Total variance explained by the principal components

1991	Total variance explained	
	Initial eigenvalues	
PC	% of variance	Cumulative %
1	24.09	24.09
2	16.68	40.77
3	10.84	51.61
4	9.40	61.01
5	8.04	69.05
6	7.28	76.33
7	6.67	83.00
8	5.98	88.98
9	4.96	93.93
10	4.15	98.09
11	1.92	100.00

Table 15.7 PCA 1991. Correlation matrix of the first three components and indicators

1991	Component		
	1	2	3
Secondary school certificate	−0.24	0.12	0.37
Unemployment rate	0.39	0.64	−0.06
Lower working competence	0.48	0.27	0.42
Elderly dependency ratio	0.88	−0.18	−0.16
Housing not of owned	−0.58	0.28	−0.11
Accommodations improper	−0.05	0.04	0.51
Overcrowding	0.12	0.65	0.28
Economic hardship	−0.24	0.75	−0.27
Hardship assistance	0.78	−0.18	−0.23
Young single-parent families	0.05	0.36	−0.59
Single-parent families elderly	0.66	0.30	0.18

Table 15.8 PCA 2001. Total variance explained by the principal components

2001	Total variance explained	
	Initial eigenvalues	
PC	% of variance	Cumulative %
1	23.01	23.01
2	15.48	38.49
3	12.00	50.49
4	9.37	59.86
5	8.84	68.70
6	7.66	76.35
7	6.51	82.86
8	5.77	88.63
9	5.14	93.77
10	4.29	98.06
11	1.94	100.00

Table 15.9 PCA 2001. Correlation matrix of the first three components and indicators

2001	Component		
	1	2	3
Secondary school certificate	−0.21	−0.44	0.65
Unemployment rate	−0.34	0.62	0.25
Lower working competence	0.10	−0.66	0.52
Elderly dependency ratio	0.87	0.18	0.03
Housing not of owned	−0.63	0.07	−0.23
Accommodations improper	−0.12	−0.10	0.13
Overcrowding	−0.04	0.37	0.39
Economic hardship	−0.44	0.50	0.40
Hardship assistance	0.83	0.01	0.01
Young single-parent families	−0.24	−0.14	−0.36
Single-parent families elderly	0.50	0.49	0.23

Table 15.10 PCA 2011. Total variance explained by the principal components

2011	Total variance explained	
	Initial eigenvalues	
PC	% of variance	Cumulative %
1	20.34	20.34
2	14.82	35.15
3	13.26	48.42
4	11.55	59.96
5	9.26	69.22
6	7.01	76.23
7	6.10	82.33
8	5.84	88.17
9	5.18	93.35
10	4.05	97.41
11	2.59	100.00

Table 15.11 PCA 2011. Correlation matrix of the first three components and indicators

2011	Component		
	1	2	3
Secondary school certificate	0.27	−0.21	0.49
Unemployment rate	−0.29	−0.45	0.43
Lower working competence	0.27	0.23	0.51
Elderly dependency ratio	0.83	0.13	−0.08
Housing not of owned	−0.35	0.61	0.02
Accommodations improper	−0.22	0.49	−0.44
Overcrowding	−0.05	0.57	0.48
Economic hardship	−0.48	−0.11	0.50
Hardship assistance	0.78	−0.07	−0.04
Young single-parent families	−0.22	0.44	0.12
Single-parent families elderly	0.51	0.40	0.31

positively correlated with the first axis are the elderly dependency ratio (0.869) and the incidence of families in hardship assistance (0.832), while the percentage of not owned homes has a negative correlation (−0.631). The second axis is characterized by a positive correlation with the unemployment rate (0.619) and a negative correlation with the incidence of employment in low-skill occupations (−0.660).

In 2011 (Tables 15.10 and 15.11), the first two components explain 35% of the total variability. The first factor is highly correlated with the elderly dependency ratio (0.833) and the incidence of families in hardship assistance (0,780). The second axis shows correlation with the incidence of housing not owned (0.610).

15.4 Conclusions and Research Perspectives

Our results show that from 1991 to 2011 there was an increase in the number of municipalities classified as "deprived" and the social environment fails to improve the discomfort induced and recorded by the material deprivation. The goodness of our results is confirmed by the PCA performed on the 11 adopted indicators. Specifically, PCA highlights that, from 1991 to 2011, we have a reduction of the variability explained by the first two factorial axes (41% in 1991, 38% of 2001 and 35% in 2011) indicating that the power of the first two principal components decreases over the decades.

However, the PCA ensures that there is a low level of dependence among the indicators of deprivation. This confirms that these indicators consistently cover, with no redundancies or repetitions, the entire multidimensional nature of the phenomenon of material and social deprivation.

However, it remains interesting to note that the first factor is always characterized by the presence of discomfort of the elderly, and the second axis (especially for 1991 and 2001) with the discomfort of employment. In particular, the first factor is characterized by high correlation with the elderly dependency ratio and with the incidence of families consisting of only elderly people in distress assistance.

We highlight that, passing from the IDM to the IAS, the state of deprivation worsens for many municipalities. This is particularly interesting considering that three of the four indicators used to switch from the IDM to the IAS are related to the aging of the population. The fact that the material deprivation worse by the inclusion of indicators related to elderly people indicates the importance of assessing policies to protect this most fragile part of the population. Indeed, the problem of an aging population regards not only Abruzzo, but also Italy.

Recent studies have focused on the identification of deprivation indices based on the fuzzy logic approach (Najjary et al. 2016; Fattore et al. 2011; Potsi et al. 2015; Betti and Verma 2007). This seems an appropriate method because the concept of deprivation is a nuanced. For this reason, it would be very interesting to extend the studies on social and material deprivation using recent advance in fuzzy theory (Maturo and Maturo 2014, 2016; Maturo and Fortuna 2016; Maturo 2016). At the same time, it could be very attractive a joint use of fuzzy indices of deprivation and recent studies on fuzzy regression models (Maturo and Hošková-Mayerová 2016; Maturo and Maturo 2013) to discover the determinants of the deprivation. Furthermore, other possible future developments could consider deprivation as a function of time and adopt a functional approach (Di Battista et al. 2016a, b, c, 2014; Ramsay and Silverman 1997) for the assessment of the disadvantaged areas.

References

Bartley, M., & Blane, D. (1994). Commentary: Appropriateness of deprivation indices must be ensured. BMJ, 309(6967), 1479–1479. doi:10.1136/bmj.309.6967.1479.

Betti, G., & Verma, V. (2007). Fuzzy measures of the incidence of relative poverty and deprivation: a multi-dimensional perspective. Statistical Methods and Applications, 17(2), 225–250. doi:10.1007/s10260-007-0062-8.

Blane, D., Townsend, P., Phillimore, P., & Beattie, A. (1989). Health and Deprivation: Inequality and the North. The British Journal of Sociology, 40(2), 344. doi:10.2307/590279.

Cadum, E., Costa, F., Biggeri, A., & Martuzzi, M. (1999). Deprivazione e mortalità: un indice di deprivazione per l'analisi delle disuguaglianze su base geografica, Epidemiologia e Prevenzione, 23, 175–187.

Carstairs, V., & Morris, R. (1989). Deprivation and health. BMJ, 299(6713), 1462–1462. doi:10.1136/bmj.299.6713.1462-a.

Di Battista, T., Fortuna, F., Maturo, F. (2014). Parametric functional analysis of variance for fish biodiversity, in: International Conference on Marine and Freshwater Environments, iMFE 2014. URL:www.scopus.com.

Di Battista, T., Fortuna, F., Maturo, F. (2016a). Environmental monitoring through functional biodiversity tools. Ecological Indicators, 60, 237–247. doi:10.1016/j.ecolind.2015.05.056.

Di Battista T., Fortuna F., Maturo F. (2016b). BioFTF: An R Package for Biodiversity Assessment with the Functional Data Analysis Approach. Ecological Indicators. doi:10.1016/j.ecolind. 2016.10.032.

Di Battista T., Fortuna F., Maturo F. (2016c). Parametric Functional Analysis of Variance for Fish Biodiversity Assessment. Journal of Environmental Informatics. doi:10.3808/jei.201600348.

Fattore, M., Brüggemann, R., & Owsiński, J. (2011). Using Poset Theory to Compare Fuzzy Multidimensional Material Deprivation Across Regions. New Perspectives in Statistical Modeling and Data Analysis, 49–56. doi:10.1007/978-3-642-11363-5_6.

Forrest, R., & Gordon, D., (1993). People and Places: a 1991 Census atlas of England, SAUS. University of Bristol.

Gordon, D. (1995). Census based deprivation indices: their weighting and validation. Journal of Epidemiology & Community Health, 49(Suppl 2), S39–S44. doi:10.1136/jech.49.suppl_2.s39.

Jarman, B. (1983). Identification of underprivileged areas. BMJ, 286(6379), 1705–1709. doi:10.1136/bmj.286.6379.1705.

Macintyre, S., Ellaway, A., & Cummins, S. (2002). Place effects on health: how can we conceptualise, operationalise and measure them? Social Science & Medicine, 55(1), 125–139. doi:10.1016/s0277-9536(01)00214-3.

Maturo, F. (2016). Dealing with randomness and vagueness in business and management sciences: the fuzzy-probabilistic approach as a tool for the study of statistical relationships between imprecise variables. Ratio Mathematica 30, 45–58.

Maturo, F., & Fortuna, F. (2016). Bell-Shaped Fuzzy Numbers Associated with the Normal Curve. Topics on Methodological and Applied Statistical Inference, 131–144. doi:10.1007/978-3-319-44093-4_13.

Maturo, A., & Maturo, F. (2013). Research in Social Sciences: Fuzzy Regression and Causal Complexity. Studies in Fuzziness and Soft Computing, 237–249. doi:10.1007/978-3-642-35635-3_18.

Maturo, A., & Maturo, F. (2016). Fuzzy Events, Fuzzy Probability and Applications in Economic and Social Sciences. Studies in Systems, Decision and Control, 223–233. doi:10.1007/978-3-319-40585-8_20.

Maturo A. & Maturo F. (2014). Finite Geometric Spaces, Steiner Systems and Cooperative Games. Analele Universitatii "Ovidius" Constanta. Seria Matematica. Vol. 22(1), pp. 189–205. ISSN: Online 1844-0835. doi: 10.2478/auom-2014-0015.

Maturo, F., & Hošková-Mayerová, Š. (2016). Fuzzy Regression Models and Alternative Operations for Economic and Social Sciences. Studies in Systems, Decision and Control, 235–247. doi:10.1007/978-3-319-40585-8_21.

Michelozzi, P., Perucci, C., Forastiere, F., Fusco, D., Ancona, A., & Dell'Orco V. (1999). Differenze sociali nella mortalità a Roma negli anni 1990–1995, Epidemiologia e Prevenzione, 23, 230-238.

Minerba, D., & Vacca, D. (2006). Gli indici di deprivazione per l'analisi delle disuguaglianze tra i comuni della Sardegna. Istituto Nazionale di statistica. http://www.istat.it/it/archivio/6727.

Najjary, Z., Saremi, H., Biglarbegian, M., & Najari, A. (2016). Identification of deprivation degrees using two models of fuzzy-clustering and fuzzy logic based on regional indices: A case study of Fars province. Cities, 58, 115–123. doi:10.1016/j.cities.2016.05.013.

Potsi, A., D'Agostino, A., Giusti, C., & Porciani, L. (2015). Childhood and capability deprivation in Italy: a multidimensional and fuzzy set approach. Qual Quant. doi:10.1007/s11135-015-0277-y.

Ramsay, J. O., & Silverman, B. W. (1997). Functional Data Analysis. Springer Series in Statistics. doi:10.1007/978-1-4757-7107-7.

Testi, A., Ivaldi, E., & Busi, A. (2005). Caratteristiche e potenzialità informative degli indici di deprivazione, Tendenze nuove, 111–124.

Chapter 16
Expected Present and Final Value of an Annuity when some Non-Central Moments of the Capitalization Factor are Unknown: Theory and an Application using R

Salvador Cruz Rambaud, Fabrizio Maturo and Ana María Sánchez Pérez

Abstract The aim of this chapter is the development of three approaches for obtaining the value of an n-payment annuity, with payments of 1 unit each, when the interest rate is random. To calculate the value of these annuities, we are going to assume that only some non-central moments of the capitalization factor are known. The first technique consists in using a tetraparametric function which depends on the arctangent function. The second expression is derived from the so-called quadratic discounting whereas the third approach is based on the approximation of the mathematical expectation of the ratio of two random variables by Mood et al. (1974). A comparison of these methodologies through an application, using the R statistical software, shows that all of them lead to different results.

Keywords Annuity · Random interest rate · Tetraparametric function · Discount factor · Mood et al. approximation

16.1 Introduction

This work aims to determine an approximate expression to obtain the present, or final, value of an annuity when the interest rate is random. In annuities assessment, fixing the interest rate has a great relevance because even small changes can result

S. Cruz Rambaud (✉) · A.M. Sánchez Pérez
Department of Economics and Business, University of Almería, Almería, Spain
e-mail: scruz@ual.es

A.M. Sánchez Pérez
e-mail: amsanchez@ual.es

F. Maturo
Department of Management and Business Administration,
University of Chieti-Pescara, Chieti, Italy
e-mail: f.maturo@unich.it

© Springer International Publishing AG 2017
Š. Hošková-Mayerová et al., *Mathematical-Statistical Models and Qualitative Theories for Economic and Social Sciences*, Studies in Systems, Decision and Control 104,
DOI 10.1007/978-3-319-54819-7_16

in major changes in the total annuity value. Thus, the determination of the interest rate value must be carried out as accurately as possible (Cruz Rambaud and Sánchez Pérez 2016; Cruz Rambaud et al. 2015).

Under the traditional approach, interest rates have been treated deterministically. In certainty contexts, the use of a single possible value for each period may be enough (Villalón et al. 2009). However, for those operations developed in uncertain environments, it is more reasonable the formulation of potential scenarios, which are subsequently reduced to one by statistical treatment (Cruz Rambaud and Valls Martínez 2002).

The determination of the interest rate value must be based on the current situation, as well as on its possible future evolution, of both the company and its environment. In this way, if prospects are unfavorable, interest rates must be higher, compared to more favorable situations, and hence the operation value is reduced as a consequence of the risk attached to it. However, in most cases, determining the interest rate of a financial operation is subject to the risk propensity/aversion of the agent to be responsible for the assessment (Suárez Suárez 2005). In this sense, the adopted interest rate would be affected by a degree of subjectivity that may over/undervalue the project.

We will consider the interest rate as a random variable which is represented as X. Therefore, the capitalization factor, $1 + i$ (Mira Navarro 2014), is also a random variable represented as U. Obviously, it is verified that $U = 1 + X$, so the relationship between means and standard deviations of both variables is:

$$\mu_U = 1 + \mu_X$$

and

$$\sigma_U = \sigma_X.$$

As a result, if X is defined in an interval $[a, b]$, U will be in the interval $[1 + a, 1 + b]$. Henceforth, when the mean and standard deviation are mentioned we will refer to the random variable U, unless otherwise specified.

In this case, the final value of an n-payment annuity, with payments of 1 unit each, made at the end of every year (annuity-immediate), valued at the rate $X = U - 1$, would be the following random variable:

$$s_{\overline{n}|U-1} = 1 + U + U^2 + \cdots + U^{n-1}. \tag{16.1}$$

Thus, its expected value is:

$$E(s_{\overline{n}|U-1}) = E(1) + E(U) + E(U^2) + \cdots + E(U^{n-1}) = 1 + \mu + \mu_2 + \cdots + \mu_{n-1}.$$

On the other hand, the final expected value of an n-payment annuity, with payments of 1 unit each, made at the beginning of every year (annuity-due), valued at the rate X, would be:

$$E(\ddot{s}_{\overline{n}|U-1}) = E(U) + E(U^2) + \cdots + E(U^n) = \mu + \mu_2 + \cdots + \mu_n, \qquad (16.2)$$

where

$$\mu_r = E(U^r)$$

is the moment of order r with respect to the origin of the random variable U. In the case that the random variable is discrete, it adopts the following expression:

$$\mu_r = E(U^r) = \sum_{i=1}^{k} p_i u_i^r, \qquad (16.3)$$

where p_i is the probability that the random variable takes the value u_i. In the continuous case, the expression of the moment of order r is:

$$\mu_r = E(U^r) = \int_{u_{min}}^{u_{max}} u f(u) du, \qquad (16.4)$$

for all values of r, being $f(u)$ the density function of the random variable U. On the other hand, the mean of order r is defined as the r-th root of the moment of order r which, in the discrete case, adopts the following expression:

$$m_r = \left(\sum_{i=1}^{k} p_i u_i^r \right)^{1/r}, \qquad (16.5)$$

whereas in the continuous case, the expression of the mean of order r is:

$$m_r = \left(\int_{u_{min}}^{u_{max}} u^r f(u) du \right)^{1/r}, \qquad (16.6)$$

for all values of r.

Below, we are going to study the limit L of the mean of order r, when r tends to $+\infty$:

$$L := \lim_{r \to +\infty} m_r. \qquad (16.7)$$

To do this, take into account that the sequence $\{m_r\}_{r=-\infty}^{+\infty}$ and, in general, the function $g(x) = m_x$, being $-\infty < x < +\infty$, is increasing since, according to the inequality of Lyapunov, for $1 < r < s$, it is verified that $[E(U^r)]^{1/r} \leq [E(U^s)]^{1/s}$. Moreover, as u_{max} (maximum value of the random variable U) is an upper bound of $g(x)$, we can deduce that m_r has a limit at infinity which will be denoted by L.

Obviously, $L \leq u_{\max}$. Let us suppose $L < u_{\max}$. In this case, there would be a u_0, such that $L < u_0 < u_{\max}$. Below, we decompose the integral which defines m_r in other two as follows:

$$m_r = \left(\int_{u_{\min}}^{u_0} u^r f(u) du + \int_{u_0}^{u_{\max}} u^r f(u) du \right)^{1/r} > \left(\int_{u_0}^{u_{\max}} u^r f(u) du \right)^{1/r} >$$

$$> \left(u_0^r \int_{u_0}^{u_{\max}} f(u) du \right)^{1/r} = u_0 k^{1/r},$$

where the density function's integral between u_0 and $u_{\max}$, $\int_{u_0}^{u_{\max}} f(u) du$, has been represented by k. Clearly, it is verified that $0 < k < 1$. So,

$$L = \lim_{r \to +\infty} m_r \geq u_0 \lim_{r \to +\infty} k^{1/r} = u_0 \cdot 1 = u_0,$$

in contradiction with the fact that $L < u_0$. Therefore, one has:

$$L = \lim_{r \to +\infty} m_r = u_{\max}. \tag{16.8}$$

Analogously, it would be shown that $l := \lim_{r \to -\infty} m_r = u_{\min}$. Consequently, the function m_x, $-\infty < x < +\infty$, has a horizontal asymptote at $y = u_{\min}$ and another one at $y = u_{\max}$, so it changes its concavity (or its convexity), which means having, at least, an inflection point.

In this chapter we will analyze the mathematical expression of the present and final expected value of an n-payment annuity, with payments of 1 unit each, made at the end/beginning of every year (annuity-immediate and annuity-due), whose calculation entails a random interest rate. Specifically, in this work these expected values are analyzed when only some non-central moments of the capitalization factor are known. In Sect. 16.2, an approach on the basis of a tetraparametric function is studied. On the other hand, in Sect. 16.3, it is developed an approach by using the so-defined quadratic discounting. In Sect. 16.4, the expression to calculate the value of an annuity is by employing the approximate formula by Mood et al. (1974). In Sect. 16.5, we present a practical example using the R statistical software. Finally, Sect. 16.6 summarizes and concludes.

16.2   The Tetraparametric Function Approach

As a result of the reasoning shown in Sect. 16.1, the curve which represents the mean of order r is an increasing function of r which can be seen in Fig. 16.1. Thus, if Q denotes the quadratic mean, or mean of order 2, H is the harmonic mean, or mean

Fig. 16.1 Graphical representation of the mean of order r (Calot 1974)

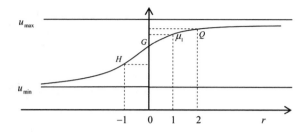

of order -1, and G is the geometric mean or mean of order 0, we have the relations shown in Fig. 16.1.

In effect, according to the Lyapunov inequality (Fisz 1963), if the moments of order r of a random variable X exist for arbitrary values of r, it is verified the following inequality:

$$\mu_r^{1/r} \leq \mu_{r+1}^{1/(r+1)};$$

(16.9)

thus, it is verified the following relationship between the corresponding means of order r and $r + 1$:

$$m_r \leq m_{r+1}.$$

(16.10)

Consequently, the curve representing the mean of order r can be fitted to the following tetraparametric function which exhibits its same shape:

$$g(r) = a \cdot \arctan(br + c) + d,$$

(16.11)

where, in a beginning, a and b are parameters greater than 0. Under these circumstances, we have:

$$\lim_{r \to +\infty} g(r) = a\frac{\pi}{2} + d = u_{max}$$

and

$$\lim_{r \to -\infty} g(r) = -a\frac{\pi}{2} + d = u_{min}.$$

By adding both equations, we deduce that $2d = u_{min} + u_{max}$, from where we can obtain the value of d:

$$d = \frac{u_{min} + u_{max}}{2}.$$

(16.12)

On the other hand, by subtracting both equations, we obtain that $\pi a = u_{max} - u_{min}$; thus, the parameter a is given by:

$$a = \frac{u_{max} - u_{min}}{\pi}.$$

(16.13)

Observe that, as expected, $a > 0$. As $g(1) = \mu_1$ and $g(2) = \mu_2^{1/2}$, one has:

$$a \cdot \arctan(b + c) + d = \mu_1 \tag{16.14}$$

and

$$a \cdot \arctan(2b + c) + d = \mu_2^{1/2}, \tag{16.15}$$

from where

$$b + c = \tan \frac{\mu_1 - d}{a} \tag{16.16}$$

and

$$2b + c = \tan \frac{\mu_2^{1/2} - d}{a}. \tag{16.17}$$

By subtracting the above two equations, we obtain:

$$b = \tan \frac{\mu_2^{1/2} - d}{a} - \tan \frac{\mu_1 - d}{a}, \tag{16.18}$$

which confirms that $b > 0$. Once determined a, b, c, and d, we can approximate the final expected value of an n-payment annuity, with payments of 1 unit each, made at the end/beginning of every year (annuity-immediate/annuity-due), valued at the rate X, by the following expressions:

$$E(s_{\overline{n}|U-1}) = 1 + m_1 + (m_2)^2 + \cdots + (m_{n-1})^{n-1} \approx 1 + \sum_{s=1}^{n-1} [a \cdot \arctan(bs + c) + d]^s \tag{16.19}$$

and

$$E(\ddot{s}_{\overline{n}|U-1}) = m_1 + (m_2)^2 + \cdots + (m_n)^n \approx \sum_{s=1}^{n} [a \cdot \arctan(bs + c) + d]^s. \tag{16.20}$$

On the other hand, the present expected value of an n-payment annuity, with payments of 1 unit each, made at the end/beginning of every year (annuity-immediate/annuity-due), valued at the rate X, respectively, would be as follows:

$$E(a_{\overline{n}|U-1}) = (m_{-1})^{-1} + (m_{-2})^{-2} + \cdots + (m_{-n})^{-n} \approx \sum_{s=1}^{n} [a \cdot \arctan(-bs + c) + d]^{-s} \tag{16.21}$$

and

$$E(\ddot{a}_{\overline{n}|U-1}) = 1 + (m_{-1})^{-1} + \cdots + (m_{-(n-1)})^{-(n-1)} \approx 1 + \sum_{s=1}^{n-1} [a \cdot \arctan(-bs + c) + d]^{-s}.$$

(16.22)

16.3   The Quadratic Discounting Approach

The variable which represents the expected present value of an n-payment annuity, with payments of 1 unit each, made at the end of every year (annuity-immediate), valued at the rate X, by using the exponential discounting is the following one:

$$a_{\overline{n}|X} = (1 + X)^{-1} + (1 + X)^{-2} + (1 + X)^{-3} + \cdots + (1 + X)^{-n}$$

(16.23)

or, equivalently, $a_{\overline{n}|X} = \frac{1-(1+X)^{-n}}{X}$. Below, we can simplify Eq. (16.23) by using the McLaurin formula. Indeed, by expanding the expression $(1 + X)^{-t}$, we have that:

$$(1 + X)^{-t} = 1 - tX + \frac{(-t)(-t-1)}{2!}X^2 - \cdots$$

(16.24)

In this sum, by removing all terms from the third one, we obtain that $(1 + X)^{-t} \approx 1 - tX + \frac{t^2+t}{2}X^2$, which means that the present value may be approximately estimated by replacing the exponential by the quadratic discounting. Once made this simplification, the present value of an n-payment annuity, with payments of 1 unit each, made at the end of every year (annuity-immediate), valued at the rate X, can be calculated by using the following approximation:

$$a_{\overline{n}|X} = \left(1 - X\frac{1^2+1}{2}X^2\right) + \left(1 - 2X + \frac{2^2+2}{2}X^2\right) + \cdots + \left(1 - nX + \frac{n^2+n}{2}X^2\right).$$

(16.25)

By applying the formulas of the sum of n first natural numbers and that of their squares, we have:

$$a_{\overline{n}|X} = n - \frac{n(n+1)}{2}X + \frac{n(n+1)}{4}\left(1 + \frac{2n+1}{3}\right)X^2.$$

Therefore,

$$E(a_{\overline{n}|X}) = n - \frac{n(n+1)}{2}E(X) + \frac{n(n+1)}{4}\left(1 + \frac{2n+1}{3}\right)E(X^2).$$

(16.26)

In the same way, the formula for the final value of an n-payment annuity, with payments of 1 unit each, made at the end of every year (annuity-immediate), valued at the rate X, after applying the aforementioned discount factor approach, would be as follows:

$$s_{\overline{n}|X} = n + \frac{(n-1)n}{2}X + \frac{(n-1)n}{4}\left(1 + \frac{2n-1}{3}\right)X^2, \tag{16.27}$$

from which

$$E(s_{\overline{n}|X}) = n + \frac{(n-1)n}{2}E(X) + \frac{(n-1)n}{4}\left(1 + \frac{2n-1}{3}\right)E(X^2). \tag{16.28}$$

In the case of an n-payment annuity, with payments of 1 unit each, made at the beginning of every year (annuity-due), valued at the rate X, the expected present and final values are given by:

$$E(\ddot{a}_{\overline{n}|X}) = n - \frac{(n-1)n}{2}E(X) + \frac{(n-1)n}{4}\left(1 + \frac{2n-1}{3}\right)E(X^2)$$

and

$$E(\ddot{s}_{\overline{n}|X}) = n + \frac{n(n+1)}{2}E(X) + \frac{n(n+1)}{4}\left(1 + \frac{2n+1}{3}\right)E(X^2). \tag{16.29}$$

16.4 The Mood et al. Approach

By using the approximate formula of the mathematical expectation of the ratio of two random variables X and Y by Mood et al. (1974), introduced as well by Rice (2006):

$$E\left(\frac{X}{Y}\right) \approx \frac{E(X)}{E(Y)} - \frac{\mathrm{cov}(X,Y)}{[E(Y)]^2} + \frac{E(X)}{[E(Y)]^3}\mathrm{var}(Y), \tag{16.30}$$

we can obtain the expression of the final expected value of an n-payment annuity, with payments of 1 unit each, made at the end of every year (annuity-immediate), valued at the rate X, as follows:

$$E(s_{\overline{n}|U-1}) = E\left(\frac{U^n - 1}{U - 1}\right) \approx$$

$$\approx \frac{E(U^n - 1)}{E(U - 1)} - \frac{\mathrm{cov}(U - 1, U^n - 1)}{[E(U - 1)]^2} + \frac{E(U^n - 1)}{[E(U - 1)]^3}\mathrm{var}(U - 1).$$

Thus, taking into account that (Fisz 1963):

$$\text{var}(U - 1) = \text{var}(U) = E(U^2) - [E(U)]^2 = \mu_2 - \mu^2$$

and

$$\text{cov}(U - 1, U^n - 1) = \text{cov}(U, U^n) = E(U^{n+1}) - E(U^n)E(U) = \mu_{n+1} - \mu_n\mu,$$

we can write

$$E(s_{\overline{n}|U-1}) \approx \frac{\mu_n - 1}{\mu - 1} - \frac{\mu_{n+1} - \mu_n\mu}{(\mu - 1)^2} + \frac{\mu_n - 1}{(\mu - 1)^3}(\mu_2 - \mu^2). \tag{16.31}$$

Analogously, the final expected value of an n-payment annuity, with payments of 1 unit each, made at the beginning of every year (annuity-due), valued at the rate X, can be deduced. Indeed,

$$E(\ddot{s}_{\overline{n}|U-1}) = E\left(\frac{U^{n-1} - U}{U - 1}\right) \approx$$

$$\approx \frac{E(U^{n-1} - U)}{E(U - 1)} - \frac{\text{cov}(U^{n+1} - U, U - 1)}{[E(U - 1)]^2} + \frac{E(U^{n+1} - U)}{[E(U - 1)]^3}\text{var}(U - 1) =$$

$$= \frac{\mu_{n+1} - \mu}{\mu - 1} - \frac{\mu_{n+2} - \mu_{n+1}\mu - \mu_2 + \mu^2}{(\mu - 1)^2} + \frac{\mu_{n+1} - \mu}{(\mu - 1)^3}(\mu_2 - \mu^2). \tag{16.32}$$

The expected present value is:

$$E(a_{\overline{n}|U-1}) = E\left(\frac{U^n - 1}{U^{n+1} - U^n}\right) \approx \frac{E(U - 1)}{E(U^{n-1} - U)} - \frac{\text{cov}(U^{n+1} - U^n, U^n - 1)}{[E(U^{n+1} - U^n)]^2} +$$

$$+ \frac{E(U^n - 1)}{[E(U^{n+1} - U^n)]^3}\text{var}(U^{n+1} - U^n) =$$

$$= \frac{\mu - 1}{\mu_{n+1} - \mu} - \frac{\mu_{2n+1} - \mu_{n+1}\mu_n - \mu_{2n} + \mu_n^2}{(\mu_{n+1} - \mu_n)^2} +$$

$$+ \frac{\mu_n - 1}{(\mu_{n+1} - \mu_n)^3}(\mu_{2n+2} - \mu_{n+1}^2 + \mu_{2n} - \mu_n^2 - 2(\mu_{2n+1} - \mu_{n+1}\mu_n)),$$

in the case of an n-payment annuity, with payments of 1 unit each, made at the end of every year (annuity-immediate), valued at the rate X, and

$$E(\ddot{a}_{\overline{n}|U-1}) = E\left(\frac{U^{n+1} - U}{U^{n+1} - U^n}\right) \approx \frac{E(U^{n+1} - U)}{E(U^{n+1} - U^n)} - \frac{\text{cov}(U^{n+1} - U^n, U^{n+1} - U)}{[E(U^{n+1} - U^n)]^2} +$$

$$+ \frac{E(U^{n+1} - U)}{[E(U^{n+1} - U^n)]^3} \, \text{var}(U^{n+1} - U^n) =$$

$$= \frac{\mu_{n+1} - \mu}{\mu_{n+1} - \mu_n} - \frac{\mu_{2n+2} - \mu_{n+1}^2 - 2\mu_{n+2} + 2\mu_{n+1}\mu + \mu_2 - \mu^2}{(\mu_{n+1} - \mu_n)^2} +$$

$$+ \frac{\mu_{n+1} - \mu}{(\mu_{n+1} - \mu_n)^3}(\mu_{2n+2} - \mu_{n+1}^2 + \mu_{2n} - \mu_n^2 - 2(\mu_{2n+1} - \mu_{n+1}\mu_n)),$$

in the case of an n-payment annuity, with payments of 1 unit each, made at the beginning of every year (annuity-due), valued at the rate X.

16.5 A Numerical Example of the Expected Final Value of an Annuity by Developing R Functions

Next, we are going to calculate the expected final value of a 6-payment annuity, with payments of 1 unit each, made at the end/beginning of every year (annuity-immediate/ annuity-due) by employing the different expressions developed in this chapter. The present value calculation has been omitted since it can be carried out similarly. Following we present some R functions to perform the proposed approaches. Because we supposed that only some non-central moments of the capitalization factor are known, the three different approaches have been applied:

- The tetraparametric function approach.
- The quadratic discounting approach.
- The Mood et al. approach.

To calculate the mean and variance, we consider the historical data of Table 16.1, specifically the monthly updates of Euribor from January 2015 to April 2016.

Before applying our functions it is necessary to load the data and install two R packages with the following codes:

```
data=c(0.298,0.255,0.212,0.180,0.165,0.163,0.167,0.161,
      0.154,0.128,0.079,0.059,0.042,-0.008,-0.012,-0.002)

install.packages("moments")
library(moments)
install.packages("labstatR")
library(labstatR)
```

Period	% Euribor in 1 year
January 2015	0.298
February 2015	0.255
March 2015	0.212
April 2015	0.180
May 2015	0.165
June 2015	0.163
July 2015	0.167
August 2015	0.161
September 2015	0.154
October 2015	0.128
November 2015	0.079
December 2015	0.059
January 2016	0.042
February 2016	−0.008
March 2016	−0.012
April 2016	−0.002

Table 16.1 Euribor (January 2015–April 2016). *Source* Bank of Spain

Following a preliminary analysis, we obtain the following mean:

$$\mu = 1.12756398,$$

whereas the calculation of the standard deviation gives a value of

$$\sigma = 0.008138746.$$

The minimum and the maximum values of the random variable U are:

$$u_{min} = a = 0.988$$

and

$$u_{max} = b = 1.298,$$

respectively.

The Tetraparametric Function Approach

To approximate the final expected value of an n-payment annuity, with payments of 1 unit each, made at the end/beginning of every year (annuity-immediate/annuity-due), valued at the rate X, we formulated two functions to reproduce Eqs. (16.19) and (16.20). The first one computes the final expected value of an n-payment annuity, with payments of 1 unit each, made at the end of every year (annuity-immediate) with random interest rates using the arctangent method:

```
FV_post_artan=function ( data , years ){
    U=1+data
   u1=mean(U)
   var=sigma2 (U)
   u2=sqrt ( var+u1 ^2)
   u_max=1+max ( data )
   u_min=1+min ( data )
   d=(u_min+u_max )/2
   a=(u_max−u_min )/ pi
   b=tan ( x =((u2−d )/ a )) − tan ( x =((u1−d )/ a ))
   c=tan ( x =((u1−d )/ a )) − b
   appo=rep (NA, years )
   s=years −1
   for ( i  in  0:s )  { appo [ i +1]=(a∗atan ( x =(b∗i+c ))+d )^ i }
   final_value=sum ( appo )
   return ( final_value )
}
```

The second function computes the final expected value of an *n*-payment annuity, with payments of 1 unit each, made at the beginning of every year (annuity-due) with random interest rates using the arctangent method:

```
FV_pre_artan=function ( data , years ){
   U=1+data
   u1=mean(U)
   var=sigma2 (U)
   u2=sqrt ( var+u1 ^2)
   u_min=1+min ( data )
   u_max=1+max ( data )
   d=(u_min+u_max )/2
   a=(u_max−u_min )/ pi
   b=tan ( x =((u2−d )/ a )) − tan ( x =((u1−d )/ a ))
   c=tan ( x =((u1−d )/ a )) − b
   appo=rep (NA, years )
   for ( i  in  1: years )  { appo [ i ]=(a∗atan ( x =(b∗i+c ))+d )^ i }
   final_value=sum ( appo )
   return ( final_value )
}
```

Using the above codes, we get the following results:

```
> FV_post_artan ( data ,6)
[1]  8.491768
> FV_pre_artan ( data ,6)
[1]  9.75462
```

The Quadratic Discounting Approach

To use Eqs. (16.28) and (16.29) for computing the final value of an n-payment annuity, with payments of 1 unit each, made at the end/beginning of every year (annuity-immediate/annuity-due), valued at the rate X, we generate the two different codes:

```
FV_post_quad=function(data,years){
   n=years
   u=mean(data)
   u2=mean(data^2)
   final_value=n+(n*(n-1)/2)*u+(n*(n-1)/4)*(1+(2*n-1)/3)*u2
   return(final_value)
}


FV_pre_quad=function(data,years){
   n=years
   u=mean(data)
   u2=mean(data^2)
   final_value=n+(n*(n+1)/2)*u+(n*(n+1)/4)*(1+(2*n-1)/3)*u2
   return(final_value)
}
```

These functions applied to our data give the following results:

```
> FV_post_quad(data,6)
[1] 8.76782
> FV_pre_quad(data,6)
[1] 9.874948
```

The Approach of Mood et al.

The last method developed in this chapter is the approach by Mood et al.; according to this perspective, we can approximate the final value of an n-payment annuity, with payments of 1 unit each, made at the end of every year (annuity-immediate), valued at the rate $X = U - 1$ Eq. (16.31), by using the following code:

```
FV_post_mood=function(data,years){
   n=years
   m=n+2
   momenti=rep(NA,m)
   U=1+data
   u=mean(U)
   for (i in 1:m) momenti[i]=moment(U,
   central = FALSE, absolute = FALSE, order =i)
   final_value=((momenti[n]-1)/(u-1))-
      ((momenti[n+1]-u*momenti[n])/((u-1)^2))+
      ((momenti[n]-1)/((u-1)^3))*
      (momenti[2]-u^2)
```

```
  return(final_value)
}
```

Analogously, the final value of an n-payment annuity, with payments of 1 unit each, made at the beginning of every year (annuity-due), valued at the rate $X = U - 1$ Eq. (16.32), may be computed by:

```
FV_pre_mood=function(data,years){
  n=years
  m=n+2
  momenti=rep(NA,m)
  U=1+data
  u=mean(U)
  for (i in 1:m) momenti[i]=moment(U,
  central = FALSE, absolute = FALSE, order =i)
  final_value =((momenti[n+1]-u)/(u-1))-
    ((momenti[n+2]-u*momenti[n+1]-momenti[2]+u^2)/
      ((u-1)^2))+
    ((momenti[n+1]-u)/((u-1)^3))*(momenti[2]-u^2)
  return(final_value)
}
```

Following our result (using data of Table 16.1):

```
> FV_post_mood(data ,6)
[1]  9.072831
> FV_pre_mood(data ,6)
[1]  10.59077
```

Our results show that the three similar give similar results. In summary, for the final value of an n-payment annuity, with payments of 1 unit each, made at the end of every year (annuity-immediate), valued at the rate $X = U - 1$, we obtain the following values:

- Tetraparametric function approach: 8.491.
- Quadratic discounting approach: 8.767.
- Mood et al. approach: 9.072.

Instead, for the final value of an n-payment annuity, with payments of 1 unit each, made at the beginning of every year (annuity-due), valued at the rate $X = U - 1$, we get:

- Tetraparametric function approach: 9.754.
- Quadratic discounting approach: 9.874.
- Mood et al. approach: 10.590.

We highlight that the final values of the Mood et al. approach are always the greatest. However, we underline that the Mood et al. approach is based on the moments; thus, the final value, computed with this last method, is strongly influenced by the distribution of the data. In this application, we assumed a normal distribution when calculating the moments of the distribution; therefore, it is reasonable to infer that if the data do not follow a normal distribution, we get very different results from those obtained with the first two methods.

To highlight this concept, we simulate interest rates following a normal distribution and repeat our test with the following code:

```
> data <-rnorm(n=365,m=0.31,sd=0.075)
> FV_post_artan(data,6)
[1]  13.18
> FV_pre_artan(data,6)
[1]  17.39
> FV_post_quad(data,6)
[1]  14.09
> FV_pre_quad(data,6)
[1]  17.33
> FV_post_mood(data,6)
[1]  13.20
> FV_pre_mood(data,6)
[1]  17.41
```

This simulation shows that, when we deal with interest rates following a normal distribution, the approaches give similar results. However, the more the interest rate are far from normality, the more the Mood et al. approach brings to results different from the tetraparametric function method.

16.6 Conclusion

In this chapter we have presented three methodologies to obtain the value of an annuity whose discount rate is a random variable. The first model is based on the curve representing the mean of order r as a tetraparametric function. On the other hand, the second model is based on the so-defined quadratic discounting, and the third one uses the approximate formula of the expected value of the ratio of two random variables. A comparison among these methodologies is presented with an R application. We considered an n-payment annuity, with payments of 1 unit each, made at the end/beginning of every year (annuity-immediate/annuity-due), valued at a random interest rate. The comparison among the developed methodologies shows that all lead to similar results.

References

Calot, G., 1974. Curso de Estadística Descriptiva. Ed. Paraninfo, Madrid.

Cruz Rambaud, S., Maturo, F., Sánchez Pérez, A.M., 2015. Approach of the value of an annuity when non-central moments of the capitalization factor are known: an R application with interest rates following normal and beta distributions. Ratio Mathematica 28, 15–30.

Cruz Rambaud, S., Sánchez Pérez, A.M., 2016. Una aproximación del valor de una renta cuando el tipo de interés es aleatorio. XXIV Jornadas de Asepuma y XII Encuentro Internacional Granada (Spain), July 7–8.

Cruz Rambaud, S., Valls Martínez, M.C., 2002. La determinación de la tasa de actualización para la valoración de empresas. Análisis Financiero 87-2, 72–85.

Fisz, M., 1963. Probability Theory and Mathematical Statistics. John Wiley and Sons, Inc, New York.

Mira Navarro, J.C., 2014. Introducción a las Operaciones Financieras. Creative Commons, http://www.miramegias.com/emodulos/fileadmin/pdfs/mof.pdf.

Mood, A.M., Graybill, F.A., Boes, D.C., 1974. Introduction to the Theory of Statistics. 3rd Ed. Boston: McGraw Hill.

Rice, J.A., 2006. Mathematical Statistics and Data Analysis. 2nd Ed. California: Duxbury Press.

Suárez Suárez, A.S., 2005. Decisiones Óptimas de Inversión y Financiación en la Empresa. 2nd Ed. Madrid, Ed. Pirámide.

Villalón, J.G., Martínez Barbeito, J., Seijas Macías, J.A., 2009. Sobre la evolución de los tantos de interés. XVII Jornadas de Asepuma y V Encuentro Internacional 17, 1–502.

Chapter 17
Assessing the Effect of Financial Crisis of Earnings Manipulation. Empirical Evidence from the Top 1,000 World Listed Companies

Francesco Paolone and Matteo Pozzoli

Abstract This chapter investigates the impact of the recent financial crisis on Earnings Manipulation (EM) by adopting the Beneish Model built up on eight financial performance indicators. The study has been conducted on the Top World Enterprises ranked by Sales Revenues in the last available fiscal year, 2013. We gathered the accounting data from financial statements of all of the companies during the crisis period (2008–2013) using *Orbis Bureau Van Dijk* database as our source and we tested the existence of EM within the Top 1,000 World Listed Companies. In doing this, we examine whether the recent financial crisis has decreased or increased the number of top companies with a high likelihood of EM. Our results show that there has been a greater probability for manipulating earnings in the first year of the global crisis: companies have had a tendency to increase creation of social wealth, in terms of generating higher profits. This would mean that the crisis has had a positive effect on handling of income by the largest companies in the world because the crisis itself has restricted the earnings manipulation policies.

Keywords Financial accounting · Earnings manipulation · Performance indicators · Transparency · Global crisis

The authors are entered in alphabetic order by their last names (Even if this chapter is the result of a joint research, the authors have contributed to it as follows: Francesco Paolone: Sects. 17.2.1, 17.2.2 and 17.3. Matteo Pozzoli: Sects. 17.4 and 17.5. Sections 17.1 and 17.6 was jointly written by the authors).

F. Paolone (✉) · M. Pozzoli
Department of Law, University of Naples "Parthenope", Naples, Italy
e-mail: francesco.paolone@uniparthenope.it

M. Pozzoli
e-mail: matteo.pozzoli@uniparthenope.it

© Springer International Publishing AG 2017
Š. Hošková-Mayerová et al., *Mathematical-Statistical Models and Qualitative Theories for Economic and Social Sciences*, Studies in Systems, Decision and Control 104,
DOI 10.1007/978-3-319-54819-7_17

17.1 Introduction

Managers could engage in account manipulation, including earnings management, to meet stakeholders' expectations therefore financial reporting results may not fairly represent firms' operations. Account manipulation can lead to inefficient capital markets (Sharpe 1964; Aboody et al. 2002; Stolowy and Breton 2004). Extant accounting research (Burgstahler and Dichev 1997; Barth et al. 2008; Wells 2001; Healy and Wahlen 1999; Badawi 2008; Richardson et al. 2001; Walker 2013; Florio 2011; Paolone and Magazzino 2014) states that executives acknowledge the importance of meeting earnings to achieve targets (i.e. loss avoidance or analysts' forecasts) as well as recognizing that earnings attainment represents a relevant motivation for accounting manipulation (Trombetta and Imperatore 2014). Stolowy and Breton (2004, p. 6–7) define account manipulation as a discretionary decision of management to make accounting choices that may affect the transfer of wealth between companies, the company and capital providers, the company and managers or managers. One form of account manipulation is Earnings Management (EM). The objective is to assess whether managers can manipulate accounts more often during a period of financial crisis than otherwise.

Studying the entire group of top world listed companies, we computed the eight indicators as defined by Beneish (1999). Beneish (1997) finds that his eight ratios, that capture financial statement distortions, provide timely assessments of the likelihood of distortions especially when considered in conjunction with management incentives. So we compute the Beneish ratios and consider management's incentive for each firm-year from 2008 to 2013. Then we grouped these observations for each year in order to assess whether companies have increased or decreased the likelihood of EM. That is, we compare the final scores across each fiscal year assuming 2008 as the year when the financial crisis became worldwide. Findings show that the first 1,000 largest companies in the world have a tendency to increase the social wealth creation, in terms of generating more profits.

This would mean that the global financial crisis has had a positive effect on the handling of the income of the largest companies in the world because the crisis itself has restricted the earnings manipulation policies (both for tax reduction and dividend policies).

Our study was conducted by adopting a well-known model of probability of accounting manipulation in order to assess the impact of the global financial crisis on the top world companies' accounts. This analysis could also be helpful to banks, other financial institutions and other lending and investing entities as it represents an additional tool useful to detect account manipulation and accounting fraud, and to reduce information asymmetry during a period of financial crisis. Finally, the results have implications for future research focusing on management incentives in concurrence with security offerings.

We assess the impact of the financial crisis on EM for the top 1,000 world listed companies ranked by Sales Revenues. We use the Beneish (1999) model based upon eight financial performance indicators to predict the likelihood of fraudulent policies of the main companies in the world. In reporting the results of our study we will begin by presenting a review of the literature on studies of EM during the financial crisis followed by an identification of the performance indicators used to determine EM probability as developed by Beneish. Then we present our empirical analyses results obtained by testing the model on the above companies and we conclude with comments on our main findings and provide suggestions for further research.

17.2 Literature Review and Hypotheses Developed

17.2.1 Prior Studies on Earnings Manipulation

EM has been defined by Schipper (1989) as an intervention in the external financial reporting process, with the intent of achieving some private advantages. Many academics have discussed the role of EM in financial reporting to adjust financial accounting data in order to mislead stakeholders about a firm's performance (Healy and Wahlen 1999; Agrawal and Chadha 2005). In this context, EM appears as an active manipulation of earnings aimed at a predetermined target (Mulford and Comiskey 2002).

According to Verona (2006), it is not possible to attribute a priori positive or negative connotations to management discretion. Discretionary actions depend on the behaviour of the executives that can be more or less professional. Pini (1991) provided a definition of discretion stating that can lead to three different approaches from which stem many conceptions of EM: perfect, ideological and, instrumental.

According to the prior literature, the "accruals" indicate the relevant adjustment, which explains the EM activities. The Accruals explain the difference between free cash flows and operating income. The accruals are determined as follows (Healy 1985; De Angelo 1986):

$$Accruals = Operating\ Income - Free\ Cash\ Flows$$

Healy (1985) and De Angelo (1986) have provided the above mentioned accrual methodology, and their research has found evidence of income manipulation in a different setting, adopting non-discretionary accruals.

Onesti and Romano (2013) investigated the areas of management discretion in impairment test accounting through three forms of EM: smooth, increase and decrease. They demonstrated the existence of EM practices in relation to the goodwill impairment test under IAS 36 (*Impairment of Assets*).

Many scholars have meticulously analyzed the relation between EM and accruals estimates driven by the advent of readily calculable EM metrics (Jones 1991; Dechow et al. 1995) and policy concerns raised by influential accounting standard setters. The relevant contribution provided by Jones (1991) is constructed on a linear regression approach which assess non-discretionary accrual factors including sales revenue and property, plant & equipment.

Prencipe et al. (2008) argued that Italian family firms are less sensitive to income-smoothing motivations in contrast to nonfamily firms, while they are similarly motivated to manage earnings for debt-covenant and leverage-related reasons. The authors assessed specific accrual (R&D capitalized cost) in which statistical tests confirmed hypothesized relationships.

Dechow et al. (1995) subsequently updated the Jones model by providing the well-known Modified Jones model, which has become one of the most widely adopted models in earnings management research. The Modified Jones model includes an adjustment to sales based on the change in the amount of receivables.

Peek et al. (2013) have recently contributed by comparing abnormal accruals across different countries. By using the two accruals estimation models, the Modified Jones model and the Dechow and Dichev model (2002), they found that accruals models exhibit considerable cross-country variation in predictive accuracy and power in detecting earnings management.

Pini (1991) pointed out that EM are defined as discretionary choices undertaken by executives in order to achieve "improper" goals. The main element on EM is the way in which managers achieve their goals by adopting discretionary actions. These actions affect the real economic and financial situation as well as the representation of corporate disclosure (Mattei 2006).

Other authors stated that EM can be achieved by using accounting methods and estimates (i.e. an accrual-based manipulation) (Bartov 1993) or by undertaking transactions that make reported income closer to some target numbers, rather than maximizing the firm's discounted expected cash flows (Roychowdhury 2006).

In addition, several studies have explored real earnings manipulation in the context of early debt retirements (Hand 1989). Other contributions (Ronen and Sadan 1981; Dye 1988; Trueman and Titman 1988) showed that Earnings Manipulation can be undertaken through asset sales.

In this context, Beneish (1999) gave his contribution by concentrating on eight financial indicators (performance ratios), and demonstrating their ability to categorize companies in two different groups: potential and non-potential earnings manipulators.

17.2.2 Prior Studies on Financial Crisis and Earnings Manipulation

According to Trombetta and Imperatore (2014), a financial crisis can be defined as a sudden or gradual interruption in the ongoing functioning of financial markets. One relevant feature of financial crises in general, is the increase of uncertainty among lenders and investors about fundamental values of assets, which leads to a greater volatility in the market prices of assets (Trombetta and Imperatore 2014).

This situation of uncertainty increases the asymmetry of information and lenders progressively lose confidence in the accuracy of the information they have about borrowers (Mishkin 1991; Gorton 2008).

Under the conditions of financial crises, financial and capital markets are more skeptical and the investors are willing to sell off their securities, sending a negative signal to the markets as well as to new potential investors who might be more reluctant to invest. They could also demand a higher Return On Investment as a consequence of the higher level of undertaken risk. Both investors and creditors might have lower propensity to invest or lend money because of the higher probability of the counterpart's default.

Many scholars have discussed the impact of financial crises on Earnings Manipulation. Kasznik and McNichols (2002) and Matsumoto (2002) have provided a significant contribution by analyzing how executives carry out earnings manipulation policies in order to attain a firm's targets and avoid, at the same time, the communication of bad earnings news to markets.

Bartov et al. (2002) described how managers manage earnings in order to alter the market's evaluation of a firm's likelihood to survive, hence reducing the average cost of capital.

Furthermore, Huijgen and Lubberink (2005) have claimed that managers are less likely to manipulate earnings in a situation of stronger litigation risk in order to reduce external exposure to litigation.

From a review of the above cited literature it is clear that several possibilities are equally likely and we can expect either more or less EM during a financial crisis. Consequently, a more in depth analysis of the relationship between EM and financial crisis on a global scale is called for in the present discussion.

To this end, we apply the original Beneish Model, also known as Manipulation Score (Beneish 1997, 1999, 2001, Beneish et al. 2013), in order to investigate whether the impact of global financial crisis on EM is positive or negative.

H_0. *There is a relationship between the global financial crisis and the Earnings Manipulation (EM) which can be demonstrated in the increase of EM that are found in accounting data in the first 1,000 largest companies by Sales Revenues. We thus expect that crisis have increased the number of largest companies with a high likelihood of EM.*

17.3 The Beneish Model

The Manipulation Score (Beneish 1997, 1999, 2001, Beneish et al. 2013) is a mathematical model which aims at identifying whether a company has a greater likelihood of managing and manipulating its earnings. The variables consist of eight financial indicators obtained from a company's financial statements and linked together within a score formula. This score shows the profile of a company as a "potential earnings manipulator". Beneish et al. (2013) suggests using the value of −1.78 as a threshold: companies who perform higher final scores are likely to have manipulated their earnings. The variables of the Model are the following (see the respective extended formulas in Appendix B):

1. DSRI (*Days Sales in Receivables Index*). It is the indicator of revenue inflation that measures the days' sales in receivables compared to the prior year. A significant increase in days' sales in receivables means a disproportionate increase in receivables relative to sales that suggests revenue inflation. The higher the increase in the DSRI the greater the likelihood that revenues and earnings are overstated.

2. GMI (*Gross Margin Index*). The decrease of Gross Margin value can be a negative signal about a company's health and future incomes. A value higher than 1 suggests a deterioration of gross margin and can force managers to manipulate earnings. To sum up, the Gross Margin is related to the change in inventories and other production that can increase the likelihood of manipulation. Thus, Beneish uses this variable which is specifically related to production costs and changes in inventory and which can cause earnings manipulation practices.

3. AQI (*Asset Quality Index*). The Asset Quality indicator is the ratio of non-current assets other than property, plant, and equipment (PPE) to total assets and measures the proportion of total assets for which future benefits are less certain. Beneish expects a positive relationship between AQI and earnings manipulation practices. The higher the value of AQI the greater the propensity of deferring and capitalizing costs in order to increase earnings.

4. SGI (*Sales Growth Index*). "If growth companies face large stock price losses at the first indication of a slowdown, they may have greater incentives than non-growth companies to manipulate earnings" (Beneish 1999, p. 27). There would be a strong positive relationship between the growth of Sales and the likelihood of EM because managers may be more incentivized to manipulate earnings.

5. DEPI (*Depreciation Index*). The DEPI measures the ratio of the depreciation rate in year t-1 to the corresponding rate in year t. If the index is greater than 1, it indicates that the tangible assets are being depreciated at a slower rate. This suggests that the firm might be revising useful asset life assumptions upwards in order to increase income. There would be a positive correlation between DEPI and the earnings manipulation.

6. SGAI (*Sales, General and Administrative Expenses Index*). This ratio shows the SGA Expenses in year t relative to the previous year. If there is a disproportioned increase in Selling, General and Administrative expenses compared to Sales Revenues, there would be a negative signal about a company's prospects. Beneish expects a strong positive association between the index and the likelihood of manipulation.

7. LVGI (*Leverage Index*). This ratio shows the Total Debt (Current and Long-term) in year t relative to the previous year. Beneish explains that LVGI was included to capture incentives in debt covenants for earnings manipulation.

8. TATA (*Total Accruals to Total Assets*). The value of Total Accruals, normalized by Total Assets, is a proxy used to assess the discretionary accounting choices undertaken by managers in order to practice manipulations. Indeed, there would be a positive correlation between Accruals and the EM.

Summing up, the eight ratios have a predictive function and focus on financial statements distortions which capture unusual accumulations in receivables (DSRI, indicative of revenue inflation), unusual growth of Sales (SGI), unusual growth of Selling, General and Administrative Expenses (SGAI), unusual capitalization and declines in depreciation (AQI and DEPI, both indicative of expense deflation), unusual propensity to borrow money (LVGI), deterioration of Gross Margin (GMI) and the extent to which reported accounting profits are supported by cash profits (TATA).

17.4 Data Collection and Sampling

The analysis has been conducted on the Top World Companies[1] ranked by Sales Revenues during the global crisis period 2008–2013 (Descriptive Statistics are in Table 17.1).

We collected financial accounting data from the Orbis *Bureau Van Dijk* database by providing firm-years observations from 2008 to 2013. We attained the percentage coverage for each year by calculating the number of available companies divided by the total number of companies. The coverages are reported in Table 17.2.

In order to look at how the trend moves, the companies are observed over the entire period of six years from 2008 to 2013.

Although there are many differences between accounting principles from country to country, we did not need to reclassify the Beneish model by adapting the financial accounting data to the different scenario. This because the Orbis database classifies financial data in a standardized manner. According to the Accounting principles, the "Selling, General and Administrative expenses" do not appear separately on financial statements and, for this reason, we use the neutral value equal to 1 for SGAI index since the Income Statement reclassification provided by the Orbis database does not indicate the Selling, General and Administrative Expenses.[2]

---

[1]We excluded companies operating in the financial sector (Banks and Financial Institutions are not included). As they have specific financial characteristics different from industrial companies. See Appendix A.

[2]If several financial accounting data are not available, Beneish (1999) assumes to keep the variable constant over the year.

Table 17.1 Descriptive Statistics

	Scores 2013	Scores 2012	Scores 2011	Scores 2010	Scores 2009	Scores 2008
Available	**722**	**716**	**711**	**695**	**662**	**641**
Missing	*278*	*284*	*289*	*305*	*338*	*359*
Mean	−1.840	−1.766	−1.787	−1.385	−1.686	−2.004
Median	−1.992	−1.969	−1.899	−1.871	−1.891	−2.226
Mode	−1.908	−2.857	−2.456	−2.547	−2.248	−2.783
St. Deviation	1.512	1.703	1.286	5.272	1.422	4.043
Min	−5.737	−5.372	−18.258	−5.206	−4.779	−46.665
Max	23.869	29.459	3.854	95.675	15.231	84.997

Table 17.2 Companies of data available

Year	2013	2012	2011	2010	2009	2008
Coverage (%)	72.20	71.60	71.10	69.50	66.20	64.10

$$\text{Manipulation Score (8M − Score)} = -4.840 + 0.920 * \text{DSRI} + 0.528 * \text{GMI}$$
$$+ 0.404 * \text{AQI} + 0,892 * \text{SGI} + 0.115 * \text{DEPI} - 0.172 * \text{SGAI}$$
$$- 0.327 * \text{LVGI} + 4.679 * \text{TATA}$$

We have adopted the original Beneish Model (8 M-Score) in order to monitor the impact of financial crisis on EM during the global financial crisis period (2008–2013). Therefore, we expect an increase in the number of top 1,000 world listed companies with a high probability of being manipulated caused by the advent of the crisis.

The eight diagnostic ratios have been computed and inserted into the M-Score formula in order to achieve the Final Score that will be later compared to the threshold of −1.78 (Beneish et al. 2013). By applying the model, it has been possible to categorize companies in two different groups: the one with a low probability of EM (companies with score below −1.78) and, the other one with a high probability (companies with score above −1.78).

17.5   Final Results

The results are reported in Table 17.3. The main points of our hypothesis were confirmed: by considering a threshold of −1.78 (Beneish et al. 2013), there is around 42% of companies with a high probability of manipulating earnings in the

Table 17.3 General Results

	2013	2012	2011	2010	2009	2008
Companies with > −1.78 High EM	303	293	323	311	300	207
Companies with < −1.78 Low EM	419	423	388	384	362	434
% High EM (%)	41.97	40.92	45.43	44.75	45.32	32.29
% Low EM (%)	58.03	59.08	54.57	55.25	54.68	67.71

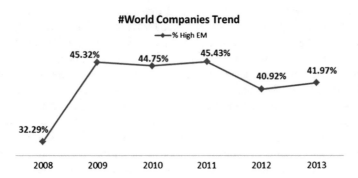

Fig. 17.1 Top 1,000 world companies trend

before-crisis period. In 2009 the percentage increases from 32.29 to 45.32% and, within the following years, we observe a steady trend with a slight decrease in 2012 (Fig. 17.1).

In the first year of global crisis, we observe that there is a greater probability for manipulating earnings; this referring to the Top 1,000 World Listed Companies that have a tendency to increase the social wealth creation, in terms of generating more profits.

This would mean that the global financial crisis might have a positive effect on the handling of the income of the largest companies in the world because the crisis itself has restricted the earnings manipulation policies (both for tax reduction and dividend policies).

As the trend in the number of manipulators shows, we have witnessed a significant rise in the number of high-risk manipulators between 2008 and 2009 as evidenced by the first year of the global financial crisis. In subsequent accounting periods the high-risk manipulations stand at percentages within the range 40–46%.

These results confirm our hypothesis that the global financial crisis, in terms of numbers, has increased the risk of manipulation in the top world listed companies. This increase remains steady along the next years up to 2013. It would be interesting to see how the trend moves after the crisis comes to an end.

These findings are relevant for financial statement users since they provide insights into the use of earnings manipulation during the global financial crisis. By taking the direction of the Beneish model into account, financial statement users get a better insight into the actual performance of a firm.

17.6 Limitations and Suggestions for Further Research

Although the study is reliable, several important limitations exist. Two important limitations should be kept in mind when applying the research findings. Most importantly, the earnings manipulation measurement model used to assess the risk of EM depends on a threshold of -1.78, fixed by Beneish in a recent study (Beneish et al. 2013). Since the findings depend on one single value and this could make the research findings less reliable. Secondly, the global financial crisis period is debatable. Since the top 1,000 world listed companies cover many different countries, the global crisis period taken into account could not be the same for all of them. The period is based on the trend in GDP worldwide showing a huge drop in 2008; consequently, we assume it overlaps with the beginning of the crisis.

A promising avenue of research would be to cluster the data in different countries as well as in different industry sectors. It would be interesting to assess the likelihood of EM occurring in manufacturing industries and financial/banking sectors. It would be useful to focus on a multiple country-setting (EU-nations as well as non EU countries) in order to analyze the impact of the crisis on EM in different contexts and compare the findings afterwards.

Furthermore, it would also be useful to consider other indicators in addition to Sales Revenues and ranking firms, for example, Operating Revenue, Net Profit, Cash & Cash Equivalents etc. as well as exploring all of the single values of Manipulation Score in order to provide an insight into specific policies which companies are carrying out.

There is considerable scope for further empirical research along the lines of the Beneish model discussed and analyzed above. However, we think it would be helpful if we could somehow find a way to consider additional performance indicators which indicate Earnings Manipulation. In the intellectually more advanced natural sciences there is a greater sense of a structuring of the national, or even international, research agenda in which key issues are approached in a systematic way.

In addition to continued empirical research based on accounting data we also need better integration between different types of research. In addition to accounting data, we need to encourage high quality surveys, more wide ranging interview studies, and a more thorough development of the theoretical foundations of accounting choices.

We also think there is scope for more laboratory-based research. The UK's Financial Reporting Council has created a Financial Reporting Lab to enable discussions between investors and preparers as well as facilitate better research links among academics, analysts, standard setters, professional bodies, company accountants, company auditors, and institutional investors.

To sum up, we need research to be more directly linked to policy-making and changing the behaviors of firms. A greater focus on identifying, disseminating, and encouraging all firms to adopt best practice may be one way to go. On the other hand, modelling average practice may not help us much here. According to Walker (2013), we may need to pay more attention to the tails of EM distribution. What causes the worst form of EM and what can we do to prevent it? What causes the best firms to deliver high quality earnings, and how can we encourage other firms to emulate them?

Appendices

Appendix A: Industry Classifications "Out of the Analysis"

64. Financial service activities, except insurance and pension funding

    641. Monetary intermediation

        6411. Central banking
        6419. Other monetary intermediation

    642. Activities of holding companies

        6420. Activities of holding companies

    643. Trusts, funds and similar financial entities

        6430. Trusts, funds and similar financial entities

    649. Other financial service activities, except insurance and pension funding

        6491. Financial leasing
        6492. Other credit granting
        6499. Other financial services*

65. Insurance, reinsurance and pension funding, except compulsory social security

    651. Insurance

        6511. Life insurance
        6512. Non-life insurance

    652. Reinsurance

        6520. Reinsurance

    653. Pension funding

        6530. Pension funding

66. Activities auxiliary to financial services and insurance activities

661. Activities auxiliary to financial services*

    6611. Administration of financial markets
    6612. Security and commodity contracts brokerage
    6619. Other activities auxiliary to financial services*

662. Activities auxiliary to insurance and pension funding

    6621. Risk and damage evaluation
    6622. Activities of insurance agents and brokers
    6629. Other activities auxiliary*

663. Fund management activities

    6630. Fund management activities

*Except insurance and pension funding.

Appendix B: Beneish's Diagnostics

Coefficient	Ratio	Formula
−4.840	Intercept	
0.920	Days Sales Receivables Index (DSRI)	$\dfrac{(Receivables_t)/(Sales_t)}{(Receivables_{t-1})/(Sales_{t-1})}$
0.528	Gross Margin Profit (GMI)	$\dfrac{(Sales_{t-1}+Cost\,of\,Goods\,Sold_{t-1})/Sales_{t-1}}{(Sales_t+Cost\,of\,Goods\,Sold_t)/Sales_t}$
0.404	Asset Quality Index (AQI)	$\dfrac{1-(Current\,Asset_t+PPE_t)/(Total\,Assets_t)}{1-(Current\,Asset_{t-1}+PPE_{t-1})/(Total\,Assets_{t-1})}$
0.892	Sales Growth Index (SGI)	$\dfrac{Sales_t}{Sales_{t-1}}$
0.115	Depreciation Index (DEPI)	$\dfrac{Depreciation_{t-1}/(Depreciation_{t-1}+PPE_{t-1})}{Depreciation_t/(Depreciation_t+PPE_t)}$
−0.172	Selling, General, Administrative Expenses Index (SGAI)	$\dfrac{(SGA\,Expenses_t)/Sales_t}{(SGA\,Expenses_{t-1})/Sales_{t-1}}$
−0.327	Leverage Index (LVGI)	$\dfrac{(LTD_t+Current\,Liabilities_t)/Total\,Assets_t}{(LTD_{t-1}+Current\,Liabilities_{t-1})/Total\,Assets_{t-1}}$
4.679	Total Accruals to Total Assets (TATA)	$\dfrac{(Cur\,Ass_t+Cash_t)-(Cur\,Liab_t-Curr\,mat\,ofLTD_t-IncomeTax_t)-Depr\&Amort_t}{Total\,Assets_t}$

Appendix C: Top 1000 Companies Divided by Country

Country	Frequence	(%)
Missing	272	
Australia	12	1.66
Austria	3	0.41
Bermuda	4	0.55
Canada	13	1.80
Cayman Islands	3	0.41
China	62	8.58
Curaçao	1	0.14
Czech Republic	1	0.14
Denmark	1	0.14
Finland	8	1.11
France	46	6.36
Germany	31	4.29
Greece	2	0.28
Hong Kong	9	1.24
Hungary	1	0.14
India	19	2.63
Ireland	7	0.97
Israel	4	0.55
Italy	4	0.55
Japan	134	18.53
Luxembourg	2	0.28
Malaysia	2	0.28
Netherlands	11	1.52
Norway	2	0.28
Panama	1	0.14
Philippines	2	0.28
Poland	2	0.28
Republic of Korea	40	5.53
Russian Federation	15	2.07
Saudi Arabia	3	0.41
Singapore	8	1.11
South Africa	1	0.14
Spain	4	0.55
Sweden	9	1.24
Switzerland	8	1.11
Taiwan	18	2.49
Thailand	5	0.69

(continued)

(continued)

Country		
	Frequence	(%)
United Arab Emirates	1	0.14
United Kingdom	29	4.01
United States of America	195	26.97
Total available	723	100.00

References

Aboody D., Hughes J., Liu J. (2002), Measuring Value Relevance in a (Possibly) Inefficient Market, *Journal of Accounting Research*, 40, 4.

Agrawal A., Chadha S. (2005), Corporate Governance and Accounting Scandals, *Journal of Law and Economics*, 48, pp. 371–406.

Badawi I.M. (2008), Preventive Measures of Corporate Accounting Frauds, presented at the 17th Annual Convention of the Global Awareness Society International, San Francisco (CA) May 2008.

Bartov E. (1993), The Timing of Asset Sales and Earnings Manipulation, *The Accounting Review*, 68, 4.

Bartov E., Givoly D., Hayn C. (2002), The rewards to meeting or beating earnings expectations, *Journal of Accounting and Economics*, 33, pp. 173–204.

Barth M.E., Landsman W.R., Lang M.H. (2008), International accounting standards and accounting quality, *Journal of Accounting Research*, 46, 3, pp. 467–498.

Beneish M.D. (1997), Detecting Gaap Violation: Implications For Assessing Earnings Management Among Firm With Extreme Financial Performance, *Journal of Accounting and Public Policy*, 16, pp. 271–309.

Beneish M.D. (1999), The Detection of Earnings Manipulation, *Financial Analysts Journal*, 55, pp. 24–36.

Beneish M.D. (2001), Earnings Management: A Perspective, *Managerial Finance*, 27, pp. 3–17.

Beneish M.D., Lee M.C., Nichols D.C. (2013), Earning Manipulation and Expected Returns, *Financial Analysts Journal*, April, pp. 57–82.

Burgstahler D., Dichev I.D. (1997), Earnings management to avoid earnings decreases and losses. *Journal of Accounting and Economics*, 24, 1, pp. 99–126.

De Angelo L. (1986), Accounting numbers as a market evaluation substitute: a study of management buyouts of public stockholders, *The Accounting Review*, 61, pp. 400–420.

Dechow P.M., Dichev I.D. (2002), The quality of accruals and earnings: the role of accrual estimation errors, *The Accounting Review*, 77, 1.

Dechow P.M., Sloan R.G., Sweeney A.P. (1995), Detecting Earnings Management, *The Accounting Review*, 70, pp. 193–225.

Dye R.A. (1988), Earnings management in an overlapping generation model, *Journal of Accounting Research*, 26, pp. 195–235.

Florio C. (2011), La verifica di *impairment* nella prospettiva delle politiche di earnings management. Profili teorici ed evidenze empiriche, *FrancoAngeli*, Milano.

Gorton G.B. (2008), The Panic of 2007, NBER Working Paper, 14358, September.

Hand J.R. (1989), Did firms undertake debt-equity swaps for an accounting paper profit or true financial gain? *The Accounting Review*, 64, pp. 587–623.

Healy P.M. (1985), Evidence on The Effect Of Bonus Schemes On Accounting Procedure And Accrual Decisions, *Journal of Accounting And Economics*, 7, pp. 85–107.

Healy P.M., Wahlen, J.M. (1999), Commentary: a review of the earnings management literature and its implications for standard setting, *Accounting Horizons*, 13, 4, pp. 365–383.

Huijgen C., Lubberink M. (2005), Earnings conservatism, litigation and contracting: the case of cross-listed firms, *Journal of Business Finance and Accounting*, 32, pp. 1275–1309.

Jones J. (1991), Earnings management during import relief investigations, *Journal of Accounting Research*, 29, 2, pp. 193–228.

Kasznik R., McNichols M. (2002), Does meeting earnings expectations matter? Evidence from analyst forecast revisions and share prices, *Journal of Accounting Research*, 40, pp. 727–759.

Matsumoto D. (2002), Management's incentives to avoid negative earnings surprises, *The Accounting Review*, 77, pp. 483–514.

Mattei M.M. (2006), *Dalle politiche di bilancio all'earnings management*, Bologna, D.u.press.

Mishkin F.S. (1991), Asymmetric information and financial crises: a historical perspective. Financial Markets and Financial Crises, *University of Chicago Press*, pp. 69–108.

Mulford C.W., Comiskey E.E. (2002), *The Financial Numbers Game: Detecting Creative Accounting Practices*, New York, John Wiley & Sons.

Onesti T., Romano M. (2013), *Earnings Management and Goodwill Accounting: Implications on Dividend Policy in Italian Listed Companies*, Global Review of Accounting and Finance, pp. 55–74.

Paolone F., Magazzino C. (2014), Earnings Manipulation among the Main Industrial Sectors: Evidence from Italy, *Economia Aziendale Online*, 4, pp. 253–261.

Peek E., Meuwissen R., Moers F., Vanstraelen A. (2013), Comparing Abnormal Accruals Estimates across Samples: An International Test, *The European Accounting Review*, 22, 3, pp. 533–572.

Pini M. (1991), *Politiche di bilancio e direzione aziendale*, Milano, Etas.

Prencipe A., Markarian G., Pozza L. (2008), Earnings Management in Family Firms: Evidence From R&D Cost Capitalization in Italy. *Family Business Review*, 21, 1, pp. 71–88.

Richardson S.A., Sloan R.G., Soliman M.T., Tuna I. (2001), Information in Accruals about the Quality of Earnings, Working paper, University of Michigan. ssrn.com/abstract = 278308.

Ronen J., Sadan S. (1981), *Smoothing Income Numbers: Objectives, Means, and Implications*, Addison-Wesley.

Roychowdhury S. (2006), Earnings Management through real activities manipulation, *Journal of Accounting and Economics*, 42, 3.

Schipper K. (1989), Commentary: earnings management, Accounting Horizons, 3, 4, pp. 91–102.

Sharpe W.F. (1964), Capital Asset Prices: A Theory of Market Equilibrium under Conditions of Risk, *The Journal of Finance*, 19, 3.

Stolowy H., Breton G. (2004), Accounts Manipulation: a literature review and proposed conceptual framework, *Review of Accounting and Finance*, 3, 1.

Trombetta M., Imperatore C. (2014), The dynamic of financial crises and its non-monotonic effects on earnings quality, *Journal of Accounting and Public Policy*, 33, pp. 205–232.

Trueman, B., Titman S. (1988), An explanation for accounting income smoothing, *Journal of Accounting Research*, 26, pp. 127–139.

Verona R. (2006), *Le politiche di bilancio. Motivazioni e riflessi economico aziendali*, Milano, Giuffrè Editore.

Walker M. (2013), How Far Can We Trust Earnings Numbers? What Research Tells Us about Earnings Management, *Accounting and Business Research*, 43, 4, pp. 445–481.

Wells J.T. (2001), *Irrational Ratios*, Journal of Accountancy, New York.

Chapter 18
Reasoning and Decision Making in Practicing Counseling

Antonio Maturo and Antonella Sciarra

Abstract Some methodologies for practicing counseling are deepen in. The aim of the counseling is in helping a person to assume his/her autonomous decisions and then rational actions for the life choices problems, coherent with own identity and objectives. The counseling procedure is considered as a dynamical decision making problem, where the awareness of alternatives and objectives and their evaluations are maieutically induced by the Counselor. After presenting some relevant practices of counseling and related decision-making procedures, this study shows the use of the mathematical theory of decisions for a formalization of the counseling methods, in order to model, clarify, and make rigorous procedures of decision. Finally it is shown that fuzzy reasoning can give a useful formal help to the task of the Counselor, because of its flexibility and closeness to human reasoning.

Keywords Counseling · Multiobjective decision making · Fuzzy reasoning · Uncertainty

Specific contributions of the two authors The work was conceived in collaboration, sharing introduction and conclusion. As for the specific contributions, paragraphs 2 and 3 are to be attributed to Dr. Sciarra and paragraphs 4 and 5 to prof. Maturo.

A. Maturo (✉)
University of Chieti-Pescara, Viale Pindaro 42, 65127 Pescara, Italy
e-mail: antomato75@gmail.com

A. Sciarra
University of Teramo, Campus Universitario di Coste Sant'Agostino
via R. Balzarini 1, 64100 Teramo, Italy
e-mail: antonella.sciarra@hotmail.com

18.1 Introduction

In Social Sciences, there are many well-established methodologies for practicing counseling (Carkuff 1972; Di Fabio 1999; May 1989; Mearns and Thorne 2006; Nanetti 2003; Perls et al. 1951; Rogers 1951, 2007; Tosi et al. 1987) on the other side in Mathematics and in Operational Research there are many patterns of Decision Making (Lindley 1985; March 1994; Maturo and Ventre 2008a, 2009a, b, c; Maturo et al. 2010; Saaty 1980; Maturo and Maturo 2013, 2014, 2017; Maturo and Fortuna 2016; Maturo 2016; Maturo and Hošková-Mayerová 2017) in condition of certainty, randomness uncertainty (Coletti and Scozzafava 2002; de Finetti 1974; Lindley 1985), and semantic uncertainty (Maturo and Ventre 2008b, 2009b; Zadeh 1975a, b). Our aim is to establish links among these theories and to look for what possible help to practicing counseling can arise by suitable formalizations of problems in terms of Decision Making theory.

The idea is that a cooperation among experts of different fields can give new points of view to face problems of persons having unease. In particular the mathematical procedures utilized to solve Decision Making problems can be very useful for the Counselor task of showing to the Client the degree of coherence between his/her objectives and his/her actions and behaviors.

An important role is played by the de Finetti subjective probability (Coletti and Scozzafava 2002; de Finetti 1974; Lindley 1985). Precisely, the coherence between objectives and actions is obtained with a thought-out assessment of probabilities to nature states and a rational aggregation of the issues associated to every alternative. Moreover this prevents a too pessimistic or a too optimistic behavior, that are consequences of not realistic evaluations of the possibility of occurrence of events.

Fuzzy reasoning and fuzzy algebra (Banon 1981; Maturo 2009; Maturo et al. 2006a, b; Sugeno 1974; Weber 1984; Yager 1986; Zadeh 1975a, b) can also give a good contribution in the counseling procedure. Fuzzy reasoning permits a quantification and a treatment of the semantic uncertainty and can help the Client to avoid extreme positions, and to consider and mediate negative and positive aspects of every situation. Fuzzy algebra (Maturo 2009; Yager 1986; Zadeh 1975a, b) permits to consider the uncertainty in all the counseling processes, and manage gradual modifications of the opinions of the Client induced by the maieutical activity of the Counselor.

18.2 Some Main Features of Counseling

Any counseling intervention pursues a general aim of development of competences and resources, needed to confront and solve actual problems that a person may meet in the his/her own path of life (Carkuff 1972; May 1989; Mearns and Thorne 2006; Nanetti 2003; Perls et al. 1951; Rogers 1951, 2007; Tosi et al. 1987).

The intervention concerns the support for the person to develop, in complete autonomy, a knowledge of self, his/her own objectives, tools, contexts, actions, strategies to be able to learn, confront and change the problems of his/her own personal and social development, problems that are perceived like criticalities to solve.

The Counselor, in his own practices, exercises an aid process that, in a maieutic way, eases, in the person, to emerge competences and the development of resources in the criticalities of the own path of life. The aim consists in helping the person in his/her choices to assume autonomous decisions and useful and rational behaviors in order to act with effectiveness and satisfaction in own contexts perceived as problematic and difficult, with the ultimate aim to reach self-realization and the wellness of the development.

This aim is realized with different methodologies and practices. We illustrate the main features of the humanistic approach elaborated by Rogers (1951, 2007) and Carkhuff (1972). This model presents the following main features, that are fundamental in practicing counseling.

The practice of counseling by Carl Rogers, called "non directive counseling", is centered on the person of the Client, in order that an autonomous and positive base force of the person emerges and operates.

Following Rogers, the Counselor accepts in the client the humanistic and existential dignity of a capable person, that can address him/her-self. The Counselor non directive role assists the person to *implementing* attitudes of *reliance, authenticity, unconditional acceptation* in the interpersonal *communicative climate* between the Counselor and the interlocutor.

The helping relation has its main aim at inducing decision making autonomy, sense of dignity and self-esteem of the person, who can experiment with a suitable climate of self-determination, assumption of responsibilities, promotion.

Non directive method is not, however, the counterpart of a professional neutrality of the Counselor, but an experience of full acceptation in the direction of the Client, to whom the Counselor practices an interesting and tolerant communication. This disposition favors the reliance of the person to change and develop him/her self toward a life more full and satisfactory, inducing in the Client the capability of self-management to recognize and get over the unease.

The empathetic technique of the Counselor consists in the attempt to understand and perceive how the client perceives and understands, to come into his/her perceptive field of meanings, where the whole situational experience finds fulfillments to realize the problem how it is lived by the Client, with just his/her words and subjective universe.

The nondirective role of the Counselor presents an empathetic and non judging communication, without prejudices, that does not proceed by analysis and clinical classification of the problems, but respecting the complete initiative of the Client, toward the representation of his/her problem in the itinerary of the interview. The basis is a continuous encouragement from the Counselor to a spontaneous expression of the Client autonomy, with the goal to help him to search for his/her true *self*. Getting over the unease, the true transformation of the situation and personality toward him/her-self, the environment and the others, is just a client's concern: the Counselor can only help him to recover the freedom to be him/her-self.

The task of the Counselor is to take care that the interviewee recovers his/her own integrity by means of the consistent perception of his/her self. The person has a conceptual representation organized by the self, like a fluid, but coherent, dynamic system of attributes and relations that the *ego* assigns to the *me* in relation to the *others*. This system is self-organizing in front of the natural and social context (Sciarra 2007). When lived experiences in the context are harmonic and congruent with the concept of self, the person reach his/her own integrity and wellness, otherwise the person is fragmented in the unease.

The existential humanistic approach of the Counselor is developed by his dispositional attitudes, that consist in following the threads of Client's discourse, in a helping relational climate of warm reception, comprehension, listening, without pre-determined schemes.

However, an operational methodology is outlined, mainly among Rogers followers, e.g. Carkhuff (1972) with *verbal reformulation* technique of the Counselor, that aims to deepen implicit meanings of the interviewee language and behavior, that give consistency and visibility to the inner attitudes of the person.

Rogers' non directive and dispositional approach and the indirectly regulated and operational one by Carkhuff are integrated on a level of greater complexity. The non directive aid of the Counselor intends to make the Client's lived experience emerges autonomously, to help him to find integrity, wellness, and coherence of his/her *self*. The indirect guide of the Counselor does not neglect to reorganize the perceptions, the attitudes, the capabilities of the person under unease, to help him/her to explore his/her own lived experience of behaviors in front of the context, discover the contradictions, and find again the consistency between objectives and effective actions.

In this way of operating, Carkhuff and his continuators elaborate a helping model, addressed to the interpersonal processes management skills of the helper towards the helpee. The operational methods of the Counselor are the verbal re-formulation of Client's words to deepen the meaning of them, and the capability to pay attention, answer, personalize, involve, explore, understand, compare, communicate and analyze impressions and evaluations... The aim of such sequential phases of the help process is to lead the person to awake to self, to have knowledge of his/her problem, to make autonomous decisions, to reorganize perceptions, behaviors and relations towards environment and persons. A progressive modification model is sketched with the aim to solve the problems of wellness in a self-regulating way.

The counseling is a professional activity for guidance, support, and development in the subject of choice ability to overcome difficult situations. Among the many fields of application, the counseling activity can be exemplified in school counseling (Maggiolini 1997).

The school is a particular form of learning and socialization community in which people with identities, expectations and roles, very different, and often conflicting, interacting in systemic equilibria often dysfunctional, developing malaise and discomfort, both in the teaching staff and in students.

So the school counseling is directed to promote a positive climate in the communication between teachers and students, improving the quality of their relationship, because often the difficulties of student learning depend on an unsatisfactory relationship with their professors, and teachers see frustrated their professional aspirations because they perceive not recognized their identity of leadership by the students.

Therefore the intervention of the school counselor focuses on decisions that can change the communicative and organizational climate of school (Quaglino 1988), helping to improve both the performance of educational success, both the relational wellbeing of the subjects (Sciarra 2007), facing and preventing aggressive behavior such as bullying and spreading a climate of balanced leadership and partnerships between teachers and students, and a rewarding communication to *mutual recognition* (Geldard and Geldard 2009).

The aims and means are in active listening development, upgrading of skills in social relations, promotion of effective and assertive communication, mastery of the problem solving method for the management of problematic situations, use of techniques for the resolution of the conflict through mediation, encouraging self-control and change with ourselves, with others, with the environment, facilitation of knowledge of self and others to understand the needs, motivations, problems, through the maieutic art of Socrates, to develop, independently, possible solutions to the discomfort.

Through the conversation help with the counselor, comes the awareness of the discomfort malaise, and mature the choice of the increase in well-being, which can be done by comparing already carried dysfunctional solutions, to be modified by activating new solutions that do not repeat the mistakes, choosing by comparison possible functional solutions to be checked.

The help offered by the conversation with the counselor (Calvo 2007), to conduct the subject from the discomfort problem to its solution, is developed in several stages. The first of six stages concerns "the welcome" that the counselor reserves to the person, paying attention to him/her without conditions, which creates mutual trust, so that the person reveal his problematic lived experience, his own limitations, but is helped to accept themselves without being judged.

Through guided conversation, the counselor and the person exchange ideas, emotions, information, reaching a profound agreement. A listening and observation skills welcomes what the person reveals with words and with the body, and the counselor exercises *mirroring* how the person communicates, verbal, para-verbal and non-verbal, to promote an empathetic exclusive acceptance.

The counselor also exercises the *reformulation* of what the person is saying, through the paraphrase that offers concise proof of a clarifying understanding and verbalization offering emotional states of subjective meanings contained in the speech.

The second phase concerns "processing" that focuses on five dimensions of understanding of the received message. First of all, the contents of which the person wants to inform the counselor, then the self-representation with which the person wants to be recognized by the counselor, then the appeal or request addressed to the counselor to think or believe and do or not do, then the ongoing relationship with counselor as perceived by the person and how he intends to change it, finally the expression of feelings and emotions that the subject expressed in the experiences mentioned.

The third phase concerns "exploration" of the problem you want to address and resolve, making conscious the subject that his feel good or feel bad depends not by events but by how the problem has been interpreted. The exploration of the problem is carried out by various analytical stages: first, the focus of the problem to help the person to grasp what is more significant than secondary aspects. Then, in order:

- the hierarchy of the problem to establish priorities relating to expectations of the person as the goal of the change,
- the identification and description of the problem as perceived by the subject,
- the evaluation of the intensity of the problem, its frequency, its duration,
- the identification of the external antecedents that were stimulus to the problem,
- identification of internal antecedents, as thoughts, images, sensations related to the problem,
- the identification of the consequent of the problem,
- understanding of the problem as perceived by the subject at body level, the emotional level, the cognitive level,
- reconstruction of the genesis and development of the problem in relation to the various influences that gave rise to it,
- the recognition of the perception of the problem, in the depths of the life of the subject,
- finally, the solution of the problem and identifying resources to define what the person wants to change, such attempts he initiated and intends to test.

The fourth phase of the conversation help develops "understanding" of the message received through various requests for clarification, exploration, investigation, etc.

The fifth phase concerns the "awareness" to be achieved by helping the person to get in touch with their emotions through various techniques such as cognitive restructuring, psychodrama etc., in order to control and modulate the identified emotions.

The sixth and final stage is the "action" that should be supported and facilitated because the subject is able to achieve his objectives of change. The subject is helped to recognize his own needs, to define his priorities, to represent the desired goal, to concretely define the steps to implement up to the goal.

The counselor encourages the subject to action with specific questions about what, who, how, why, where, when, will be realized. In this way, the counselor helps the person to develop action plans to follow step by step, setting deadlines, the beginning and the end of the action, taking care of even small positive reinforcements, avoiding negative reinforcement that discourage, identifying the precise resources required.

18.3 Counseling and Decision Making

The fulcrum around which revolve the different practices of counseling is represented by the theme of the decisions. In fact counseling is a facilitating path for the subject to take autonomous decisions in problematic situations with the help of a counselor.

It's a real indirect training decisions for themselves resolve their problems, getting the counselor's help to walk along all stages of the decision-making process.

The main steps in the process relate to the following fields:

- awareness of personal goals, means, motives, expectations, preferences:
- define the terms of the situation in which you are located;
- clearly formulate the problem to be addressed;
- explore the alternatives of possible solutions;
- consider the consequences of each possible choice;
- weigh the conditions of risk and uncertainty of the environment;
- to consider the criteria that guide the choice;
- assess, with the criteria, the advantages and disadvantages of alternative choices being compared;
- collect from all previous stages the information necessary to deliberate and reach the decision for final action.

The counseling is a true learning the stages of decision making (March 1994) and the path of problem solving (Pennati 2005) with the preparatory help and indirect guidance of a counselor.

The counseling is the set of relational practices of facilitation and support put in place by the counselor to prepare the final decision to action, the choice of which, however, falls on the actor's entire autonomous responsibility.

The decision process is divided into two parts: the deliberation and choice.

The deliberation concerns the way in which they are exploring all possible alternatives and factors to consider for a reasonable decision while choosing regards the last stage in which among the alternatives we come to a final decision for action.

In both phases, of deliberation and choice, it operates the indirect guidance of the open dialogue with the counselor. This is maieutic, kind of Socratic dialogue: a set of questions and answers, objections and replies, information and details exchanged between the counselor and the subject, so that the latter can draw within himself, with the indirect help of the counselor, the choices for settle with its own independent decisions his difficulties to manage change his life paths.

It is inner decisions that can transform the problematic situation from the malaise to the welfare, not as a diagnostic therapy for diseases, but as a transformative care with which the counselor helps the person the best choices for her own life transitions. The person's life transitions affect a choice of a job or a course of study, a partner or friends, a residence or a financial investment, an expense or a conflict, among other options representing difficulties overcome, problems to solve, significant changes to manage.

Take for example the decisions of a young graduate inherent in the difficult choice of university course of studies to be undertaken. The counselor cannot prescribe to the subject a solution, but it must help him to help himself, to find his own decision, accompanying him to explore all the complex decision-making variables to solve his problem.

Because the young man may not be aware of his own goals, then the counselor will help to deepen maieutically, within himself, the awareness of his propensities, aptitudes, vocations, expectations, to connect the choice of university course of studies with the creation of objectives and motivations for his life aspirations.

Because the young man cannot know the wide range of features present in various universities, then the counselor will orient to acquire information of the various existing curricula, their characteristics, their employment opportunities, cultural resources present, the quality of life of the host cities, international relations, the costs involved, future work benefits etc.

The counselor will facilitate in the young man the clear wording of the decision problem, making him analyze, compared to his expectations, the various alternative courses of study, leading him to consider the possible consequences of each choice, considering the consequences in the light of the preference criteria, the objectives which aims to achieve, considering the conditions of risk due to the uncertainty of the environment, on which the subject can only make predictions with the calculation of probabilities, because rarely operates under certainty. The final choice is made by developing rational calculation for the most advantageous and appropriate resolution taking into account all the elements before involved.

The outlined process configures the rational decision model. This model considers, with respect to a problem to be solved, more action alternatives; of each alternative it assesses the consequences, takes into account the state of the environment, which may be under certainty, risk, uncertainty, and establishes selection criteria based on preferred values.

The person with the help of the counselor, chooses among the possible alternatives not only based on the state estimated the environment, not only based on the consequences, but according adopted preference criteria, ranging from the search for maximum profit, in search of maximum socially distributed utility, in search of maximum personal security value, the search for the minimum environmental damage, searching for the prevention of the risk for unwanted consequences, etc.

In addition to the logic model of the consequences, in which the rational decision maker operates with the most useful choices compared to the objectives, there is also a model based on the logic of appropriateness, which describes a decision process based on the logic of the values to be respected and not on the consequences useful. In this case the decision maker operates for the needs of realization of his own identity and cultural belonging which impose to choose taking into account the social rules and values to be respected.

These two models represent the distinction already made by Weber (1922) on rational actions that can be oriented on the one hand to achieve a utilitarian purpose, the other to gain the respect of rules and social values. In the first case it is the utilitarian decision-pattern, with the aim of homo economicus who decides optimizing the means to the purpose, in the second case it is the homo sociologicus normative decision model, value-oriented, who chooses the action more adequate according to his own value-beliefs concerning the identity of his cultural socialization.

The decision, moreover, is not only a free subjective choice between alternatives of action justified on the basis of inclinations, beliefs, preferences, expectations of the decision-maker, but also take into account the objective state of the environment that can influence and even change the result of the chosen action.

The decision maker, and with him the counselor who helps and guides, must understand and consider, in the subjective calculation of the action most suitable, even the state of the environment which may present objective conditions of certainty, probability, uncertainty, offering a panorama of constraints and possibilities each time different, influential on rational deliberation of the final subjective choice, under a criterion in whose light evaluate the most appropriate decision.

There is no absolute decision, but only a decision of situational conditioning action (Boudon 1979), of which the subject has to estimate, among the possible choices, the pros and cons according to the objectives, the states of the environment, the consequences, and the evaluation criteria.

For example, if before going out I can decide among the alternatives to take or not to take an umbrella, I'll have to estimate the probability of the state of the environment in which it can rain or no rain. If I choose to take an umbrella, between the consequences in some cases will have an advantage, in other cases it will have a useless burden.

I will have to consider these elements of choice in the light of a criterion, rational and/or based on the values (which may be, e.g., the regulatory and cultural priorities of maximum prevention and caution, or the useful economic priorities of the minimum weight and minimum space or one of their mix), which will direct the most appropriate decision.

It must be recognized that the decision maker, and the counselor that facilitates the choices freely process the decision, but the choice is conditioned by the situation in the sense of situational logic of Popper (Sciarra 2006), for which an individual performs a rational choice when he acts in a manner adapted to the situation in which he is located. In addition the decision maker, and the counselor, operate in a state of bounded rationality in the sense of Simon (1982), according to which an individual is never provided with complete knowledge, but has partial knowledge that can lead to errors, so do not look for the optimal solution but as satisfactory, taking into account, among other variables, the interaction with others.

The counseling as facilitation process offered to a person in order to develop skills and competences to "help to help themselves" in the development of autonomous decisions, is a real empowerment (Piccardo 1995) to render less latent and liberate all the energy, cognitive resources, the values, emotions, latent in the subject, to help him solve the difficulties of his life choices. Counseling is also a training on concrete cases of a problem solving, offered to the subject as a guide to the main rational and motivational passages, for the decision. Counseling, moreover, is the formation of a habit to take into account the environment that we face, according to a situational logic which relates the adequacy of the action to the given situation.

The process of problem solving involves the following major moments that the counselor will offer, as a path of facilitation, to the subject who decides: awareness of a choice problem, conceptual and visual representation of the elements of the problem with breaking down into simple sub-problems, definition of situational logic of the problem in terms of clarification of rational and adaptive relationship between objectives and environment. The subject will take into account the possibilities and constraints, ends and means, risks and certainties, to formulate clear and distinct alternative hypotheses even with brainstorming.

The subject must verify the solutions by trial and error even in the mental simulation, finally he has to compare the possible solutions for the purpose of choosing the best solution, both by the economic logic of utility, either from the normative logic of membership values, to land, in terms of a reasonably motivated decision, the final decision which solves the problem of choice.

When an individual decision not only depends on the personal goals that deal with a particular state of nature, but is influenced by other people or is co-determined by the decisions of others, then it presents a case of interdependence between strategic players, with situations that may be formalized by the mathematics of game theory (Colombo 2003).

Then the counselor can support the decisions of the subject by helping him to imagine others' moves, simulating response strategies, to implement the gaming patterns that have proven successful. The simulation game strategic decisions is possible if the counselor encourages identification with the diversity of points of view of others, in their objectives, situations, resources, preferences, criteria, making possible to reconstruct the calculation of others' choices and decisions, so to understand and anticipate the moves in his own game plan.

18.4   A Mathematical Modeling for Counseling

Some mathematical models for Counseling can be built starting from the patterns of Decision Making and Game theories.

The first task of the Counselor is to obtain that, in a given instant t of the counseling procedure, the Client is aware of a set A_t of own alternatives or strategies and a set O_t of own objectives and is persuaded to look for the strategies that are coherent with own objectives and to follow one of these strategies, dropping out the incoherent alternatives.

The sets A_t and O_t can change in the counseling process. The Counselor can obtain, maieutically, that the Client be conscious of other alternatives and objectives, realizes that some alternatives before considered are not feasible, or some objectives that seemed important in a moment of emotion, actually are not important for his/her, and so can be deleted from the list of the objectives.

A second task of the Counselor is to facilitate the Client in founding the relations between his/her own objectives and each of his/her possible behaviors, in a coherent and realistic vision of his/her wishes, tools, constraints, and possibilities; then the Client is assisted in elaborating his/her evaluation of the situation with consequent framework of expectations, perspectives of the possible alternatives, i.e. to understand how actions are tied with his/her wishes.

The Counselor, using Ars Maieutica, and establishing together with the interlocutor a relation that allows to attune his/her techniques to the emotional framework and the cognitive objectives of the Client, has to get that the client himself builds his/her answers and decides the own proper social actions.

Of course, the formalization should be understood "as needed" and anyway it should be a light, an ideal target, a far reference point, that gives the Client the logic framework of his/her thoughts, in order that he/she avoids dispersions, incoherencies between wishes and effects of his/her actions. A complete clarification of the objectives or a complete knowledge of the alternatives is neither possible, nor desirable; indeed, an excess of detail increases the complexity and loses sight of main logical thread.

The help of decision theory, in the construction of the frame of the expectations consequent to each action, is important. The person to be oriented must be aware of what he/she gets, or the frame of the possible outcomes related with any alternative and each his/her objective. Decision theory helps to collect these data in order that the subject be oriented toward the actions that are more coherent with his/her wishes.

Decisions may develop in certainty conditions or randomly. In other words an action may give rise to just one or several possible consequences. The Counselor must help the Client to find out the possible consequences of his/her actions and set choice criteria neither too optimist, nor too pessimist. To this aim, a crucial help is given by subjective probability, based on the coherence of the opinions. In fact, subjective probability allows to assess a coherent probability distribution to the outcomes of an action.

Let $A = \{a_1, a_2, ..., a_m\}$ be the set of alternatives, $S = \{s_1, s_2, ..., s_r\}$ the set of nature states, i.e. a set of not impossible events, pairwise disjoint, and such that their union is the certain event. In the classical Decision Making problem formalization, the existence of the utility matrix $U = (u_{ih})$ is assumed, where u_{ih} is a real number that represents the utility for the Decision Maker if he choose the alternative a_i and the event s_h happens.

The pessimistic point of view leads to assume as score of a_i the minimum, respect to h, of the numbers u_{ih}, the optimistic one considers the maximum of such numbers.

In (de Finetti 1974; Lindley 1985) it is proved that a coherent point of view leads to obtain the score of every a_i in two steps. Firstly, a subjective probability assessment $\{p_1, p_2, \ldots, p_r\}$ to the set of events $\{s_1, s_2, \ldots, s_r\}$ is given. After, the score of a_i is assumed to be equal to the *prevision* of a_i, given by the formula

$$P(a_i) = p_1 u_{i1} + p_2 u_{i2} + \ldots + p_r u_{ir}. \tag{18.1}$$

In (Maturo and Ventre 2008b, 2009b) fuzzy extensions of formula (18.1) are considered, starting by two different points of view.

Of course, in practicing counseling, the Counselor must obtain probabilities p_h and utilities u_{ih}, in a maieutical way, by the Client. The coherent synthesis of the opinions of the Client is the prevision (18.1).

The rational behavior is considering as scores of objectives a coherent synthesis of elements of information or opinions.

Lindley (1985) claims that, if the utility matrix $U = (u_{ih})$ is given, only the prevision is a coherent synthesis. But, in general, it is very difficult to obtain the matrix U. In general, the maximum result of the activity of the Counselor is to lead the Client to give a classification of pairs (a_i, s_h), from the most preferable to the least desirable, i.e. a preorder relation is obtained.

In order to obtain numerical scores, a useful procedure is Saaty's AHP (Maturo and Ventre 2009c; Saaty 2007). This process is based on questions that the Counselor proposes to the Client, with the aim to get measures of the pairwise comparisons of the desirability of a set of objects (the pairs (a_i, s_h) in our case).

If there are many objectives, and $O = \{o_1, o_2, \ldots, o_n\}$ is the set of objectives, then for every objective o_j, there is a different utility matrix $U^j = (u^j_{ih})$ and then, for every alternative a_i and objective o_j, the prevision of a_i with respect to o_j is the real number

$$P^j(a_i) = p_1 u^j_{i1} + p_2 u^j_{i2} + \ldots + p_r u^j_{ir}. \tag{18.2}$$

An important task of the Counselor is to help the Client to be aware of own objectives. Of course, every person has a very high number of objectives in her/his life, and from the ars maieutica of the Counselor the most important objectives of the Client must emerge and only they are to be considered in the mathematical Decision Making model.

Moreover, the set $O = \{o_1, o_2, \ldots, o_n\}$ of the relevant objectives of the model must be classified by the Client. The ideal situation is that the Client finds a rational way to associate to every objective o_j a positive real number w_j that measures the importance that the Client attributes to the objective o_j.

Also for the weights of the objectives a suitable procedure is given by the AHP of Saaty (Maturo and Ventre 2009c; Saaty 2007). For every pair (o_{j1}, o_{j2}) of objectives, the Counselor, with a set of questions, does the Client say what is the one preferred or that the objectives are equally preferred. In the first case the Counselor must obtain by the client an integer number belonging to the interval

[2, 9] that measures to what extent the most preferred objective is more important for the Client than the last preferred.

From the responses obtained, with the AHP procedure, a vector $w = (w_1, w_2, \ldots, w_n)$ of weights of objectives is obtained, where w_j is a positive real number expressing the importance that the Client gives to the objective o_j, and the following normalization condition is satisfied:

$$w_1 + w_2 + \ldots + w_n = 1. \tag{18.3}$$

Before to implement the mathematical procedure to obtain the vector w a verification of the coherence of the responses is necessary. In particular the transitivity of the preferences must be verified. On the contrary, the Counselor, with a patient procedure, must propose the questions in a different form in order to avoid incoherence.

When the previsions $P^j(a_i)$ and the weigths w_j are obtained, a rational measure of the opportunity of the action a_i by the Client is given by the following score

$$s(a_i) = w_1 P^1(a_i) + w_2 P^2(a_i) + \ldots + w_n P^n(a_i). \tag{18.4}$$

The greater is the number $s(a_i)$, the more agreeable is the action a_i. We emphasize that this conclusion is only a coherent consequence of the opinions of the Client and the assumption of particular mathematical procedures (e.g., the consideration of formula of prevision or formula (18.4) to aggregate information on weights and previsions).

Obtaining scores $s(a_i)$ is important mainly as an help to the Decision Making, without a claim to be definitive and not modifiable preference measure of the opportunity of the Client actions.

18.5 Fuzzy Reasoning for Counseling

Fuzzy reasoning and arithmetic may help. They permit a gradual procedure in the changes of points of view, gradual and dynamical attribution of the degrees of importance to the objectives, management of semantic and emotional uncertainty.

Furthermore, the consequences of an action cannot be, in general, defined in a sharp way; rather it is opportune that the Client does not renounce to his/her doubts in favor of a choice that is rash, premature and of doubtful effectiveness. From this point of view, fuzzy logic and linguistic variables may be effective, in that they are expressed as imprecise numbers, but plastic and gradually modifiable numbers as far as the opinions of the Client become more clear.

The fuzzy extensions of the decision theory are able to gather the various elements and shows a clear framework, in a language that is close to the human language, of the path that goes from the awareness of the own proper expectations to the coherent action.

In particular, the utility matrix $U = (u_{ih})$ considered in the previous section is replaced by a matrix $U^* = (u_{ih}^*)$ where every u_{ih}^* is a fuzzy number that expresses a value of a linguistic variable (Zadeh 1975a, b). Moreover, the probability assessment is replaced by an assessment of fuzzy probabilities $\{p_1^*, p_2^*, ..., p_r^*\}$ to the set of events $S = \{s_1, s_2, ..., s_r\}$, where every p_h^* is a fuzzy number with support contained in the interval $[0, 1]$.

By considering the Zadeh's extensions (Yager 1986; Zadeh 1975a, b) of the usual addition and multiplication to fuzzy numbers, or alternative fuzzy operations (Maturo 2009), for every alternative a_i, we can introduce the *fuzzy prevision* (Maturo and Ventre 2008b, 2009b) by means of the formula

$$P^*(a_i) = p_1^* u_{i1}^* + p_2^* u_{i2}^* + ... + p_r^* u_{ir}^*. \tag{18.5}$$

If the importance of the objectives is expressed by values of a linguistic variable, then also the weights of objectives are fuzzy numbers w_j^*, $j = 1, 2, ..., n$. Then formula (18.4) is replaced by the more general

$$s^*(a_i) = w_1^* P^{*1}(a_i) + w_2^* P^{*2}(a_i) + ... + w_n^* P^{*n}(a_i), \tag{18.6}$$

where the addition is the extension of the usual addition with the Zadeh extension principle and the multiplication is an approximation of the Zadeh multiplication, built with the aim to preserve the shape of the class of the considered fuzzy numbers (Maturo 2009).

Unlike numbers $s(a_i)$, in general the fuzzy numbers $s^*(a_i)$ are not totally ordered. This can appear a drawback by a mathematical point of view, but, on the contrary, it is an advantage in the practice of the counseling, as the fuzzy number $s^*(a_i)$ contains, in its core and in its support, the history of the uncertainty of the opinions of the Client and then it is a more realistic global representation and it is a measure of the opportunities of his/her choices, more coherent with his/her opinions.

18.6 Conclusions

From previous sections it seems natural the conclusion that an interaction between the maieutical ability and capability to find strategies of the Counselor and the power of the mathematical models for Decision Making can be very useful to solve unease problems e to show to the Client a clear vision of the consequence of her/his possible actions.

The illusory certainties, obtained by questionable assumptions, must be replaced by controlled uncertainties. The tools to take into account the uncertainties and control their consequences in all the counseling procedures are the probabilistic and fuzzy reasoning. In particular, before any action it is necessary to have some information about the facility of occurrence of nature states and this leads to consider the de Finetti subjective probability. More in general, the uncertainty on the assessments of these probabilities can be controlled by considering fuzzy subjective probabilities expressed by fuzzy numbers.

The utility of an action with respect to an objective is often very doubtful. Such uncertainty can be controlled by measuring the utilities with fuzzy numbers.

In the fuzzy ambit, the aggregation of utilities and subjective probabilities associated to every possible action of the Client is made with the tools of the fuzzy algebra and the fuzzy reasoning. They permit to obtain, as a final score of every action, a fuzzy number that provides not only a measure of the validity of this action, but also contains a résumé of the whole history of doubts and uncertainties on the process of evaluation.

A treatment of the aggregation of the previsions and the weights of objectives more sophisticated than the one considered in the previous Sec. takes into account also the logical relations among the objectives (Maturo and Ventre 2009b; Maturo et al. 2006a, b, 2010). A generalization of the utilities is obtained by utilizing fuzzy measures decomposable with respect to a t-conorm $\oplus$ (Banon 1981; Sugeno 1974; Weber 1984). From such a viewpoint a fuzzy prevision can be defined, in which the addition is replaced by the operation $\oplus$. Applications of these theories to Decision Making may be found in (Maturo and Ventre 2009b; Maturo et al. 2006a, b, 2010).

However, as a final result of the application of the mathematical model, the Client obtains a vector of fuzzy (in particular crisp) numbers $s^* = (s^*(a_1), s^*(a_2), \ldots, s^*(a_m))$, where $s^*(a_i)$ measures the advisability of the action a_i, taking into account all the opinions and doubts expressed by the Client and the end of the maieutic work of the Counselor.

The vector s^* is an important reference point for Counselor and Client, a basis for understanding the consequences of their future activities, interactions, strategies and actions.

References

Banon G.: Distinction between several subsets of fuzzy measures. Int. J. Fuzzy Sets and Systems 5, 291–305, (1981).
Boudon R.: La logique du social, Hachette, Paris, (1979).
Calvo V.: Il colloquio di counseling, Il Mulino, Bologna, (2007).
Carkhuff R.R.: The Art of Helping. Amherst. MA: HRD Press, (1972).
Coletti G., Scozzafava R.: Probabilistic Logic in a Coherent Setting. Kluwer Academic Publishers, Dordrecht, (2002).
Colombo F., Introduzione alla teoria dei giochi, Carocci, Roma, (2003).
de Finetti B.: Theory of Probability. J. Wiley, New York, (1974).
Di Fabio A.: Counseling. Dalla teoria all'applicazione, Giunti, Firenze, (1999).
Geldard K., Geldard D: Il counseling agli adolescenti, Erickson, Trento, (2009).
Lindley D.V.. Making Decisions. John Wiley & Sons, London, (1985).
Maggiolini A.: Counseling a scuola, Franco Angeli, Milano, (1997).
March J.G.: A primer on Decision Making. How Decision Happen. The Free Press, New York, (1994).

Maturo, A, Ventre, A.G.S.: Aggregation and consensus in multiobjective and multiperson decision making. International Journal of Uncertainty, Fuzziness and Knowledge-Based Systems, Vol. 17, No 4, 491–499, (2009).

Maturo, A, Ventre, A.G.S.: An Application of the Analytic Hierarchy Process to Enhancing Consensus in Multiagent Decision Making. In: ISAHP2009, Proceedings of the International Symposium on the Analytic Hierarchy Process for Multicriteria Decision Making, July 29-August 1, 2009, University of Pittsburg, Pittsburgh, Pennsylvania, paper 48, 1–12, (2009).

Maturo, A, Ventre, A.G.S.: Fuzzy Previsions and Applications to Social Sciences. In: Kroupa T. and Vejnarová J. (eds.). Proceedings of the 8th Workshop on Uncertainty Processing (Wupes'09) Liblice, Czech Rep. September 19–23, 2009, pp. 167–175, (2009).

Maturo, A, Ventre, A.G.S.: Models for Consensus in Multiperson Decision Making. In: NAFIPS 2008 Conference Proceedings. Regular Papers 50014. IEEE Press, New York, (2008).

Maturo, A., Maturo, F.: Finite Geometric Spaces, Steiner Systems and Cooperative Games. Analele Universitatii "Ovidius" Constanta. Seria Matematica. Vol. 22(1), pp. 189–205 (2014). ISSN: Online 1844–0835. doi:10.2478/auom-2014-0015.

Maturo, A., Maturo, F.: Fuzzy Events, Fuzzy Probability and Applications in Economic and Social Sciences. Springer International Publishing, Cham. pp. 223–233 (2017). URL: http://dx.doi.org/10.1007/978-3-319-40585-8_20, doi:10.1007/978-3-319-40585-8_20.

Maturo, A., Maturo, F.: Research in Social Sciences: Fuzzy Regression and Causal Complexity. Springer Berlin Heidelberg, Berlin, Heidelberg. pp. 237–249 (2013). URL: http://dx.doi.org/10.1007/978-3-642-35635-3_18, doi:10.1007/978-3-642-35635-3_18.

Maturo, A., Squillante, M., and Ventre A.G.S.: Consistency for assessments of uncertainty evaluations in non-additive settings. In: Amenta, P., D'Ambra, L., Squillante, M., Ventre, A.G.S. (eds.) Metodi, modelli e tecnologie dell'informazione a supporto delle decisioni, pp. 75–88. Franco Angeli, Milano, (2006).

Maturo, A., Squillante, M., and Ventre A.G.S.: Consistency for nonadditive measures: analytical and algebraic methods. In: Reusch, B. (ed.) Computational Intelligence, Theory and Applications, pp. 29–40. Springer, Berlin, (2006).

Maturo, A., Squillante, M., and Ventre A.G.S.: Decision Making, Fuzzy Measures, and Hyperstructures, Advances and Applications in Statistical Sciences, Volume 2, No. 2, 233–253, (2010).

Maturo, A., Ventre, A.G.S.: On Some Extensions of the de Finetti Coherent Prevision in a Fuzzy Ambit. Journal of Basic Science 4, No. 1, 95—103, (2008).

Maturo, A.: Alternative Fuzzy Operations and Applications to Social Sciences. International Journal of Intelligent Systems, Vol. 24, pp. 1243—1264, (2009).

Maturo, F., Fortuna, F.: Bell-Shaped Fuzzy Numbers Associated with the Normal Curve. Springer International Publishing, Cham. pp. 131–144 (2016). URL: http://dx.doi.org/10.1007/978-3-319-44093-4 13, doi:10.1007/978-3-319-44093-4 13.

Maturo, F., Hošková-Mayerová, S.: Fuzzy Regression Models and Alternative Operations for Economic and Social Sciences. Springer International Publishing, Cham. pp. 235–247 (2017). URL: http://dx.doi.org/10.1007/978-3-319-40585-8 21, doi:10.1007/978-3-319-40585-8 21.

Maturo, F.: Dealing with randomness and vagueness in business and management sciences: the fuzzy probabilistic approach as a tool for the study of statistical relationships between imprecise variables. Ratio Mathematica 30, 45–58 (2016).

May R.: The art of Counseling. Human Horizons Series. Souvenir Press, New York, (1989).

Mearns D., Thorne B.: Counseling centrato sulla persona, Erickson, Trento, (2006).

Nanetti F.: Il Counseling: modelli a confronto. Pluralismo teorico e pratico. Quattroventi, Urbino, (2003).

Pennati A.: Risolvere problemi, dentro e fuori dalle organizzazioni. Una guida al problem solving metodologico, Franco Angeli, Milano, (2005).

Perls F., Hefferline R.F., Goodman P.: Gestalt therapy. Julian Press, New York, (1951).

Piccardo C.: Empowerment, Cortina, Milano, (1995).

Quaglino G.P.: I climi organizzativi, Il Mulino, Bologna, (1988).

Rogers C.R.: Client-Centered Therapy: Its Current Practice, Implications, and Theory, Houghton Mifflin, Boston, (1951).

Rogers, C. R.: Counseling and psychotherapy, Rogers Press, Denton, Texas, (2007).

Saaty T.L.: The Analytic Hierarchy Process. McGraw-Hill, New York, (1980).

Sciarra E.: Karl Popper e l'epistemologia delle scienze storico-sociali, Libreria Universitaria Editrice, Chieti, (2006).

Sciarra E.: Paradigmi e metodi di ricerca sulla socializzazione autorganizzante. Edizioni Scientifiche Sigraf, Pescara. Italy, (2007).

Simon H.A.: Models of bounded rationality, Mit Press, Cambridge (Mass.), (1982).

Sugeno M.: Theory of fuzzy integral and its applications, Ph. D. Thesis, Tokyo, (1974).

Tosi D.J., Leclair S.W., Peters H.J., Murphy M.A.: Theories and applications of counseling. Charles Thomas publisher, Springfield, (1987).

Weber M.: Wirtschaft und Gesellschaft, Mohr, Tübingen, (1922).

Weber S.: Decomposable measures and integrals for Archimedean t-conorms. J. Math. Anal. Appl. 101 (1), 114—138, (1984).

Yager R.: A characterization of the extension principle. Fuzzy Sets Syst;18:205—217, (1986).

Zadeh L.: The concept of a linguistic variable and its application to approximate reasoning, Inf Sci 1975; 8, Part I:199–249, Part 2: 301—357, (1975).

Zadeh L.: The concept of a linguistic variable and its applications to approximate reasoning, Part III. Inf Sci; 9: 43—80, (1975).

Chapter 19
Risks Associated with Reality: How Society Views the Current Wave of Migration; One Common Problem—Two Different Solutions

Ana Vallejo Andrada, Šárka Hošková-Mayerová, Josef Krahulec and José Luis Sarasola Sanchez-Serrano

Abstract This chapter is dealing with migration problems in general, in particular with the immigration, and is covering two territories, Andalusia (Spain) and the Czech Republic. The problem is described in a pre-case study, which covers results concerning citizens' approach to an urgent social topic, i.e., migration and immigration and risks related to these questions. First, there is given a summary about the history of migrations in both regions; next, the current situation in those regions is characterized; after that, the questionnaire was prepared with the idea of how people feel this phenomenon, and survey was made. Finally, based on the results obtained, possible risks are presented and some strategies how to deal with inconvenient situation, which might arise, are suggested. Since the pre-case study showed highly different approaches of both nationalities, the authors concluded to continue this study, expand the number of respondents so that results obtained later on could be considered significant.

A. Vallejo Andrada · J.L. Sarasola Sanchez-Serrano
Faculty of Social Science, Pablo de Olavide University, Sevilla, Spain
e-mail: Ana-va-94@hotmail.com

J.L. Sarasola Sanchez-Serrano
e-mail: Jlsarsan@upo.es

Š. Hošková-Mayerová (✉)
Department of Mathematics and Physics, University of Defence,
Brno, Czech Republic
e-mail: sarka.mayerova@unob.cz

J. Krahulec
Department of Emergency, University of Defence, Brno, Czech Republic
e-mail: josef.krahulec@unob.cz

© Springer International Publishing AG 2017   283
Š. Hošková-Mayerová et al., *Mathematical-Statistical Models and Qualitative Theories for Economic and Social Sciences*, Studies in Systems, Decision and Control 104,
DOI 10.1007/978-3-319-54819-7_19

19.1 Introduction

The migration phenomenon has become one of the hottest issues nowadays, however, migrations have been known since the time the first humans populated the planet.

"The scientific research located the first humans in Africa; from this area they migrated and started to populate the whole planet. In the most part of human existence, the nomadism and not the sedentary lifestyle have to be a characteristic part of human life" (ACCEM 2008, p. 6).

It is perfectly understandable that the debate about the current influx of migrants to Europe is precarious and polarized. Migration has always been one of the main drivers of human history; and that is why it has never been seen as contradictory to the phenomenon of both positive and negative. Large waves of migration in the past were often accompanied by the collapse of a functioning order. On the other hand, migration was often at the mythical beginning of the national history.

19.2 Two Different Points of View on Migration: Andalusia and the Czech Republic

In this part of research we are going to discuss the migration phenomenon in two territories: Andalusia and the Czech Republic; the idea is to emphasise possible consequences of this process and try to suggest possible solutions.

Since respondents answers are affected by their societies and culture as well as by historical events, the Czech culture cannot be the same as the Andalusian one; therefore, possible answers to the same problem in both societies will be different, however the problem is the same but not the society.

For that reason, before discussing the migration nowadays, it is necessary to make a summary about the history of migration in general and in both territories in detail. In addition, a survey of the population of both zones will be made to know how people feel this phenomenon and what relevant measures could be taken (Rosicka et al. 2008; Rosicka 2004).

19.2.1 Brief Migration History

Migration is as old as humankind itself, and always has a crucial influence on the further historical development. Leaving aside the assessment of the current situation, we can remind seven largest and most significant migrations in human history (Vojáček 2015).

- **Biblical Exodus**

Legendary escape of Jews led by Prophet Moses from Egyptian captivity occurred in the 13th century B.C and it is still considered the oldest historical migration wave.

- **Moving nations:** *it led the Czechs to Central Europe!*

Perhaps, the largest wave of 'moving of nations' occurred in late antiquity and the Middle Ages; it was the largest migration of all time, which permanently transformed the map of Europe and North Africa: it was caused mostly by predatory raids of nomadic barbarian tribes on the decaying Roman Empire. However, a significant role was also played by boosted European population, climate change and the transformation of the existing way of life. The foundations of many states were being laid during the centuries-long process. West Slavic tribes came to Central Europe within this mass migration: later, the current Czech nation could form of them. However, this 'moving of nations' spell disaster and irretrievable extinction for the Romans (Gavriluță 2016).

- **The fall of the Byzantine Empire: speeding up the onset of the Renaissance**

The conquest of Constantinople by the Turks in 1543 marks the end of the Byzantine Empire, which was the last remnant of the ancient Roman Empire and causes extensive emigration of Byzantine scholars to western countries, especially in contemporary Italy. These scholars bring with them knowledge of Greek as well as ancient Greek philosophy and culture: that existing medieval society had about all only vague dim idea. Escaped intellectuals contribute in a significant way to the birth of the Renaissance.

- **The expulsion of Jews and Muslims from Spain: the King impoverishes his own country!**

After the unification of Spain at the end of the 15th century, the Catholic King Ferdinand of Aragon strives to persuade all subjects to exercise his faith. Therefore the Spanish Inquisition is established and it is primarily focused on local Jews and Muslims. In spring of 1492 this fact led to expulsion of all Jews from the Iberian Peninsula and several hundreds of thousands of people are leaving the territory. They find refuge in the Ottoman Empire where the Sultan Bajazet II receives them with open arms. He is well aware how such a strong group of population can be beneficial economically. *"It is foolish, indeed, to call Ferdinand a wise ruler,"* says Sultan. *"It is he who impoverished his own country and enriched mine!"* This opinion is shared by today's historians as well. Neither the confiscated property can outweigh the loss caused by the *Spanish Edict of expulsion!*

- **The settlement of the New World:** *it was caused by hunger, unwanted faith and desire for wealth!*

The discovery by America by Christopher Columbus in October 1492 triggers off a massive migration of people in Europe on a new continent. In addition to adventurers and greedy colonizers, those are also unwanted religious groups in Europe: all of them are looking overseas for a safe place to practise their faith. At other times, the massive migration results from tragic events such as Great Famine in Ireland in the mid-19th century: about two million Irish go to the United States during that period. The arrival of a number of capable and brave people gradually makes the United States the most advanced world superpower. However, on the other hand, the indigenous Indian population is almost exterminated due to immigrants!

- **Leaving Protestants from Bohemia during the Thirty-Year War:** *the Teacher of Nations flees as well!*

The 30 year War is responsible for one of the major migration waves taking place within the Bohemian territory. The victory of the Catholic troops led by Habsburg Emperor Ferdinand II in the Battle of White Mountain induces Czech Protestant Estates to escape after 1620, particularly to the Netherlands, Sweden, and German-speaking countries. Thus, from the Bohemian kingdom, considerable amount of artists and scholars leave, e.g. a Baroque painter Karel Skreta, or Jan Amos Komensky, a philosopher and writer known as the "Teacher of Nations".

- **World War II:** *it expels 60 million people from homes!*

The Second World War is responsible for the largest migration in history in terms of number of refugees. The most pessimistic estimates state that 60 million people had to leave their homes: those were mainly Jews and fleeing abroad was often the only way not to end up in a gas chamber.

Even this very brief historical survey shows that migration had and always has its pros and cons. (Richtermocová 2016)

19.2.2 Andalusian Historical Social Migration Context

In terms of Andalusian historical social migration context, in the following section, some data explaining the result of the surveys will be added and further analyzed.

The book "La identidad del pueblo andaluz" characterizes many facts, such as the geographic position: "Andalucía has a position between continent and oceans, also being the only territory with France which has a coastline on both the Atlantic and Mediterranean, and it is the closest part in Europe to Africa" (Cano et al. 2001, p. 17).

This situation presents the fact that many cultures have inhabited Andalusia, for example, Phoenicians, Greeks, Tartessians, Romans and Arabs because of evidently constant commercial relations with other countries.

Due to that fact, Andalusian people are used to keep in touch with other cultures, both sporadically and permanently, especially with the Muslim religion.

Year	2015	2014	2013	2012	2011	2010	2009	2008
Number of immigrants per 1,000 residents	5.08	4.78	4.43	5.01	5.82	5.91	6.41	9.21

Fig. 19.1 Basic demographic indicators, Andalusia

This fact has created "a peculiar phenomenon of cultural addition/assimilation/synthesis", the result of which has characterised Andalusia as a mix of cultures; if we study Andalusia history it shows us how Andalusia culture takes in and integrates other cultures (Cano et al. 2001, p. 19).

We would also like to highlight that in general, Spain had been for many decades an "immigrant exporter country" as well as a reception country: in particular, it was Andalusia.

Consequently, we can say that Andalusia population is used to living with other nationalities in its homeland both temporarily and long-term.

19.2.3 Current Situation in Andalusia

We are going to describe the current immigration situation in Andalusia; therefore, the percentage of immigrants in Andalucía during the period 2008–2015 will be analyzed as well as nationalities of these immigrants in order to create the complex Andalusia characteristics.

According to the "Instituto General de Estadísiticas" data in the graph below in 2015, in Andalusia there were 5.08 immigrants per 1,000 residents in contrast to 2008: that time there were 9.21 immigrants per 1,000 inhabitants. The decline of immigrants started in 2009 and corresponds with the beginning of the serious impact of the economic crisis (Fig. 19.1).

In the next graph, we can see how the number of immigrants began to decline in 2009; it might be as a consequence of the economic crisis; there was a short recovery in the last year (Fig. 19.2).

Currently "The III Integral Plan for the immigration in Andalucía Horizonte 2016"[1] came in force: its principal goal is to promote the integration of immigrants in all aspects of life as well as to recover more information about migration.

According to "La Consejería de Justicia e Interior de la Junta de Andalucía",[2] nowadays, there are 192 associations working with immigrants distributed as follows: Almería (78), Cádiz (8), Córdoba (6), Granada (30), Huelva (13), Jaén (3), Málaga (25), and Sevilla (29) (Fig. 19.3).

---

[1]http://www.juntadeandalucia.es/organismos/justiciaeinterior/areas/politicas-migratorias/planes-inmigracion.html.

[2]http://www.juntadeandalucia.es/justiciaeinterior/opencms/portal/Justicia/ContenidosEspecificos/Asociaciones/BancoDatos/asociaciones?entrada=destinatarios&destinatarios=7.

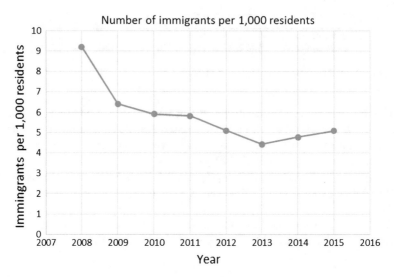

Fig. 19.2 Number of immigrants per 1,000 residents

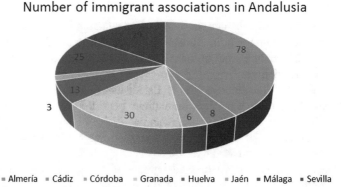

Fig. 19.3 Immigrant associations in Andalusia

19.3   Migration in the Czech Republic

Migration is a phenomenon, which is connected with human civilization from time immemorial. Under the term migration we can understand some movement of quantity of people from one country to the other and also some movement within one country. From the general point of view, migration is a natural effect, when people are moving (migrating) from the poorer country into the richer country, e.g., from desert countries into countries with water sources and rich flora and fauna. From the current point of view, the migration is comprehended a bit differently.

Currently, the item migration is very frequently connected with illegal migration. Events of last days but also events from previous time caused that quantity of

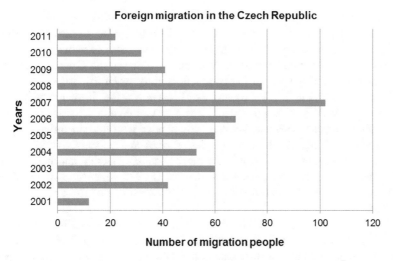

Fig. 19.4 Migration in the Czech Republic in 2001–2011. *Source* Migrace v České republice (2012)

people from the North Africa and Middle East started to move. Destruction of political systems and successive political destabilisation, civil unrests or economical consequences were the most important reasons why thousands of people started to move towards Europe with the vision of new and better life. Nowadays, the Czech Republic is especially a transit country for people moving to the Germany or Sweden.

Migration of people could be viewed in the conditions of the Czech Republic as a new phenomenon; however, the reality looks different. At the time of Czechoslovakia, there was also migration: for example, a German community had around 3 million of people; there were also minorities of Hungarians, Russians, Poles or Jews. After disintegration of Czechoslovakia and opening borders it was possible to start speaking about new dimension of migration. According the Czech Statistical Office, by the Ministry of Interior of the Czech Republic in 2010, in the Czech Republic there were registered 426,749 foreigners. Most often them were people from the Ukraine (128,636 people; 30%), then from Slovakia (71,392 people; 17%), Vietnam (60,931 people; 14%), Russia (31,037 people; 7%) or Poland (18,572 people; 4.4%) [1] (Fig. 19.4).

19.3.1 The Presence of the Vietnamese Community in the Czech Republic—An Excursion into the Past

As already mentioned, the Vietnamese community is the third largest in the Czech Republic. A considerable number of Vietnamese immigrants in the Czech Republic compared to other Central European countries results primarily from immigration-friendly measures in the past and the positive experience of the Vietnamese before 1989. Many students, who had studied in Czechoslovakia before 1989, achieved great success in both the public and private sectors.

To understand the community of Vietnamese living in the Czech Republic, we have to return to the fifties of the twentieth century when the first citizens of the Democratic Republic of Vietnam (DRV) came to the former Czech-Slovak Republic (CSR). Videlicet, an agreement on economic and scientific cooperation was signed between CSR and DRV; in compliance with that document, since the late fifties, hundreds of students, trainees, stayees and personnel (later on) were coming to be educated primarily in the mechanical engineering and light industry. After that, many other agreements followed until 1985 when the number of Vietnamese citizens in Czechoslovakia started decreasing due to due to political decisions.

With the change in political, legal and social atmosphere after 1989, the situation of Vietnamese citizens in Czechoslovakia had to change as well. Many of them had to return home and those who remained had to change the purpose and manner of their stay: instead of employment contracts based on bilateral agreements, they Vietnamese started applying for residence in the Czechoslovak Federative Republic (CSFR) because of the family reunification or business together with Czech citizens.

Already in the late nineties, the Czech Republic, a new wave of Vietnamese immigrants is coming to the Czech Republic: they differ from the previous ones in exclusively economical motivation. A large part of them come from rural regions and cities of the former North Vietnam, mainly from poor provinces.

However, this community does not represent almost any problem in the Czech Republic as well as Slovak citizens or immigrants from Poland. The question of Russian and Ukrainian immigrants is somewhat problematic, especially with regard to the risks associated with Russian or Ukrainian-language mafia. However, the most pressing issue for the Czech Republic residents consists in Romany migration and all the problems and risks associated with it (Analysis 2011; Speranza 2016).

The issue of Romany minority in the Czech Republic is a sore problem, which, among others, is interlinked with the long history of persecution and harassment of Romanies in the Czech Republic and Slovakia territory since the Middle Ages and the recent large migration of Slovak Romanies in the Czech Republic after the breakup of the federation.

The alarm was triggered in 2000 when almost 8010 Slovak citizens (in the vast majority Romanies) submitted the application for asylum. In 2002, the total number of applications from Slovakia reached 843 (the third position in the ranking after the Ukraine and Vietnam), and in the following year the number jumped to 1,055 (Czech statistical Office 2011). Why did a wave of migration raise so suddenly? The answer is simple: the Slovak Romanies discovered a wonderful way to exploit "the Czech welfare state".

19.3.1.1   Little History—The Romanies in the Czech and Slovak Countries

According to Emilia Horvathova, a significant Slovak gypsyologist, the Romanies were first discovered in the Bohemian countries and Slovakia in the 13th century along with the crusade returning from the Holy Land at the period when the Hungarian territory experienced the raid of the Tatars in 1241: the Romanies were fleeing en masse to Bohemia to be rescued from raiders. The first actual statement about the Romanies in Bohemian countries dates back to 1399: those were musicians, blacksmiths or even soldiers in the service of Hungarian kings. At that period, the coexistence with the Romany population did not cause almost any troubles to the original population of the Czech basin. The problems began after the defeat of the Hungarian army by the Turks at Mohacs on August 28, 1526: in fear of the Turks and their possible "spies", the Romanies began to be persecuted. In 1538, the King Ferdinand I Habsburg issued the first act restricting the movement of the Romanies, and his followers, Leopold I and Joseph I continued this restrictive approach. Maria Theresa prohibited to use the expression "Romany" and "Gypsy" and ordered to call the Romanies only "new citizens" or "new Hungarians". Prohibitions and persecutions alternated with attempts to assimilate ethnic groups: however, none of them met with any particular interest by common Romany population. The Second World War brought far the worst persecution of the Romany population: During the Romany holocaust (known as Pojarmos), about 8,000 Romanies were killed in Czechia and about 1,000 were killed in Slovakia. The post-war census of 1947 states that 101,000 Romanies were living in Czechoslovakia: 84,438 were living in Slovakia and the rest in Czechia. Their number was steadily increasing due to high birth-rates, and after the breakup of the federation in 1993, the number of the Romanies in Slovakia was estimated to be 400,000; in the Czech Republic the number was estimated to be 150,000 inhabitants (Nečas and Miklušáková 2002).

19.4   Analyses of Andalusia and Czech Surveys

Why do the Czechs respond significantly differently from the Andalusia citizens to the questions 7–10? It will not be easy to find the answer to this question; nevertheless, it might be rooted in the history of both nations with different historical backgrounds and experience.

19.4.1   Methodology

This part will be devoted to analysing the surveys, which had been conducted by 400 citizens within each of both territories, Andalusia and the Czech Republic. The questions were written in the official language of each country so that everybody could understand properly. Further, questions are presented in English. The surveys consisted of 10 questions; four of them were general introductory questions, such as sex, age, economical situations and education.

Questionnaire:

1. Are you a male or female?

   Male Female

2. What is your age?

   18 to 24 25 to 34 35 to 44 45 to 54 55 to 64 65 to 74 75 or older

3. What is your economical situation?

   Lower class Working class Middle class Upper class

4. What is the highest level of education you have completed?

   Basic studies High school College University

5. Have you ever heard about migrations?

   Yes No

6. How much do you consider you know about migrations?

   I do not know anything about it I know very litle

   I know something I know a lot about it

7. What do you think about migrations?

   It is something positive It is something negative I do not know

8. How much do you consider immigrants have affected your life? $1 < 5$

   1 2 3 4 5

9. Which of these words do you consider are more related with immigration? (you can choose more than one)

   Multicultural phenomenon Terrorist Integration Criminality

   Economic problems Economic benefits Lower unemployment

   Social benefits Tolerance Danger

10. How do you consider the impact of immigration in your country

    Positive Negative Neutral

Further, they had to answer two questions about the migrations in terms of a general aspect; "have you ever heard about the word migration?" and they should have judged "how much do you know about migration?"

Finally, there were four specific questions about immigration related to the personal point of view of the respondent (saying if it is positive or negative); respondents were selecting from the list of expressions, which was/were more related with migrations; they had to mark 1–5 how much has immigration affected their life, and, evaluate the impact in their homeland.

19.4.2   Results

Resulting from the first block of questions related to the personal information, we cannot find a big difference between Andalusia and the Czech Republic; however, in the Andalusia survey, we can find much wider range of age than in the Czech Republic: therefore, the level of studies may result in more varieties as well. (See question no. 1, 2, 3 and 4.)

Considering the second block of questions related with migrations, in both regions, the high percentage of respondents knows the word migrations, and a high percentage of Czechs ranges between "I know something" and "I know a lot about it" (91.77%); Andalusia respondents (73.81%) say they have this knowledge about migrations. (See question no. 5 and 6.)

However, when analyzing the last block of questions related with immigration, we can find bigger differences.

When we ask the question: "How do you consider migration?", 48.3% of Andalusia people do not know how to consider it and 41.31% think it is something positive; on the contrary 22.95% of the Czech citizens do not know how to consider it and 68.8% of them think it is something negative. (See question no. 7.)

As to the impact on the life, we are able to find similar results; however, when we ask them to match the word immigration with a list of expressions, we can see that the majority of Andalusia people match it with the words multicultural phenomena at first, then with integration and tolerance; on the other hand, the Czech people underline expressions multicultural phenomena, terrorism, criminality and danger. (See question no. 8 and 9.)

The last question related with the impact of immigration on the countries, we can find again diverse results between both territories: the majority of Andalusians (43.62%) consider it has had a neutral impact and 36.99% consider it has had a positive impact in their countries, the majority of Czechs consider it has had a negative impact (73.77%), and 19.67% expresses a neutral impact. (See question no. 10.)

19.4.3 Discussion

First of all we should start characterizing two different expressions: racism and xenophobia. The former means that we do not like the immigrants, but we do not attack them; however, the latter means that we take some actions against immigrants.

People start to be racist when they have to compete for the resources, when they consider the government to be more in favour of the immigrants and more against inland population. How to tackle such situations? We believe, we should work with the immigrants and with the population of the country.

As becomes evident from the results, both territories have a completely different point of view on immigrants; for that reason, the work process in both territories should be different.

In our opinion, we should spend more time on the integration advert campaign, which should involve both Czech people as well as immigrants: the idea should consist in promoting the integrations in all aspects of life; in Andalusia, we should invest more resources in occupational campaigns.

19.5 Conclusion

The answers, in particular to questions 7, 9, and 10 being surveyed in both regions were different; therefore, the authors of this pre-case study concluded to continue studying these above-mentioned problems: in particular, to gain much larger sample of respondents, not only in terms of the total number. It should primarily be a representative number of respondents in each gender, age and education categories as well as in essential pairs of these categories so that correlation between individual categories and their relation to immigrants could be examined (Tomei 2016).

The authors of the study believe that sufficiently detailed knowledge in this field may help not only to understand the approach and attitude of the examined regions population towards immigrants but this knowledge can also help to prevent potential risks in terms of social tensions, social unrest or other undesirable effects associated with current and future migration waves.

Based on the information collected and its contradiction, the authors would like to refine the draft measures, which could be taken and performed by administration as well as by other stakeholder charities.

The authors are intending to provide the case study results to selected charity organizations for their use.

Acknowledgements The second author was supported within the project for "Development of basic and applied research" developed in the long term by the departments of theoretical and applied bases of the FMT UoD (Project code: "VYZKUMFVT" (DZRO K-217)) supported by the Ministry of Defence of the Czech Republic.

Supplement

Question 1: Are you male or female?
(Responded: 400 | Respond refused: 2)

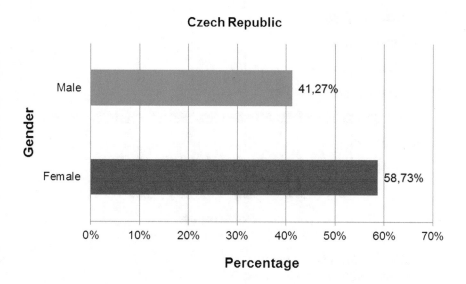

Question 1: Are you male or female?
(Responded: 401 | Respond refused: 1)

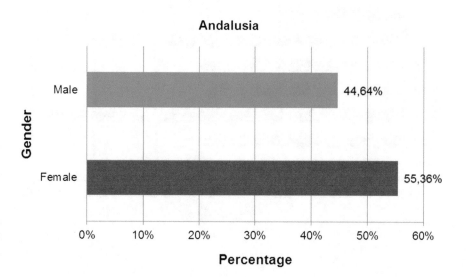

Question 2: How old are you?
(Responded: 399 | Respond refused: 3)

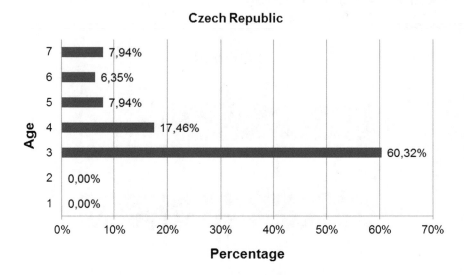

Question 2: How old are you?
(Responded: 400 | Respond refused: 2)

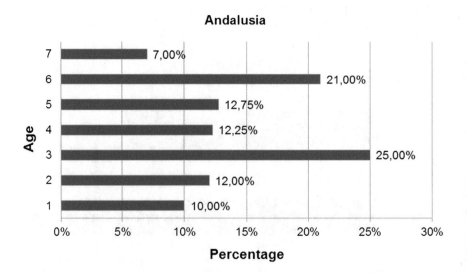

Question 3: What is your economical situation?
(Responded: 401 | Respond refused: 1)

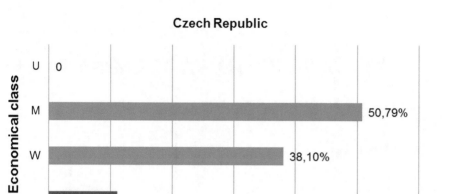

Question 3: What is your economical situation?
(Responded: 398 | Respond refused: 4)

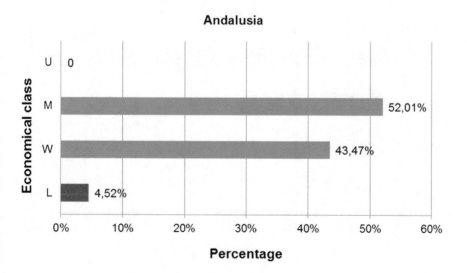

Question 4: What is the highest level of education you have completed?
(Responded: 394 | Respond refused: 8)

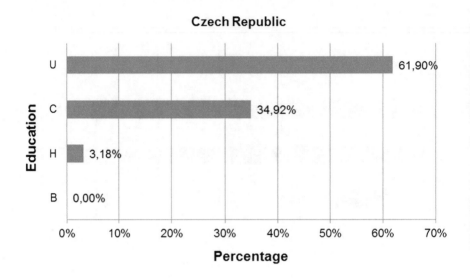

Question 4: What is the highest level of education you have completed?
(Responded: 399 | Respond refused: 3)

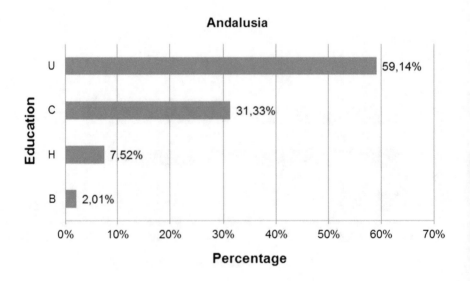

Question 5: Have you ever heard about migrations?
(Responded: 398 | Respond refused: 4)

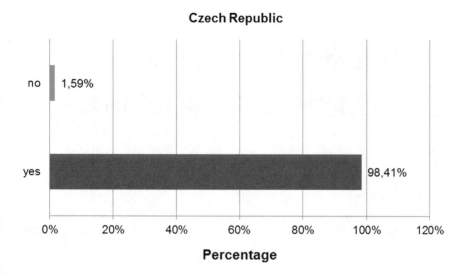

Question 5: Have you ever heard about migrations?
(Responded: 400 | Respond refused: 2)

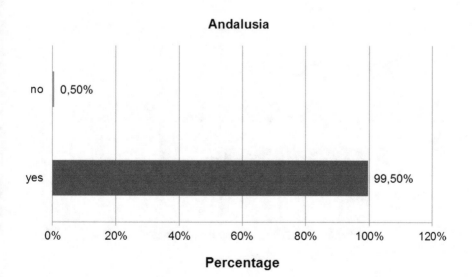

Question 6: How much do you consider you know about migrations?
(Responded: 397 | Respond refused: 5)

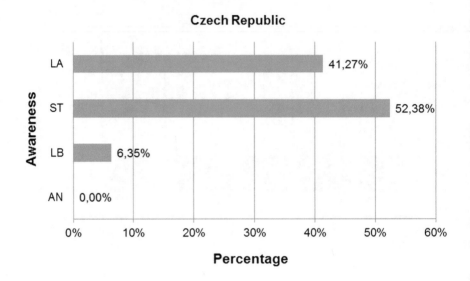

Question 6: How much do you consider you know about migrations?
(Responded: 399 | Respond refused: 3)

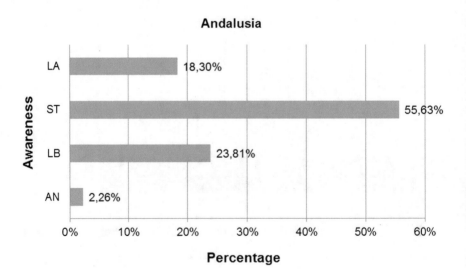

Question 7: What do you think about migrations?
(Responded: 400 | Respond refused: 2)

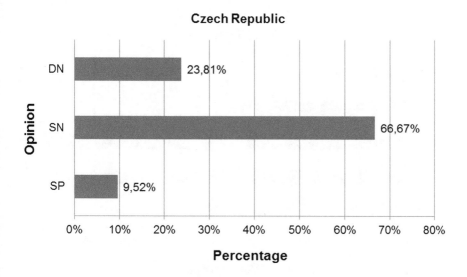

Czech Republic

Question 7: What do you think about migrations?
(Responded: 399 | Respond refused: 3)

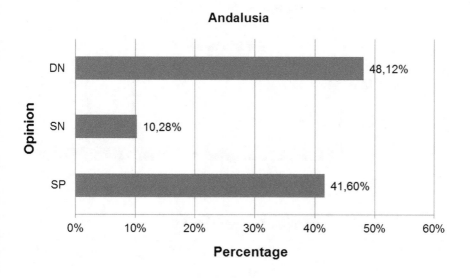

Andalusia

Question 8: How much do you consider immigrants have affected your life? 1<5

(Responded: 399 | Respond refused: 3)

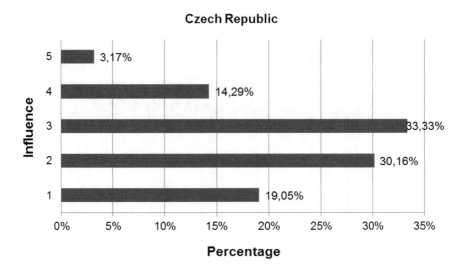

Question 8: How much do you consider immigrants have affected your life? 1<5

(Responded: 397 | Respond refused: 5)

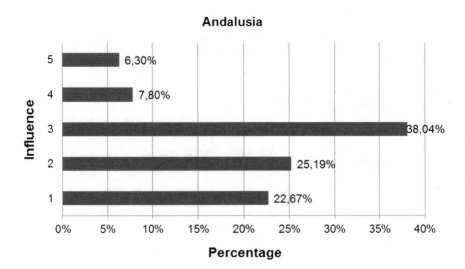

Question 9: Which of these words do you consider are more related with immigration? (you can choose more than one)
(Responded: 399 | Respond refused: 3)

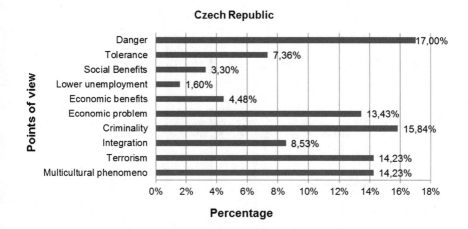

Question 9: Which of these words do you consider are more related with immigration? (you can choose more than one)
(Responded: 399 | Respond refused: 3)

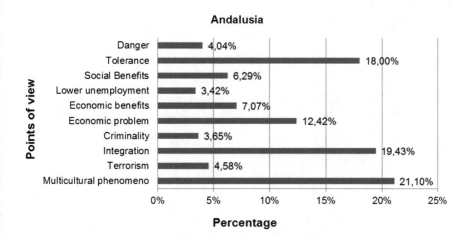

Question 10: How do you consider the impact of immigration in your country?

(Responded: 399 | Respond refused: 3)

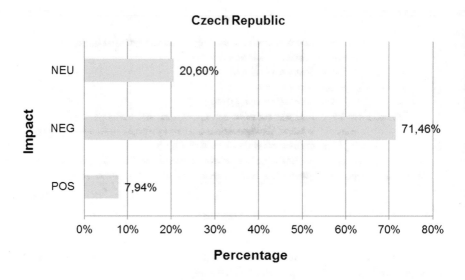

Question 10: How do you consider the impact of immigration in your country?

(Responded: 394 | Respond refused: 8)

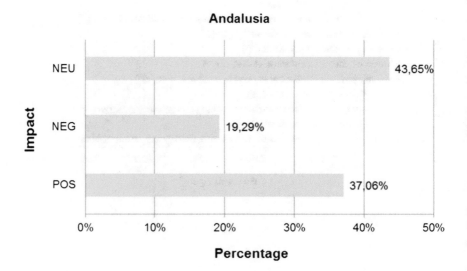

References

ACCEM, (2008) Historia de las migraciones, España como emisor y receptor de inmigrantes, ISBN 978-84-691-1789-7.

Analysis: Bye in the Czech Republic: Romany migration from Slovakia in the Czech Republic, [online 05-12-2011]. Available online at: http://www.demografie.info/?cz_detail_clanku&artclID=789.

Cano, G., Cazorla, J., Cruces, C., Delgado, C., Escalera, J., Lacomba, J.A., Moreno, I., Ropero, M. (2001) La Identidad del Pueblo Andaluz, Defensor del Pueblo Andaluz. ISBN 84-89549-51-6.

Gavriluță, N., (2016) Religious Beliefs and Superstitions in Contemporary Romania. A Socio-Anthropological Perspective, Studies in Systems, Decision and Control, Vol. 66, Maturo (Eds.), Recent Trends in Social Systems: Quantitative Theories and Quantitative Models, 978-3-319-40583-4, 3–9.

Migrace v České republice. (2012) In: *Český statistický úřad* [online]. Praha: Český statistický úřad, 2012 [online 07-26-2016]. Available online at: https://www.czso.cz/documents/10180/20567025/104035-12k01.pdf/91f72d10-2f63-484b-991d-89a1d8135758?version=1.0.

Nečas, C., Miklušáková, M. (2002). Historie Romů na území České republiky, Český rozhlas, http://romove.radio.cz/cz/clanek/18785.

Richtermocová, T. (2016) Migrace v České Republice, její klady a zápory. In: DOCPLAYER [online 07-26-2016]. Praha: docplayer.cz, 2016. Available online at: http://docplayer.cz/56443-Migrace-v-ceske-republice-jeji-klady-a-zapory.html.

Rosicka, Z. (2004) Cultural Diversity and Multinational collaboration. *International Conference proceedings "Crisis Managment in Europe – Problems and Perspectives"*. Polcje, Slovenia, 2004, 24–28.

Rosicka, Z., Benes, L., Fleissig, P. (2008) Vita in societate secura. Monograph. Univerzita Pardubice, 2008. ISBN 978-80-7395-117-7.

Speranza, S., (2016) Public Values and Social Communication, Studies in Systems, Decision and Control, Vol. 66, Maturo (Eds.), Recent Trends in Social Systems: Quantitative Theories and Quantitative Models, 978-3-319-40583-4, 107–126.

Tomei, G. (2016) When Statistics Are Moved by Words. Biopolitic of International Migration Flows in Contemporary Italy, Studies in Systems, Decision and Control, Vol. 66, Maturo (Eds.), Recent Trends in Social Systems: Quantitative Theories and Quantitative Models, 978-3-319-40583-4, 23–30.

Vojáček, L., (2015) 7 largest migrations in history. The eternal chase for a better future! (2016) [online 03-09-2016]. Available online at: http://epochaplus.cz/?p=12561.

Part II
Recent Trends in Qualitative Theories for Economic and Social Sciences

Chapter 20
Sociability and Dependence

José Luis Sarasola Sánchez-Serrano, Ana Vallejo Andrada
and Alberto Sarasola Fernandez

Abstract In this article, we are going to talk about the social aspect of elderly
people in urban areas, especially about disabled elderly people and how disability
affects their socialization. With the aim to cover this topic we have done research
which involves a survey, an interview and a discussion group, with the idea to
collect all relevant information, we have compared the results of these three techniques and made a complete evaluation of these findings.

Keywords Social work · Elderly people · Urban areas · Disabled people ·
Sociability

20.1 Introduction

This research has analysed the elderly people society in urban areas in Sevilla city,
Andalucía, Spain. Nowadays this research is vital because of the movement of
people between rural zones to urban areas as well as the fact that life expectancy has
increased.

Consequently, it was necessary to analyse the variability of the situations of
senior citizens especially focusing on dependency aspects.

It is crucial to understand the new social relations that can appear in this context
and create new ways to integrate this collective with these new necessities.

J.L. Sarasola Sánchez-Serrano (✉) · A. Vallejo Andrada · A. Sarasola Fernandez
Social Sciences Departament, Pablo de Olavide University, Sevilla, Spain
e-mail: Jlsarsan@upo.es

A. Vallejo Andrada
e-mail: ana-va-94@hotmail.com

A. Sarasola Fernandez
e-mail: Alberto-sgs@hotmail.com

© Springer International Publishing AG 2017   309
Š. Hošková-Mayerová et al., *Mathematical-Statistical Models and Qualitative Theories
for Economic and Social Sciences*, Studies in Systems, Decision and Control 104,
DOI 10.1007/978-3-319-54819-7_20

For this reason, this research is focused on the complete stages of when the person retires to the moment of dependency. We considerer this stage of the life cycle like a continuum which has its own stages.

The first aging signs normally start when the person is fifty-five years old, with the emergence of corporal changes and the deterioration of sensory and perceptive functions.

The second stage of aging is related with retirement, and it is characterized by the loss of a social interaction roll within society. This change is less visible than physical changes.

The third stage begins when the elderly start to lose loved ones, especially their spouse.

The fourth stage comes when activities get reduced, illness appears, and the restricting processes commence. (Docel Fernandez and Gutierrez Barbarrusa 2006:38)

As well as also having to take into account the different degrees of dependence that could affect the population, as these degrees would affect the previous stages. In Andalucía, Spain we differentiate between:

1. Low Dependency: We regard this degree of dependency to be when the person needs help to perform some basic, everyday activities from time to time.
2. Moderate Dependency: We consider this degree of dependency to be when the person needs help to perform a wide range of everyday basic activities, but does not need complete supervision to be able to do them.
3. Severe Dependency: We deem this degree of dependency to be when the person needs help to perform all the everyday basic, menial activities, because he or she has lost their physical, mental, intellectual or sensory capacity, so he or she needs somebody's complete care to live (Sempere Navarro and Cavas Martínez 2007:140–141; Hoskova-Mayerova 2017).

According to this information, we decided to start an investigation, in conjunction with Pablo de Olavide University Teachers (who are also in the research group Research in Social Sciences and Social Politics (PAI SEJ-452)), to elaborate a study about the sociability of old people in urban area and its relation with the degrees of dependency using qualitative and quantitative techniques.

Highlighting within the study's framework the differences between the elderly inhabitants' sociable areas in Seville City.

And research contributions in order to improve the old people sociable areas. Symbolising the crucial relationship between sociability and integration of seniors in the urban area, as well as the benefits the urban areas can have if it integrates old people, as well as analysing the relationship between the different old people's degree of dependency and their sociability techniques (Malagón et al. 2006; Gubiani 2017).

20.2   Hypothesis

In Seville, there are a wide range of sociable possibilities for senior citizens in relation to the architectural planning of the city areas dedicated to this work. For this reason, the old peoples' sociability will be different depending on the characteristics of the area where they live.

The higher the level of dependency old people have, the higher the degree in gaining access to suitable social areas. Because social areas are not normally adapted for old people with high dependency.

When we are talking about the different activities as well as the time people dedicate to them we have to take into consideration there are different generations of old people so the activities and the time will be different depending on their ages and characteristics.

The more money and free time the elderly have, the greater the variety of activities they are able to access.

In relation to an old persons' knowledge of the different resources available, the most popular are health centre resources and local Medical staff.

20.3   Research Techniques

The research techniques which were used in the study were:

1. Surveys: The type of survey we decided to use is a closed survey with the intention of gathering specific information from the general public.
2. Interviews: The people who were interviewed were selected from the surveys, taking into account the ones who had more information in relation to the research topic; which again targeted the public in general.
3. Discussion group: This technique provided us with more qualitative and subjective information which was crucial to the study.
4. Participant observation: We used this technique as a means of verifying the findings of our surveys, interviews and discussion groups. This also involved the direct observation of the selected urban zones of our research.
5. Map creation and zoning urban areas: This technique consisted of map creation and resource analysis.
6. Others techniques

20.4   Methodology

With the idea to achieve the objectives of the investigation, this research has been divided into three different phases:

1. First phase: Research preparation, during this time the professionals were focused in defining the objectives and preparing all the necessary elements to carry out the research

   • *Key concept definition*: This part was carried out by the research group members, and was the basis of the researcher's theoretical part.
   • *Measurement tool elaboration and designed:* This part consisted in all measurement tools (the quantitative and qualitative) elaboration and design that were going to be used in the research as well as the necessary support and instruments to create a research protocol. We can highlight:

      – A previous analytical test about the social areas
      – Records about the social areas
      – Interview and survey scripts
      – Discussion group scripts
      – Staff Selection
      – Social area locations
      – Previous social areas analytical test
      – Date test analysis

   • *Re-structure the survey object and urban areas*: This part includes all the possible necessary changes as a result of the first phase conclusions (Tomei, 2017).

2. Second Phase: Object analysis
   This is the most extensive phase of the research, and is divided into these different activities:

   • Zoning of urban areas: In this part the professional team's aim is to create a complete map of the urban areas where they are going to work, indentifying which areas are accessible to old people.
   • Carrying out the closed interviews and the surveys described in the First Phase
   • Arranging and analysing the closed interviews and survey data.
   • Carrying out the open interviews.
   • Analysing the open interview data.
   • Starting the discussion group.
   • Analysing the discussion group data.
   • Carrying out the participant observation.

3. Third Phase: Research conclusion
   This is the last phase, in this phase we are going to create the research conclusion in relation to the hypothesis and creating a final report, this part will be composed by these different actions:

- Establishing the final research conclusions
- Writing and working out the final research report: taking into account the previous phase's findings, as well as the information they recovered during the complete investigation process.
- Data presentation and distribution: This part will involve the publications and distribution of the research results using these instruments:
  - Publishing the full research
  - Organizing seminars in Pablo de Olavide University.

20.5   Results

Survey results

(A) General information

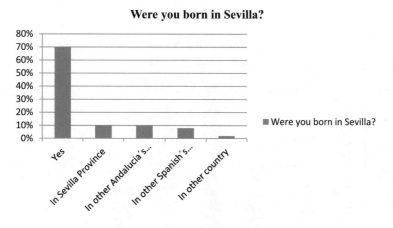

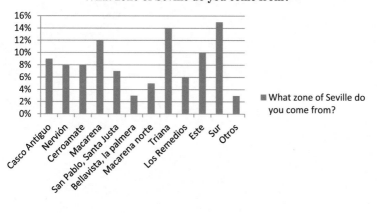

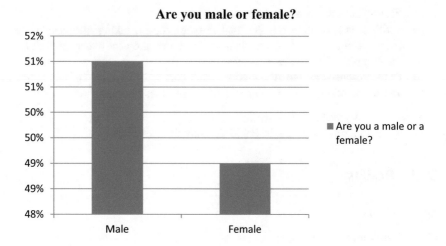

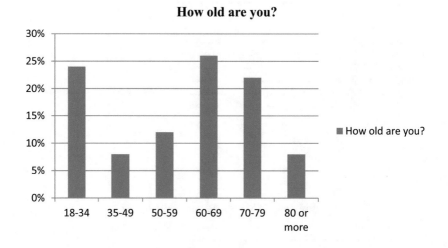

What is your civil status?

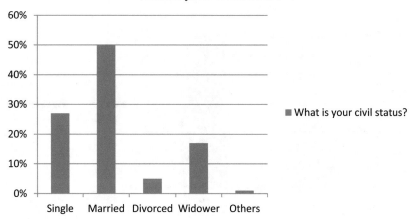

Academic Experience

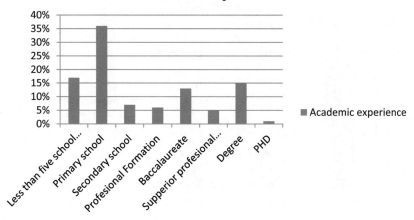

Do you do any voluntary activities?

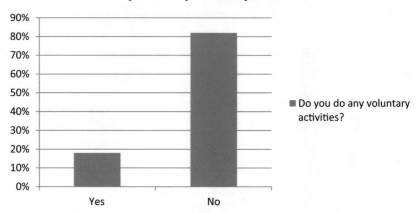

Where do you normally do activities with people of your own age?

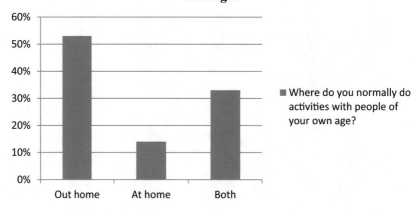

(B) Free time information

Do you do these activities alone or in a group?

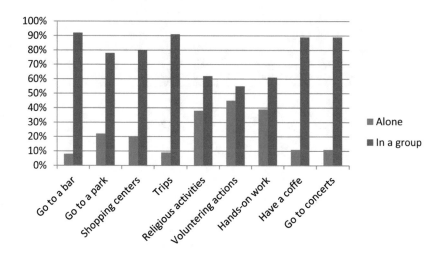

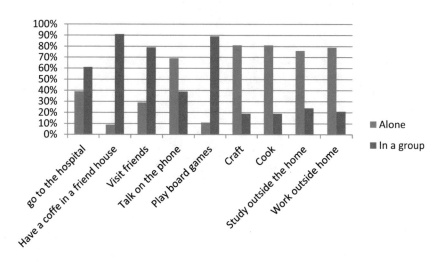

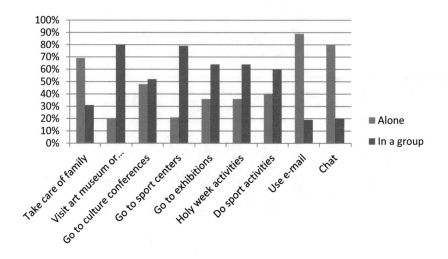

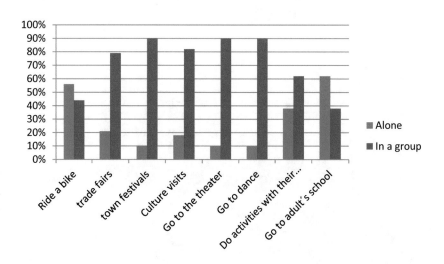

(C) Difficulties dependence can cause

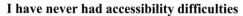

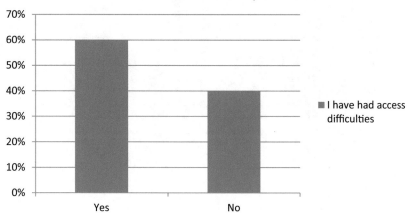

I have never had accessibility difficulties

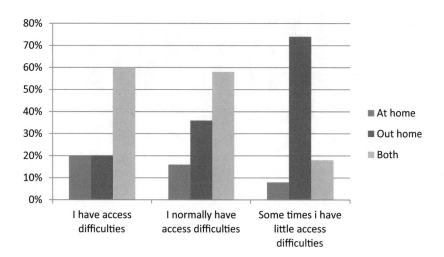

Do you have accessibility difficulties in these situations?

Do you have accessibility difficulties in these places?

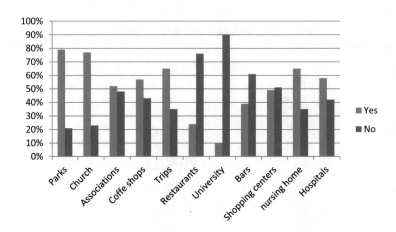

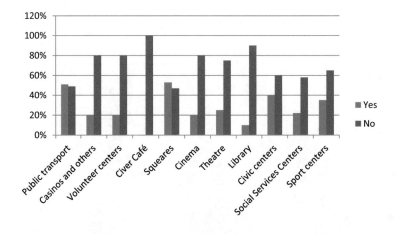

(D) Possible resources, how much people know about them and how much they
     use them.

How much do you know about these resource, do you use them?

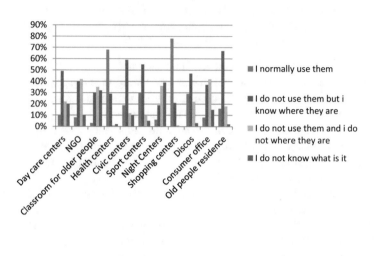

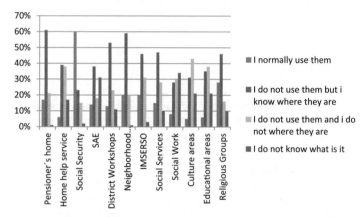

   In order to make the result section more understandable, we are going to divide it
into different sections with relation to the different topics.

(A) General sociable areas for elderly citizens
     In general terms, we can say the senior's characteristics` in urban areas pre-
     pared for them are:

   • Dependent people do not normally use urban areas prepared for the older
     generation.
   • Women have a greater percentage of dependence than men.
   • There is an important presence of people younger than 60 years of age.

- There is a difference between men and women and how they use these areas, due to the fact that women normally go as groups whereas men normally go alone.
- Elderly people normally create groups with people of the same age as them in the more common sociable areas, we call them "Generational Groups".
- In relation with the hypothesis, the more elderly participate in their society, the more integrated they will become, this is verified.

(B) About the activities elderly citizens take part in.

The less common for elderly citizens are: Going to the park, having coffee, visiting exhibitions, visiting museums and/or monuments, going to the theatre, doing voluntary schemes, playing board games, riding a bike, going to a trade fair, dancing, attending workshops, attending adult education centres, going to town festivals, visiting family, attending concerts, going to cultural conferences, going to the cinema, enjoying cultural visits, using e-mail.

(C) In relation to dependency and the possible limitations

The respondents have stated that they find difficulties accessing sociable areas both indoor and outdoor.

And the general perception the population have, is that disabled elderly people have a disadvantage in comparison with the rest of the general population when they want to gain access to the sociable areas.

20.6   Conclusion

To sum up we can say that despite Seville having plenty of sociable areas designed for old people, it looks like not everybody has the same access opportunities to these areas, both factors which can contribute to this are: The area where they live and the degree of dependency they have.

As we can see in our results dependent people have access problems both outdoors and indoors, so this confirms our hypothesis "The higher the level of dependency old people have, the higher the degree in gaining access to suitable social areas. Because social areas are not normally adapted for old people with high dependency."

The area where they live is also directly related to their economic situation because the more money they have the better area they live in, and the better the area is, the better the facilities are for senior citizens included.

Consequently, although we do not have enough data to confirm our hypothesis "The more money and free time the elderly have, the greater the variety of activities they are able to access" at least in the accommodation factor we could see a possible relationship between the money they have and the greater the variety of activities they are able to access, whereas, we do not have any information about their free time so we cannot suggest a connection between it and the previous item.

About the resources they have and how much they used them, as we can see in the result part, our previous hypothesis was wrong, elderly people know and used social security centres as much as health care centres.

References

Docel Fernández, L.V. and Gutierrez Barbarrusa, T. (2006): *La sobrecarga de las cuidadoras de personas dependientes*. Tirant Lo Blanch. Valencia.

Gubiani, D. (et al.), (2017) *From Trajectory Modeling to Social Habits and Behaviors Analysis*, Decision and Control, Vol. 66, Maturo (Eds.), Recent Trends in Social Systems: Quantitative Theories and Quantitative Models 371–385.

Hošková-Mayerová, Š. (2017) *Education and Training in Crisis Management*. In: The European Proceedings of Social & Behavioural Sciences EpSBS, Volume XVI. Future Academy, 2017, p. 849–856.

Malagón, J.L; Barrera, E; and Sarasola, J.L. (2006). *Análisis de la sociabilidad de las personas mayores en el medio urbano su relación con el grado de pependencia*. Registro territorial de la propiedad intelectual de Andalucía. Sevilla.

Sempere Navarro, A.V.; y Cavas Martínez, F. (2007): *Ley de Dependencia. Estudio de la Ley 39/2006 de Promoción de la Autonomía Personal y Atención a las personas en situación de dependencia*. Aranzadi, S.A. Navarra.

Tomei, G. (2017) *When Statistics Are Moved by Words. Biopolitic of International Migration Flows in Contemporary Italy*, Decision and Control, Vol. 66, Maturo (Eds.), Recent Trends in Social Systems: Quantitative Theories and Quantitative Models, 23–30.

Chapter 21
School Institutions as Complex Systemic Organizations: A Multidisciplinary Perspective

Grazia Angeloni

Abstract The chapter that follows, tries to highlight the reasons why educational institutions should be considered complex systemic organizations. The multidisciplinary approach suggested tends to make use, on one hand, of different lenses, in order to appreciate the organizational phenomenon taken into consideration, on the other, it is an effort to join different epistemes for practical purposes. It is itself the expression of the contemporary man's attitude toward complexity. Since this is showed through circles of relations and interactions among the component parts of phenomena and accidents, the suggestion is to develop a cognitive method to stay with it side by side, in order to manage it, without being entrapped by it. This possibility lies in a new system of thought, critical and systemic that acknowledges what is happening and the same problematic nature of reality in terms of dynamism and probability, instead of linearity and consequentiality.

Keywords School organizations · Complexity · Systemic thought · Multidisciplinary perspective

There is no place devoted to the formal and purposeful process of education that cannot be identified with school. An articulated, never concluded and always improvable intentional process puts in such a place the human being as the pivotal resource, meant both as an immaterial entity and as an individual and social being to whom the teaching activity is addressed. As a matter of fact, school was born with a peculiar intention, with a well determined goal that is to promote and develop the human being's personality, through the acquisition of free, critical, systematic learning (Molinari et al. 2000). The pursuit of this goal sees the contribution of a plenty of actors whose functions are the same as their purpose, both matched with the pupils' education. Those actors are primarily the teachers. Their actions are not only intentional, that is, deliberately goal oriented for their achievement, but also formalized, i.e. adjusted on the basis of national and local

G. Angeloni (✉)
University "G.d'Annunzio" Chieti-Pescara, Chieti, Italy
e-mail: graziaangeloni@unich.it

© Springer International Publishing AG 2017   325
Š. Hošková-Mayerová et al., *Mathematical-Statistical Models and Qualitative Theories for Economic and Social Sciences*, Studies in Systems, Decision and Control 104, DOI 10.1007/978-3-319-54819-7_21

curricula where the different fields of knowledge are arranged, working loads set in terms of subjects, knowledge, skills and times are well scheduled, in other words defined with accuracy both on a national and local level; where teaching activities underlie a reasoned choice of approaches, methods and teaching/learning strategies.

School is therefore a social construction made up of people who share the same goal which is among other things directed, in the exercise of their duties and functions, to ensure and safeguard each and every one's social right to be educated. Pupils are thus education right holder.

But neither a group of people who is going to pursue a common purpose seems not to satisfy the definition of organization, nor the teachers of a school who set their pupils' education as a common goal to achieve, through a series of formal activities, can recall an organization in its deepest meaning. To understand the concept of "organizational phenomenon" may be advisable to move to another context of reference. As Ferrante and Zan (1998) argued, a group of passengers traveling by train and having formalized its willingness to reach a common destination, upon payment of a ticket cannot be named organization. It's up to us the duty, therefore, to complete the statement in order to have a clear vision of the features of the social construction that is our object of study. Chester Barnard calls organization "*a set of activities or forces of two or more people, consciously coordinated*", and yet Ferrante and Zan (1998) believe that "*organization, than any other social construction, is a form of collective action based on repeated processes of differentiation and integration generally stable and intentional*". The first definition is further developed by the author through a story, fairly known in the organizational literature as the boulder parable. Let us imagine to travel by a car. Suddenly we see a boulder that prevents us from going ahead. A first instinctive response would induce us to get out and give the necessary pressure on the obstacle to its removal. But an accurate reflection, makes us realize to be powerless and need the help of other people to achieve the goal. And yet this is not enough, that is, it is not sufficient to think that anyone who releases his strength on the object can actually move it. To be effective, the joint efforts of more people have also to be coordinated, i.e.to be released on the obstacle intentionally and with order, in the same sense or direction and at the same time. The organization therefore exists in the first place as a cooperative system: "a set of activities or forces of two or more people".

It seems appropriate right now, to clarify the terminology used, to avoid misunderstanding about the term "cooperation". We could define it as the will to work with others—from the Latin *cum-operari*—in a "consciously coordinated"way, i.e. giving order—*cum-ordinatio*—to action, through strategies, rules and procedures. According Calvani and Rotta (2000) while cooperation presupposes a relationship of symmetry among the agents who come into structured and finalized contact, collaboration, which at the lexical level is often used as a synonym of the first term, refers to a greater extent to a supportive and reciprocal relationship among those who interact. Cooperation if well regulated and governed can become itself coordination and can be assumed as a mechanism for integration to bring a diversified reality into a whole. We came then to the interpretation of the second definition of organization presented to us by Ferrante and Zan, in which differentiation and integration are the key words. The concept of differentiation refers to the division of labor, that is, functions, tasks, activities, on one hand, on the other it refers to specialization. Arguably, the more a person repeats over the time certain work performances linked with his own function and role, the more he acquires procedural knowledge about the tasks related to this function, rather than on the other ones. Through differentiation every organization of people, is transformed from an undifferentiated universe, into a system of roles. This system, as a matter of fact, tends to stabilize and stands still over the time and may be challenged only in case of critical events or particular dysfunctions which affect the organization as a whole. Differentiating means then to establish "who is doing what". If we go further and assume other discipline as lenses to interpret events and facts, we can find that the organizational literature gives the word organization a double meaning: (1) an institution founded on the division of labor and skills; (2) the way in which the entity is organized (Rebora 2001; Bonazzi 2002), with obvious reference to its internal arrangement, therefore its structure.

Thus, in the school system the differentiation of work and tasks takes shape and is embodied in the functions and roles performed by each professional category. To make a short list of them, in order to show the differences as for each function and activity we may say, starting from the top of the organization, that the school principal is by law responsible to ensure the management of the entire institution legally represented by him/her. He/she otherwise is responsible for the service results, in terms of outcomes, as well as for the financial and instrumental resources management. He/she has also autonomous powers of direction, coordination and human resources development, in compliance with the governing bodies. He/she organizes school activities inspired by the criteria of fairness, effectiveness and educational efficiency. After him/her, the teachers are involved in teaching their disciplines, and in functional activities to teaching, which are expressed in planning their educational interventions, leading educational research, updating their knowledge, in the assessment and evaluation necessary for an effective performance, in relationships with pupils and families and with all the school staff. Moving forward, the director of administrative services directs and manages the school administrative services on the principal's behalf. The administrative staff performs administrative functions, accounting, management and operational related activities, in cooperation with the school principal, the teaching staff, the school governing bodies, the families. The support staff fulfills general tasks and functions of supervision and pupils accommodation. They keep the hygienic conditions relating to the school premises as well, working in cooperation with the teachers and the administrative director.

The system of roles in school, however, defined by the Italian regulations, creates a particular organizational structure that can be described through a top-down organization: the school principal, the teachers, the director of the administrative services, his/her staff, the school cooperators. However it does not seem appropriate here, to deepen the concept of the organizational structure, for the whole topic's sake. The process of differentiation, also expresses on one side the need for a minimal rationalization of the organizational processes, on the other the acquisition of a reasonable assurance that each person can be assigned a task to perform (Spaltro and Piscicelli 2002). Needless to say that the fulfillment of a given function requires an adequate system of knowledge, skills and competencies necessary for the implementation of the above mentioned tasks. Therefore differentiation processes and a good system of roles should always be accompanied by appropriate coherent process of personnel selection (Ferrante and Zan 1998). There are, in these terms no exception to this principle as far as regards the school principal's cooperators and the teachers appointed by the Board of Teachers whose functions are "instrumental" to the school educational plan. They have to possess, as a matter of fact, a body of knowledge, abilities and skills before they have been appointed for such a role, by a selection procedure based on coherent criteria referred to each specific task. Nevertheless the organization, cannot be considered and appreciated as a whole if differentiation is not accompanied by integration, if unity, in other words does not take the place of division. Now one question arises on who is able to bring back to a whole the complex system of diversified and specialized tasks. A traditional approach, therefore rational to the study of organizations, as for the process of coordination, entrusts the hierarchy of authority, giving it power. According to such a perspective the "authority" should seem the only one capable of integrating, recompose each function within a unified framework directly targetted to organizational purposes. This view seems still to be accepted in the school of autonomy. In fact, one of the tasks of the school principal is, after all, the coordination of the institution into a whole. A more modernistic perspective instead disengages the concept of integration from the "chain of command", attributing it to each organizational actor. In both the cases, integration is a process that makes use of some mechanisms in order to produce its effects. They may be of different nature: rules and procedures, through which one determines the criteria for "how to do something," and that should provide a coherent collective behavior; technologies and work plans that set the different tasks into a logical and sequential continuum and determine, among other things, the time within which carrying them out; the strategies to whom any actor can inspire his way of acting, but also the cross-level communication among positions and operating units and among them with hierarchy which is a powerful tool for integration. It is not, ultimately, to be underestimated the role played by the values, symbols, the shared tacit assumptions that make up the culture of the organization and can act either as the real glue or as a disaggregating force among the human resources who provide school with its own identity, building its culture and acting within it.

The processes of integration and coordination take place primarily in those formal school structures, crystallized in some decades ago regulations and called by them "governing bodies": having a technical nature, such as the teachers' board, the class, interclass or intersection council, or a political purpose: the school board.

The teachers' board is, as a matter of fact, an organization in itself, as a multipersonal entity, composed of the school principal and the teachers. It carries out a legally significant action, fully appropriate to the aims of the educational system. Moreover, the same Latin word *collegium*, which indicated in the Roman society the plenty of the associative phenomena, perfectly sums up the definition of organization as a group of people who, through cooperation and order strives to the pursuit of common goals. All the school governing bodies have also their own structure, their own organizational chart, i.e. a system of roles within which each member performs his/her function, according to a precise juridical asset. Some examples can be found in the chairman, the vice president and the secretary. A sort of structure within the structure, as if it were, an organization within the organization. But we know that the organizational phenomenon does not recall the mere organizational structure, not only it. It is in the first place a social construction made up of a group of actors, a group who overcomes the limitations of each component, giving place, thanks to the human beings' efforts who deliberately act in it, to a new collective social entity called therefore collegial. Collegiality then becomes the result of a cooperative action endowed with the feature of willingness; it becomes interaction that is increased whenever the will to share, and by it the process of negotiation of meaning and significance occurs. To be put under discussion, to arrive at a vision sharing, is that same reality which is linked with the educational performance, goal of the entire school organization. Collegiality can become, in this sense, the expression of a stronger group unity, the true process of synthesis and identity construction for a professional category, in our case the teachers' one.

Schools are therefore organizations, but as Romei says "*the organization is the means that allows us, if not to rule, at least not to be completely overwhelmed by complexity; to deal with complexity as active players who wish to move into it with a leading role*". Nowadays the dominant paradigm is in fact precisely the one related to complexity. We live in a complex world, relations among men are complex too. Going on and on we could attribute the adjective complex to any event or reality. As already pointed by Gaston Bachelard in his book *The new scientific spirit* published in 1951, in nature there are no simple phenomena, as each phenomenon is a web of relations. But basically what does the complexity mean and how it is showed and again how to deal with it? Complex is the intrinsic property of a phenomenon irreducible to any description or explanation perceived as such by an observer (Maturana and Varela 1988), which is to say "*a qualitative property of phenomena that refers to a combination of plurality and independence at the same time, without being reucible to any analytical explanation*". If, then, it is the man who gives the phenomena, according to his perception the attribute of complex, "*ultimately things in themselves are not be simple or complex: it is the observer and his own system of observation who find them as such*", we must agree that complexity is not a structural property of the reality, but it is a way to see it, it is the man's attempt to understand reality. (Ceruti, Morin (1985); Prygogine and Stengers (1985); Von Foerster 1985; Lanzara 1993; Weick 1993, 1997). Since knowledge implies a building process by human beings, the reality where the man acts and that he knows is the one that he merely builds and recreates time after time, using, to say it with Romei, the "*magic wand of representation*". Again, the just mentioned author claims that the human being is complex by nature. It follows that complexity of the world is nothing but the projection of such a human quality. Organizations that, as we have argued so far, are composed solely of people, can be defined, for a kind of reflexive property, complex phenomena and schools are not the exception. Among other things it is the organization itself, according to the classical theory and particularly for authors such as Max Weber and Fredrick Taylor the human being's attempt to cope with complexity in rational terms, opposing this an apparatus of explicit rules, to reduce it and control it. However, if the purpose of human organizations is to avoid the "termite hill", paradoxically men create more and more complex organizations, increasingly similar to it (Kundera 1988). In addition, reducing any organization to prescriptive rules and principles, would acknowledge man as solely reason, disregarding the value of other qualities such as emotions which also play a key role in the processes of human interaction and knowledge building. As Morin argued, the human being is not just a *homo sapiens*, but a *homo demens*, an emotionally insane man, full of illusions, myths, madness, affection and instinct. The classical perspective, actually adds to complexity a new paradigm, that of rationality or better rationalization that would claim to reduce the complex phenomenon into its component parts, to be placed selectively under a sort of rank analysis, forgetting, therefore, to focus on the relations among those parts inextricably bound, "interlaced"—complexus means precisely to be interlaced, tied up—that shape the phenomenon as a whole. Moreover rationality and rationalization are not the same thing, rather they are diametrically opposed. In fact rationalization *"puts at the very forst place logical consistency, trying to dissolve the empirical elements, to remove them, to reject what does not comply with the rules, thus falling into dogmatism (…). It is not an accident that Freud used the term "rationalization" to describe this neurotic and/or psychotic trend where the subject is trapped in a closed system of explanations, without any connection with reality, albeit with his own logic. Somehow the big difference between rationality and rationalization is that the first recalls openness, the other just closure, the closing of the system itself"*. It seems that a reductionist approach to complexity, that seeks to order hierarchically decomposing the multiplicity of a phenomenon observed in the individual parts, divided in turn into small units and that tries to give explanation, from hierarchically highest level to the lowest, is to be avoided since unproductive. Similarly a descriptive approach seems counterproductive, if we take for granted the definition of complexity. *"The description collects empirical knowledge on the many shapes a phenomenon assumes when it is observed from different points of view, although with no correlation to a unitary pattern"*. The descriptive approach would not give reason of the complex nature of an event or phenomenon. Otherwise, a morphological approach would seem more appropriate to highlight *"the shapes that phenomenic complexity assumes in front of the observer. It is to go beyond the mere description of the complexity of a phenomenon, showing how the multiple dimensions that characterize the dynamic evolution are related to each other by generating forms of non-determination. The shape of complexity for a phenomenon is thus the shape which displays its discontinuity in a particular domain of observation, depending on the refinement values assumed in its characterizing dimensions"*. Indeterminacy and discontinuity refer to the assumption that one cannot give any explanation of the phenomenon in terms of predictability, abolishing complexity, but they assume a peculiar meaning:

"to make clear the irreducibility of the phenomenon to whatever straightforward explanation, showing, qualitatively, what is the shape taken by the phenomena in front of an observer".

The character of dynamism and uncertainty, or rather of possibility connotes, as part of the genetic approach (Maturana and Varela 1988) the relationships among the various components of a phenomenon which despite their relationship according to a perturbation–compensation pattern, retain their specific autonomy. Following this perspective, any explanation of the event or complex phenomenon, starting from its original multiplicity is carried out in terms of possibility. The benefit of this approach is that it does not pretend to be summed up in a complete and comprehensive formula, being linked to a partial and local dimension: the focus is just on some elements that characterize the autonomy of the components and on some of their ways to interact. It is true that living in a complex world means to be equipped with a "method to go on with complexity" (Romei 2000), a multidi-mensional approach that recognizes, as already noted by Bocchi and Ceruti that there is not a privileged access road to complexity but rather multiple pathways, since it is showed as never unique, but in multiple facts. An approach such as that suggested by Romei which uses continuous investigation as a leading technique to move between clarity and darkness, between certainty and uncertainty on the complex facts. And through the human being's reflective investigation, a human being nestled in the problematic nature of experience, starting to grab the first units of certainty, to define the lines of coherent actions. *"We must learn to sail in an ocean of uncertainties through islands of certainty."* A method, in his features, distinctly deweyan that starting from the problematic nature of the experience— which by its nature is a complex phenomenon, as it encompasses the whole of the phenomena of nature, objects, events, thought, conscience, in their mutual relations and transactions—places the human being in a doubtful state of mind. The man, through the observation phase comes to conceive some suggestions the most grounded of them, as checked in progress, will prove hypothetical solutions. And if not enough, in the unbroken chain of action, the man as reflective and thoughtful human being is called to acknowledge the temporary nature of some achieved certainties, that are just problematic situations able to generate suggestions for solutions to be put under a continuous checking process. Therefore defining an organization as a complex phenomenon means giving reason to the multiple shapes it can take in the observer's eyes. Not by chance the famous study of Gareth Morgan, published in a volume gives an account of the multiplicity and variety of shapes in which the organization can be perceived, depending on whether you focus on certain peculiar and recurring aspects, featured by the character of variety, autonomy and relation. The organization then becomes, as suggested by Romei, the paradigm of human behavior in front of complexity, a paradigm to be considered in the kuhnian meaning: as a well defined framework of exploration and scientific research, and as a set of tools for analysis and development necessary for research itself.

Conceiving the school as a complex organizational phenomenon puts us as observers, paraphrasing Bocchi and Ceruti, in front of a challenge that involves the ability of waiving the myth of certainty, completeness, exhaustiveness, order. A challenge, which requires a kind of *metanoia* that is, a change of mentality and mental models. It also requires the unconditional adherence to a "Fifth Discipline": a systemic thought that does not use straightness, but roundness, that does not lead to the perception of reality through straight lines, but which reveals itself as circles of influences, thus taking into account the system of interrelations, instead of linear chains of cause and effect. Moreover the systemic thought acknowledges the *unitas multiplex* principle necessary for addressing complexity in terms of dynamism and probability.

References

Bonazzi G., *Come studiare le organizzazioni*, Il Mulino, Bologna, 2002.

Calvani A., Rotta M., *Fare formazione in internet*. Erickson, Trento, 2000.

Ferrante M., Zan S., *Il fenomeno organizzativo*, La Nuova Italia Scientifica, Roma, 1994, (II edizione, Carocci, Roma, 1998).

Kundera M., *L'arte del romanzo*, Adelphi, Milano, 1988.

Lanzara G.F., *Capacità negativa. Competenza progettuale e modelli di intervento nelle organizzazioni*, Il Mulino, Bologna, 1993.

Maturana H., Varela F., *Autopoiesis and Cognition: the realization of the living*, Reidl, London, 1980, (It.trans. Autopoiesi e cognizione, Marsilio, Venezia, 1988).

Molinari L., Gabrielli M., *La professione docente nella scuola dell'autonomia. Funzione, stato giuridico, responsabilità*, Anicia, Roma, 2000.

Morin E., *Le vie della complessità*, in Bocchi G., Ceruti M. (a cura di), *La sfida della complessità*, Feltrinelli, Milano, 1985, pp. 49–60.

Prygogine L., Stengers I., *Order out of chaos*, Falming, London, 1985.

Rebora G., *Manuale di organizzazione aziendale*, Carocci, Roma, 2001.

Romei P., *L'organizzazione come trama. Fondamenti per la conoscenza e lo studio dei fenomeni organizzativi*, Cedam, Padova, 2000.

Spaltro E., de Vito Piscicelli P., *Psicologia per le organizzazioni. Teoria e pratica del comportamento organizzativo*, Carocci, Roma, 2002.

Von Foerster H., *Cibernetica ed epistemologia: storie e prospettive*, in Bocchi G., Ceruti M. (edited by), *La sfida della complessità*, Feltrinelli, Milano, 1985, pp. 112–140.

Weick K.E., *Sensemaking in organizations*, Sage, London, 1995, (It.trans.*Senso e significato nell'organizzazione. Alla ricerca delle ambiguità e delle contraddizioni nei processi organizzativi*, Raffaello Cortina, Milano, 1997).

Weick K.E., *The social psychology of organizing*, Random House, New York, 1969, (It.trans. *Organizzare. La psicologia sociale dei processi organizzativi*, Isedi, Milano, 1993).

Chapter 22
Advanced Technologies for Social Communication: Methods and Techniques in Online Learning

Roberto Salvatori

Abstract The speed of change and the spread of ICT (Information and communications technology) in Education puts the entire educational system in the position of having to constantly think about new methods and redesign teaching models adapted to a globalized and interconnected society, where knowledge is distributed, easily accessible and constantly updated. The massive introduction of technology in learning environments has not given the expected improvement in teaching, in fact, there has been a misuse of technology with a deterioration in school performance. Education must therefore aim to develop certain skills to enable a student to work independently usefully and effectively. The new systems and teaching models must be designed so as to shift the focus on the student. It is therefore necessary to work on different directions that are represented by the intrinsic motivation, extrinsic motivation, curiosity, favorable environment and distributed cognition.

Keywords New technology in education · Innovative technology · Information and communications technology · Massive open online courses · Online learning · Connectivism · Intrinsic motivation · Extrinsic motivation

22.1 Introduction

The next future will certainly be characterized by the presence of innovative teaching since in the last decade there has been a strong influence and pervasiveness of innovative technologies in society and in our daily lives. Already in 2007, a survey showed that 88% of students aged between 14 and 19 years of age, participated actively in "blog" and "Forum", to socialize and share information. The speed of change and the spread of new information and communication technology

R. Salvatori (✉)
University of Teramo Studies—Communication Science, Teramo, Italy
e-mail: rsalvatori@unite.it

R. Salvatori
"G. D'Annunzio" University of Chieti-Pescara, Chieti, Italy

© Springer International Publishing AG 2017
Š. Hošková-Mayerová et al., *Mathematical-Statistical Models and Qualitative Theories for Economic and Social Sciences*, Studies in Systems, Decision and Control 104,
DOI 10.1007/978-3-319-54819-7_22

puts the entire educational system in the position of having to think about new methods and redesign teaching models adapted to a globalized and interconnected society, where knowledge is distributed, easily accessible and constantly updated. Some experts in the field of education had speculated that the massive entry of new technologies in education would produce and lead to the necessary educational innovation. (Hoskova-Mayerova 2017). Based on these beliefs, different government institutions have invested heavily to equip several schools with innovative teaching. It appeared that the MIWB (multimedia interactive whiteboards, tablet, wireless devices to connect to the Internet) has spread the practice of ICT in education (Information and Communication Technology).

Unfortunately, the expected improvement in education has not occurred and, indeed, there has been a widespread misuse of the so-called innovative technologies and IWB (interactive whiteboards) used in "compatibility mode" to the traditional blackboard. It was then verified that a interactive educational technology does not guarantee an interactive lesson, reaching the absurd in which "you can do interactive lessons with traditional blackboard and not interactive lessons with interactive whiteboard" (Fierli 2010).

Research conducted on the use of media and new technologies in education seem to show that the use of the media does not improve the teaching and in some cases it makes teaching even worse. Analyzing the vast amount of research carried out recently in this field, it is concluded that the most important aspect is not on the type of media or its use in teaching, but is based on the quality of teaching. At first, and according to studies conducted by the government agency English BECTA (British Educational Communications and Technology Association) in primary and secondary schools, the interactive whiteboard (IWB) have improved teaching and learning.

Later it was found that the studies were methodologically questionable, since they were the result of observations or interviews with teachers and students. A probable cause that has misled researchers BECTA has probably originated by the wave of enthusiasm in the entry of new technologies in education certainly contributing to a temporary improvement of learning. A subsequent study made by Steve Higgins and collaborators, correlational (new technologies and school positive results) showed that the novelty of the new technology produces a temporary improvement. The research was carried out by comparing British schools with IWB and without IWB. In a first stage there was a distinct advantage of schools with IWB while in the long run worsened the performance, bringing schools without IWB to achieve better results. A great job of research in this area was completed by Hattie (2009) by analyzing nearly 52.000 studies on millions of students. The results showed, that, even with minimum differences, new technologies appear to be less effective than the traditional methods. This leads to the logical conclusion that the tool does not guarantee success in training and teaching, and the presence of advanced technologies in the fields of training does not ensure the quality of teaching. (Hošková, Mayerová and Rosická 2015; Hošková-Mayerová and Potůček 2016).

22.2 Reflections Didactic-Pedagogic in the Digital Age

It should certainly be paid attention to the introduction of new technologies in education. The teacher must have solid skills in pedagogical-didactic because the traditional lecture is reduced and encourages greater student autonomy. Education must therefore aim to develop certain skills to enable a student to work independently, usefully and effectively. A useful classification was obtained by David Rumelhart and Donald Norman (Rumelhart and Norman 1978), which identifies the necessary ability that the student must have for a scientific approach (be able of abstraction, be able to think critically, be capable of scientific imagination, be able to transfer, be able to treat information mathematically, be able to do analysis and synthesis, adequately consider the technical languages, tap into the knowledge to fill gaps, awareness in distinction between common sense and science).

From the perspective of pedagogical-didactic, it is unthinkable to adopt only traditional methods and patterns, ignoring the possibilities, inherent in new technologies, to facilitate the predisposition for learning which is the starting point or essential condition for effective learning and durability. A very interesting reflection of John Vailati, more than a century ago, (May 1906) shows with amazing foresight, the critical point of the current education system. It reads, in fact "educated men, teachers, students of pedagogy that would reject the proposal to engage with terror, if only for a week, to attend three conferences a day, one after the other…" especially on issues of great interest to them; yet they "do not see the absurdity of teaching, that force the students from 10 to 18 years old to remain nailed on average for 5 h, during all their years, to school, as if there were no other systems or means to achieve the same results" (Vailati, 1906).

Without obviously exaggerating, it would be desirable to insert in teaching today, moments of reflection and study, supported effectively by the new technologies, MOOC (Massive Open Online Courses), OER (Open Educational Resource) and other models of online learning, in which the student is left free to explore, experiment and learn in the great network of knowledge represented by the web with the help of new communication technologies. In this regard, it is enlightening reflection of Jerome Bruner, with regard to education and teaching machine programmed: "… it is clear that the machine is not intended to replace the teacher; indeed it may lead to the request for a greater number of teachers, and best teachers, when the heaviest part of the teaching is entrusted to the mechanical means. Neither seems a justified fear that the machine can dehumanize the educational process more than they do books: the program for such machines can be educational staff like a book, it can be composed with the spirit or be stupid, it may be enjoyable or boring as an arid exercise" (Bruner 1960). It therefore appears necessary that the world of education analyzes problems and compares them to current generations, keeping in mind the pervasiveness of new technologies that have influenced and changed the forms of communication, leisure, study and work, (Facci et al. 2013).

22.3   Online Learning Environments and Open Educational Resources

One thing to keep in mind, as noted by Gonzalez (2004) for the "half-life time of knowledge" regarding the length of time between the information obtained and its obsolescence. Half of the information known today was unknown a decade ago. The amount of information in the world has increased dramatically and, according to studies by the American Society of Training and Documentation (ASTD), it doubles every eighteen months. Our society is liquid and complex, characterized by the dynamics of the information ephemeral life in which the mind has to constantly rework the information to fit the new contexts.

To combat this phenomenon many educational organizations have adopted systems of e-learning at a distance and special methods for the dissemination of information.

Even in Europe there are many innovative initiatives in terms of learning and open educational resources with a number of projects proposed by the European Commission.

In particular on September 2013 it started an ambitious project called "Open Education Europe" to ensure shared access and to open the educational resources (OER) in Europe. The goal is to share open educational resources in Europe, even in different languages, with students, teachers and researchers and to investigate (via a portal in this area of research) issues relating to new technologies in teaching, through the consultation of research, events and experiences in place as well as the reading of the major magazines worldwide.

Today you can take a free course at Harvard, Yale or Stanford without the need to travel to the United States by taking advantage of the system MOOC (Massive open online courses) simply registering in the specific area and benefit from the educational materials made available to members (thousands of videos, handouts, tests, projects, experiments, video-lessons).

The primary issue is on the teaching model to follow. It should, in fact, point out that in this context are found obsolete and inadequate teaching methods (educational technology and multiple network resources) for the digital environment.

It would be desirable to set up specific learning environments that take into account the group dynamics that are created outside of the specific content acquisition through the use of forums, the community in general and resource sharing through social networks. So be careful, as indicated by Calvani, to adapt technology to the needs of students and assist the design and preparation of learning environments, suggesting suitable mediators (technical devices, regulatory, human or otherwise) for careful expression, personal growth, social and cooperative construction of knowledge (Calvani 2008).

The Teaching today, on the contrary, considers the student like a container to be filled in with information by focusing all the attention on the teacher who is the keeper of knowledge and from whom exclusively take the basics. Despite the innovative forces of modern pedagogy, learning is almost always centered on the "theme" and not "the problem", so learning attempts more or less braves remains based on the relationship between teacher and student and not the technical of "communicating vessels".

Whereas the modern teaching enhances the student putting him in the centre, as a live person with the desire to learn, learning to learn for life long learning (Trentin 2011). The spread of the improvised teaching practices leads the student to the false belief that the objective has been achieved instead it is necessary to re-elaborate the knowledge, personalize it, integrate the information learned.

22.4   Semplessità

According to Berthoz, contemporary society characterized by excessive propensity to the complexity, helped to develop in humans a special ability: the "semplessità" that, as reported by the author "is this set of solutions found by living organisms that, despite the complexity of the natural processes, the brain can prepare the act and anticipate consequences". The concept of "semplessità" is the decrypting the complexity through the application of specific rules that simplify the thinking and behavior. With this technique, man can successfully face complexity in its different forms.

Studies carried out by Berthoz (2011) lead to a finding that the mental processes performed by humans to solve problems during its evolution, greatly affect higher cognitive functions (memory, reasoning, creativity, etc.). He says that "without a doubt every time that our brain anticipates action appears a different state. And as the perception is always simulation of an action in the world, conscious perception is always an anticipation of some event that will occur in the world, regardless of the fact that the event is produced by the subject percipient or not". The semplessità requires a process made up of several factors among which we can cite the inhibition, selection, connection, imagination to achieve a specific purpose.

Some of these tools can be applied for teaching such as:

(1) Speed (the assumption of possible development of an action);
(2) The separation of functions and modularity (simplification or temporal differentiation);
(3) Memory (comparison between past and present);
(4) Reliability (more redundant solutions assessing their consistency).

Some of these principles have been applied to a new teaching model created by center for Research on Education to the Media Information and Technology (CREMIT) directed by Rivoltella, through episodes of Learning Located that develop into three distinct phases: first an anticipatory time (useful to provide adequate stimuli); Later operative (troubleshooting with the technique of problem-solving); and the last step ristrutturativo (pointing and reviewing what has been developed through the two previous times). With the intention of looking for a suitable model for teaching, strongly influenced by technology and by network, George Siemens has formulated a theory of learning based on the new educational needs and by noting the various changes that have occurred with the introduction of new technologies and in general, the special connections and relationships in the digital age. His theory, called "Connectivism", is the combination of the main theories of learning with the integration of recent issues raised by the new technologies of communication. As claimed by Bruner, knowledge is "make sense" that is a creative decoding personal to understanding the reality. Seymour Papert said, "the goal is to teach in a way that offers the greatest learning with a minimum of teaching. (…) The other necessary fundamental change reflects an African proverb: If a man is hungry you can give him a fish, but even better is to give him a fishing line and teach him to fish". In the search of information on the network you need to know the place where to draw the information but, for an effective learning, you need to have the critical thinking skills to analyze the information and find the knowledge of the possible implications of the formal and informal aspects that can be created.

22.5   Connectivism

The method "connectivist" of Siemens (2006), compared to behaviorism, cognitivism and constructivism, examines the specific learning processes of the digital society through "the exploration of forms of connective knowledge and learning mechanisms are the magnifying glass through which study trends of learning and knowledge".

The connectivism was born under "the pedagogical thrust" of philosophers and educators of previous generations Piaget, Dewey, Papert and her introduction of tune with the digital society.

All theories have links with the experiences and knowledge of the past, the connectivism is related to socio-cultural theories of Vygotsky, "on affordances" of James Gibson, Wittgenstein understanding negotiated Postman technology that embodies an ideology Papert on the importance of practice, Bruner for self-efficacy in learning, on the collective intelligence of Pierre Levy, Marshall McLuhan for the relationship with the new technologies.

The Cognitivism applies the same principles of the of the networks theories in terms of knowledge and learning processes, including technologies as an integral part of the knowledge and cognitive processes.

The Connectivism highlights the fluidity of knowledge and of information on the network as well as the speed and dynamic content, and seeks, through the network, to give a logic sense and coherence to the various information and knowledge present in an extremely fragmentary and disjointed. Several educators and industry experts have criticized the model of Siemens finding little rigorous part of learning as there is a predominance of technologies at the expense of the educational aspects.

Calvani is of the opinion that "a transfer of wild Connectivism to school can lead one to believe that it is enough to put the students in the network to produce knowledge, consolidating widespread stereotype, that most technologies are used, in any manner you face, the better for learning".

From these considerations it is clear that there is a certain interest in the teaching methods to be employed in a digital society, characterized by the changing sources of information, in order to obtain a profitable and stable learning.

In order to improve teaching methods with new technologies must act primarily on the motivational profile of the students working on intrinsic motivation (curiosity and the need for competence) (Rosická and Hošková-Mayerová 2014a, b).

22.6   The Curiosity

As emphasized by Berlyne (1960) in front of the placing of conflicting information with those in our possession, it is activated a mechanism for the reduction of this uncertainty. In a research carried out by Berlyne and Frommer (1966), it was noted that two similar stories but with different characters attracted the curiosity of the students in a very different way, benefiting the stories with characters having fanciful strange names.

It should be emphasized that the uncertainty not always pushes to look for information, leading to a total disinterest motivational when the level of uncertainty is too high or too low. Therefore, there is an optimal level of stimulation (Hunt 1965), where curiosity has a regulatory role and through an appropriate dosage, both qualitative and quantitative information in inputs. The relationship between the conflicting information and the activation of mind is well known since ancient times and has always been considered the essential element in the scientific and philosophical investigation. Plato in the Theaetetus (155d) has Socrates say (Bianchi and Di Giovanni 2007) that the philosophy does not begin if not with wonder. Aristotle in the Metaphysics (928B) states that through the wonder men have started down the road of philosophizing. Comte says that the sciences respond to the "fundamental need to feel our intelligence to know the laws of the phenomena and to feel how this need is deep and imperious, just think for a moment of the psychological effects of wonder". Popper states that "every problem comes from the discovery of an apparent contradiction between what we know and what we believe one facts". As we can see in Plato, Aristotle, Comte and Popper, have the same common denominator: curiosity. Curiosity is therefore a very important lever of social influence, should not be underestimated, particularly in the use of training in general but especially in the online learning systems type MOOCs or other training methods.

22.7 The Motivation

As it is clear from the international literature (White 1959; Hunt 1963; Pittman et al. 1982; Deci and Ryan 1985; Vallerand et al. 1993) motivation can be further split into extrinsic and intrinsic factors.

Extrinsic factors are all those actions carried out for its own benefit or advantage, external type and fueled by environmental factors, while the intrinsic elements are supported by the acts or behaviors carried out for the pleasure and satisfaction obtained "feeds" the factors internal to the person. So it is possible to infer that external factors must be continually sought and encouraged through appropriate reinforcements while the internal factors are "self-supporting".

As experienced in three years 2010/11–2012/2013 (Salvatori 2015) and as also supported by Maehr (1976), the permanent motivation (continuing motivation) should be used in all training program particularly in the post-graduate, as it stimulates the motivation even after finishing the course of study.

In training through systems e-learning or systems MOOCs act it surely the extrinsic motivations as there is a possibility of career enhancement, obtaining the qualification, social relations, ways of improving working conditions, etc. while is it certainly necessary to support and encourage the intrinsic motivations through the method of research (Suchman 1962) or Socratic method (Collins 1982). Fall into intrinsic motivation (Bianchi and Di Giovanni 2007) all the recreational-cognitive (curiosity and the need for competence) and realistic-social (need for affiliation and need for achievement). The students primarily enroll for a self-interest dictated by environmental reinforcements (extrinsic motivation), for the materials, forms of interaction (intrinsic motivation—curiosity) and to test themselves (intrinsic motivation—need for competence) to relate to other students (intrinsic motivation—need for affiliation) and to succeed and realize their projects (intrinsic motivation—need for achievement).

In a system of online learning the motivation is crucial because as evidenced by many studies and research in the field, the "dispersion of the students" is really high (can reach over 80%).

They can be very useful the following aspects (Bianchi and Di Giovanni 2007):
Keep alive the curiosity (through unusual information);
Provide feedback to students;
Goal setting (guide the student toward calibrated objectives pursued);
Check the attributions of success and failure;
Check implicit theories of intelligence (to increase the pressure to study).

22.8 Conclusions

The new systems and teaching models must be designed so as to move the focus on the student, and especially, for using it in an appropriate manner in the network, more and more dynamic and interactive, full of services and information, in order to achieve top involvement of the students, their re-elaborative ability and learning.

We need to work on different directions that are represented by intrinsic and extrinsic motivation, by curiosity, by a favorable environment and distributed cognition. This teaching mode therefore requires a great effort by both the teachers and students, but it is essential to prepare the next generation to deal independently with the complexity of contemporary problems and preparing them favorably to interact with others within a community in which the value represents an element of growth for the benefit of the distributed intelligence.

References

Bauman Z., Vita liquida, Laterza, Roma, 2008.
Bauman Z., La società dell'incertezza. Bologna, Il Mulino, 1999.
Berlyne D.E., Conflict arousal and curiosity. McGraw-Hill, New York, 1960.
Berlyne D.E., Frommer F.D., Some determinants on the incidence and content of children's questions. Child D., 1966.
Berthoz. A., La semplessità. Codice edizioni, Torino, 2011.
Bruner J., La ricerca del significato. Per una psicologia culturale, Bollati Boringhieri, Torino, 1992.
Bruner J., The Process of Education, Cambridge, Harvard University Press, 1960.
Bianchi A. P. Di Giovanni, E. Di Giovanni, Empowerment. Che cosa vuol dire? Really New Minds, 2015.
Calvani A. (2008) Connetivism: New paradigm or fascinating pout-pourri?, Je-LKS, pp. 247–252, 2008.
Collins W.A. (1982) Cognitive processing in television viewing, in D. Pearle all. Television and Behavior: ten years of scientific progress and implications for the eighties, Washington D.C., Us Government Printing Office.
Di Giovanni P., Psicologia della comunicazione, Zanichelli, 2007.
Deci E.L., Ryan R.M. (1985) Intrinsic motivation and self-determination in human behavior, New York, Plenum Press, 1985.
Facci M., Valorzi S., Berti M., Generazione Cloud, Erickson, Trento, 2013.
Fierli M., (2010) Le tecnologie nella scuola: che cosa si dice e che cosa succede davvero, Rivista educational 2.0, RCS libri.
Gonzalez C., The Role of Blended Learning in the World of Technology, http://www.unt.edu/benchmarks/archives/2004/september04/eis.htm (Retrieved 4/04/2016).
Hoškova-Mayerová Š. and Potůček R., (2016) Qualitative and Quantitative Evaluation of the Entrance, Draft Tests from Mathematics Studies in Systems, Recent Trends in Social Systems: Quantitative Theories and Quantitative Models, Decision and Control, Vol. 66, Maturo (Eds.), 53–64.
Hoškova-Mayerová, Š., Rosická, Z. (2015). E-Learning Pros and Cons: Active Learning Culture? Procedia - Social and Behavioral Sciences, 191, no. June 2015, pp. 958–962.

Hošková-Mayerová, Š. (2017) Education and Training in Crisis Management. In: *The European Proceedings of Social & Behavioural Sciences EpSBS, Volume XVI.* Future Academy, 2017, p. 849–856.

Hattie J. Visible learning: a synthesis of over 800 meta analyses relating to achievement. London-New York: Routledge, 2009.

Hendry L.B., Kloep M. (2002) Lifespan development. Resources, challengers and risks. Thompson Learning: London; trad. it. Lo sviluppo nel ciclo di vita. Bologna: Il Mulino, 2003.

Higgins S., Falzon C., Hall I., Moseley D., Smith F., Smith H., Wall K 'Embedding ICT in the literacy and numeracy strategies: Final report.', Project Report. University of Newcastle, upon Tyne, Newcastle, 2005.

Hunt J. Intrinsic motivation and its role in psychological development. In D. Levine, University of Nebraska Press, 1965.

Hunt J., 1963, Motivation inherent in information processing and action. In O.J. Harvey (ed.) Motivation and socila interaction, Ronald Press, New York, 1963.

Maehr, M.L. (1976), Continuing motivation: An analysis of a seldom considered educational outcome. Review of Educational Research, 46(3), 443–462, 1976.

Mariani L., Saper apprendere. Atteggiamenti, motivazioni, stili e strategie per insegnare a imparare, Edizioni libreriauniversitaria.it, Padova, 2010.

Papert S., I bambini e il computer. Milano, Rizzoli, 1994.

Pettenati M.C., Cicognini M.E., Guerin E., Mangione M.R. Personal Knowledge management skills for lifelong-learners 2.0, Hershey, PA, IGI http://www.igi-global.com.

Pittman T.S., Boggiano A.K., Ruble D.N., 1982 Intrinsic and extrinsic motivational orientations: Interactive effects of reward, competence feedback and task complexity. In J. Levine e M. Wang Eds Teacher and student perceptions: implications for learning. Hillsdale (NJ), Erlbaum, 1982.

Rosická, Z, Hošková-Mayerová, Š. (2014a) Motivation to Study and Work with Talented Students. *Procedia - Social and Behavioral Sciences*, 114, no. February/2014, p. 234–238.

Rosická, Z, Hošková-Mayerová, Š. (2014b) Efficient Methods at Combined Study Programs. *Procedia - Social and Behavioral Sciences*, 131, no. May/2014, p. 135–139.

Rumelhart D.E., Norman D.A. Accretion, tuning and restructuring, Three modes of learning, in Cotton J.W., Klatzky R. (eds.) Semantic factors in cognition. Erlbaum:Hillsdale (NJ), 1978.

Rumelhart D.E., Norman D.A. Analogical processes in learning, in Anderson J.R. (ed.) Cognitive skills and their acquisition. Hillsdale (NJ): Erlbaum, 1981.

Salvatori R., Tecnologie e-learning e mutamenti sociali, Edizioni Complexity, 2015.

Siemens G., Knowing Knowledge, http://ltc.umanitoba.ca/wikis/KnowingKnowledge/index.php/Main_Page, 2006.

Suchman, 1962, in The Block Scheduling, J. Allen Queen, 153–155, 2009 Corwin Press, 2009.

Trentin B, La didattica in pericolo, Rivista Educazione e scuola, 2011.

Vailati G., Rivista di Psicologia Applicata, pp. 160–166, Bologna, 1906.

Vallerand R. J., Pelletier L.G., Blais M.R., Briere N.M., Sencal C., Vallieres E.F., 1993, On the Assessment of intrinsic, extrinsic, and a motivation in education: Evidence on the concurrent and construct validity of the Academic Motivation Scale. Educational and Psychological Measurement, 53, 159–172, 1993.

White R., Motivation reconsidered: The concept of competence. Psychological Review, 66, 297–233, 1959.

Chapter 23
Social Distance as an Interpretation of a Territory

Valentina Savini

Abstract This contribution deals with the concept of social distance. The physical and spatial aspects of social distance identified by Simmel disappeared from the theoretical setting when the issue was analyzed in America by the Chicago School. The approaches that combine the physical and relational dimensions, based on Italian and English sociologists of the last few years, have to be preferred. The main objective of the research on social distance is to go beyond Bauman's paradigm of liquidity—without groping to reconstruct the social complexity—and think of society as the physical and symbolic place that comes to life and reproduces the social structure. The conclusions that can be drawn from the survey introduced in this chapter relate to (a) the perception of social distance with regard to some significant variables; (b) the social distance as propensity to get closer or farther away the other, especially the local Gypsies and foreigners in general; (c) the construction of complex indexes (index of innovation/traditionalism and index of social background) which allows to study the composition of the social structure of Pescara and the attitude that people seem to manifest.

Keywords Social distance · Inclusion · Exclusion · Prejudice · Social space

23.1 Introduction

Social distance is a classic concept of sociology and it is closely related to that of space. The space, which has always fascinated many sociologists, today is particularly interesting because of two aspects. On one hand the new means of communication and transport seem to erase distances, and on the other hand there is a close relationship between human activities and spaces. Our social practices contribute to the configuration of the spaces. If it is true that it depends on our ways to shape the

V. Savini (✉)
Dipartimento di Economia Aziendale, Università degli studi G. d'Annunzio
Chieti-Pescara, Viale Pindaro 42, 65127 Pescara, Italy
e-mail: valentina.svn@gmail.com

© Springer International Publishing AG 2017   343
Š. Hošková-Mayerová et al., *Mathematical-Statistical Models and Qualitative Theories for Economic and Social Sciences*, Studies in Systems, Decision and Control 104,
DOI 10.1007/978-3-319-54819-7_23

space around us and, as in a kind of dialectical relationship, it is also true that the socially structured spaces contribute to reproduce certain human practices. When the individual moves within a space he/she must confront the material dimension, given by the physicality of the objects that compose it, and the affective dimension that emerges from the lived in a given environment. The relational capacity of the space is such that it is simplistic to indicate a single term that surrounds us. Living in space is to act and interact, relate to others and pursuing goals. We can talk about places and distances to indicate different aspects of the same object of study.

Over the years social distance has been studied several times from different points of view. There are two main approaches to the study of social distance: the American one developed during the Twenties, when studies of the Chicago School concerned the issues of ethnic, linguistic and religious affiliation, and differences between white and black cultures are also seen in their ways of organizing the space around them; the second approach is that of the late nineteenth century when European thought was focused on distances generated by the construction of systems of inequality of historical-philosophical-economic matrix.[1]

Many years have passed without significant sociological contributions to the study of social space and social distance. We can only recall the recent work of some English scholars and an Italian research group to which this survey is inspired.

This chapter aims to describe the evolution of the approaches to the study of the space and the social distance concept and to summarize the main achievements of the author's Ph.D. thesis: an empirical research conducted in 2014 in Italy, specifically in the city of Pescara.[2]

23.2 Sociological Definition of Social Distance

23.2.1 Georg Simmel's Sociology

The literature agrees in considering Georg Simmel the first and the most comprehensive theorist of social distance (Simmel 1908). This is because he was also the

---

[1]European scholars had no doubts about the existence of a class system: the main theoretical references are Karl Marx and Max Weber who studied the concepts of class stratification and mobility. The differentiation of individuals and their classification class was under their position or status ascribed and depending on the mode of production prevailing. The European social structure was always dual: oppressor/oppressed, owners/workers, white-collar/blue-collar workers. Inequality had economic foundations—according to Marx (1921)—or was considered as a phenomenon of the distribution of political and economic power—according to Weber (1934).

[2]Pescara is the major city of the Abruzzo region of Italy. Located on the Adriatic coast at the mouth of the Aterno-Pescara River, the present-day municipality was formed in 1927 joining the municipalities of Pescara, the part of the city to the south of the river, and Castellammare Adriatico, the part of the city to the north of the river. The surrounding area was formed into the province of Pescara.

theoretician of social space: space is an element through which the incessant flow of daily life crystallizes in social forms (such as institutions, social relations, artistic works, etc., in other words, culture). But space is not a shape in itself: it produces social forms. Spatial forms are those social relations that materialize in space.

According to Simmel, the space as a form has five basic qualities. The first is the exclusivity, namely the fact that each point in space exists only for those accesses to it. Based on this element, Simmel defines two pure types of social formations: the spatial formations, for example, the State and the over-spatial formations such as the Church. Among them is an intermediate phenomenon, which is the city that has its own space, but the importance of a city often goes beyond its geographical borders. The second characteristic is given by the boundaries. The boundaries for a social group is like a frame for a work of art because defining space helps to make sense to the content. It delimits, fixes and informs the processes and relationships that live within the group, defining its membership. The principle of fixing stresses the importance of having a landmark in space around which stabilize the social life —individuals and roles—in order to ensure the continuity of the society. Other elements that characterize the space, important for the purposes of this research are the proximity/distance and the mobility. The categories of closeness and remoteness are probably more related to space and greatly affect the social relations. To be near or to be far spatially—or physically—it involves creating different social relations: the proximity is tied hand in glove with the perception and the senses; the distance, on the contrary, is related to intellectual or mental processes. There seems to be a real sociology of the eye, the smell and the hearing, explicated in detail in the *Excursus on the sociology of the senses*, in which each sense is recognized by the capacity to promote the association with other individuals or, on the contrary, the ability to separate people.

However, the distance, regarding the complete absence of stimuli, repulsions and attractions, assumes the prevalence of intellectual processes. A relationship can survive the physical distance of its members only under certain conditions. In fact the minimum requirement is that members of the association share a very high cultural level, therefore the abstraction allows us to relate with others even if not physically present here and now; in the case of no or slight intellectuality, the spatial distance adds social distance. It occurs also that closeness and emotional distance travel separately: their link disappears if the parties involved are separated by different cultural levels—this is because the intellectuality interposes men—or if they live in the big city; the outcome is always the same: indifference and exclusion against those people who, despite being physically close, are emotionally distant. In the first case, the indifference is caused by intellectuality that, excluding the impulsiveness of sensory response, manifests itself in a cold and detached behavior; but, in the case of the modern city, the multitude of contacts with innumerable people, or the excess of stimuli that characterize city life, causes the spacing, which becomes a mere instrument of protection.[3]

---

[3]Generating in this way what the author calls *"blasé attitude"*.

The last quality of the space is mobility that is the ability of men to move from one place to another. Simmel associates the study of mobility, considered the main quality of modernity—like incessant flow of life, to the idea of movement, the phenomenon of nomadism, by comparing sedentary society with those that migrate and using as criteria of comparison the need for differentiation of social experience content. Mobility also relates to the theme of proximity because in the *Excursus on the foreigner* changes the way to understand the closeness because of the stranger who shows this aspect, because he is not part of the group from the start—but he arrived later—he is both near and far: member of the group but at the same time not "native". This implies that to the stranger is given a different social personality, because he/she is capable of being in the group and at the same time distant from it, so it can be "objective".

In the formal sociology of Simmel "social distance plays a dual role, both descriptive and epistemological. Simmel thinks the society as a whole, the outcome of distancing processes in which social and spatial factors are intertwined to give a community of individuals a particular form and a particular order in the relationships and interactions between groups and subjects" (Cesareo 2007: 11). Therefore, Simmel considers the social distance as a structural and multidimensional phenomenon because distancing processes are determined by social and spatial factors, that intersecting themselves, they give life to a community of individuals, which in turn relating themselves activate the phenomena of physical/spatial and social approaching and distancing.

23.2.2 *Empirical Contributions from the Other Side of the Ocean*

Contrary to other authors such as Marx and Weber, Simmel has never started a real paradigm. This probably because he differed from the philosophical tradition of his time. But he strongly influenced other scholars, such as Robert Park, one of the founding fathers of the Chicago School. In the crossing of the Atlantic Ocean, the complex simmelian theory, able to stand out among all due to the interpenetration of the physical, symbolic and geometric dimensions of space, lost some essential features. In the operational translation of the concept the Chigagoans left out an important element that is the ability of co-production between social distance and urban context (Park 1925; Bogardus 1925a, b). They lost sight of the "constructivist" vision of Simmel and with Emory Bogardus the theoretical bases passed to be more "psychologist".

Emory Bogardus is remembered for the construction of the famous social distance scale. According to this author, social distance could be defined as the lack of sympathy and understanding that continues to exist even when the physical distances have been eliminated, as for example in the city, where people live in close proximity to each other. The main significance of the social distance lies in its connection with the maintenance of a status: once you reach a certain status, this will be hardly abandoned, people will fight hard to keep what they earn. Despite the spatial proximity, in the typically American metropolitan city of the Twenties, there were huge social distances, found in *"The existence of boys' predatory gangs, of high juvenile-delinquency rates, and of crime waves in cities is an index of social distance. Race riots are chiefly urban phenomena revealing social distance. Descriptions of the large city as the "lonesomest spot anywhere", or as "the most unsocial place in the world", are expressions of social distance"* (Bogardus 1926: 40–41).

After a very long process of trial the final scale took these appearances:

1. "Would marry
2. Would have as regular friends
3. Would work beside in an office
4. Would have several families in my neighborhood
5. Would have merely as speaking acquaintances
6. Would have lie outside my neighborhood
7. *Would have live outside my country"* (Bogardus 1933: 269).

During the Twenties, Bogardus was in charge of the studies on ethnic coexistence and therefore the concept of distance was closer to (ethnic) prejudice than the one formulated by Simmel. This approach assumed that the processes of distancing depended on those a priori that characterize every culture, that is, the processes of categorization. Simmel believed that the social distance depended on the categorization process: the type of relationship that we want to establish with someone will depend on the categories, these a priori, in which we put the individual, through which we automatically attribute him a certain identity. This implies that the researcher should consider the socio-cultural connotation of the company—in the sense of community—or group of which he will study the process of distancing. Bogardus instead neglected the role of knowledge and its social construction in the process of creation of social distance. He disregarded the categories necessary to the process of the other's categorization, from which depend the processes of spacing/approaching, since it forms itself at the level of the social construction of knowledge.

Contemporary research try to take on the original concept of social distance, taking care not to neglect any of the elements considered essential, that is relational and spatial (in the sense of organization of social life in space and its management). In one of the latest studies on the recovery of the real simmelian sense of the concept, the distinction between psychologist and structural meaning is expressed in these terms:

At one extreme, Bogardus (1925) emphasizes subjective social distance and refers to the social approval or prestige of various social groups, as measured by the perceived level of intimacy (neighbourliness, friendship, marriage and so on) that respondents would find acceptable with individuals from different national, ethnic and religious groups. Similarly, Park distinguishes the degree of intimacy that occurs between individuals from different social groups: 'The degree of... intimacy measures the influence which each has over the other.' (Park 1950: 257). [...]

At the other, structural, extreme, Sorokin (1927) emphasizes not social distance, but social space. However, the two concepts are complementary; distance implies a space of some kind, while a space is defined by the objects that are located in it and the distances between them. These different perspectives open up the possibility of a more precise and consistent account of individual action and social structure, in which actual relationships of intimacy are used to measure the degree of social distance between individuals and groups and thus of the social space in which the distances occur. Far from being no more than a metaphor, interaction distance can be empirically identified by mapping relations of proximity (Bottero and Prandy 2003: 179).

The main objective of all present researches on social distance is to move beyond the liquidity paradigm of Bauman (2000)—without trying to reconstruct the social complexity—and consider the society as a physical and symbolic place that comes to life and reproduces the social structure (Cesareo 2007; Bichi 2007; Frudà 2007; D'Amato 2009; Fantozzi and La Spina 2010; Pascuzzi 2010; Tacchi 2010; Coco 2011, Hošková-Mayerová and Rosická 2015).

23.3 The Empirical Research

The interest in the analysis of social distance in the territory of Pescara moves from the major social and structural changes which were brought by the imbalance of industry that have greatly modified the city. Pescara, despite not being the regional capital, is the largest city of Abruzzo and also the one with the highest concentration of population in urban areas. The city has undergone a massive urbanization and has experienced a strong process of socio-economic development since world war II. It expands, seamless, over a vast territory that borders the boundaries of the other provinces, and that presents an heterogeneous social structure. The situation, which looks fluid and various in demographic and socio-economic terms, has produced often phenomena of social anomie and has always been a source of conflicts that are increasingly taking on, extreme forms of intolerance and sometimes of real social deviancy.

23.3.1   Methodology

The Italian sociologist Gianni Statera wrote that two fundamental principles can be or should be drawn from the empirical research of the past and that they are at the basis of any investigation: "(1) A wide-ranging research project requires the integration of descriptive dimension with the explaining one of the research drawing;

(2) consequently, the adoption of a single detection technique (experiment, or systematic observation, or interview) is insufficient. Resulting usually necessary to resort to differentiate and coordinate tools" (Statera 1990: 73).

In other words, research that aspires to be broad and comprehensive should always connect the descriptive dimension to that for explanation of the research design and, therefore, use different detection instruments but connected together. Normally at this end we resort to what is usually called "triangulation", that is the use of three different levels of analysis that in the specific case of social distance were: research background, informal interviews with privileged witnesses and finally the collection of quantitative data through a standardized questionnaire.

The search for social background was carried away from the existing literature on the subject and on those related: the contributions of the Chicago School, from Park to Bogardus, then gradually more recent contributions to the current issue of the concept of social distance, as it has been considered by the national Italian survey "the social distance in Italian urban areas" (Cesareo 2007; Bichi 2007; Frudà 2007; D'Amato 2009; Fantozzi and La Spina 2010; Pascuzzi 2010; Tacchi 2010; Coco 2011). Besides we considered the principles of social distance and those related to it, such as space and social mobility. The city finally has been studied both from the theoretical point of view and as a description of the socio-environmental characteristics of the place where the phenomenon studied comes to life.

We used informal interviews with key informants to collect part of these "background" information. They were local politician, intellectuals and academics. With each of these experts we have tried to deepen the knowledge of some topics related to research: the history of the territory on which the city of Pescara has arisen; the evolution in terms of urban development of the city; socio-economic characteristics of the main neighborhoods, or more suitable areas of residence; the main features of the phenomena related to the presence of Gypsies and foreigners.

The testimonies collected helped to frame the problem and to define accordingly the dimensional areas that constitute the phenomenon of social distance, calibrating it to the city of Pescara, the survey scenario. Besides, whether it be a research mainly explanatory or mainly descriptive, their common denominator is the identification of the dimensions that make up the problem that needs to be explained. In fact, as Statera writes: "the problem of conceptualization is crucial not only from the logical point of view, but also with regard to what is perhaps the most productive task of the social researcher: measurement" (Statera 1990: 86).

The term "measurement" is used by Lazarsfeld and Statera: even if we do not achieve an objective measure, that is a "number", in sociology a classification as well may mark an important step forward in the analysis of complex social phenomena.

The methodology proposed by Lazarsfeld, through some logical steps, makes starting from conceptualization of a complex and abstract problem to reach its concrete measurement. The social distance has been subjected to this procedure of simplification. In a first phase we identified the dimensions and then the indicators that allowed us to give form and substance to the social distance in Pescara.

In addition to basic sociological data, the dimensions identified are the social distance acted, the social distance experienced and perceived social distance; the problem area concerning neighborhood issues; prejudice is the dimensional area for measuring any stereotype against the Gypsies, foreigners and a list of figures, or categories of persons, disadvantaged such as ex-prisoners, ex-alcoholics, ex-drug addicts. Finally, the area of social personality includes the variables that pertain to cultural consumption, values, religion, political orientation and the future vision for their lives and the world.

23.3.2 Questionnaire

In order to investigate the processes of approaching and distancing of Pescara citizens, we administered a structured questionnaire to a sample built on purpose. Structure and contents of the questionnaire had traced the dimensions previously identified. Indeed, the questionnaire consists of many dimensional areas (Lazarsfeld and Barton 1951) such as socio-demographic data, social distance (perceived, acted and experienced), stereotypes and social personality. In the first area, called "the basic sociological data", we collected data on variables like gender, year of birth to verify the age, place of birth with the province, marital status, level of education, employment typology and employment status, level of education and employment of the father and mother of the respondent; we also asked about children and the condition of cohabitation. In the second problematic area we turn to detect the three dimensions of social distance: perceived, acted and experienced. The first type refers to the distance or proximity of respondents express against a number of subjects, for each of which were required to give a score from 1 (highest proximity) to 10 (maximum distance). Then, to simplify the reading of the data, the scores were added together in order to return only two modalities: near-far. The acted distance is the distance we place between ourselves and others. To detect we used two questions: the first attempts to capture the data by querying the subject of how to behave towards that person you do not want to have anything to do with and the second question is asked to give a reason. The same thing happens for the experienced distance, that is the perception that the person interviewed has to be turned away from certain people and we asked again to indicate the supposed motives.

The dimension "barriers and social distances" aimed to evaluate whether there are cultural, economic, value, social and political barriers in the surveyed city. We referred to the perception that individuals have of certain obstacles to get along with everyone.

The indicators that follow relate to the dimension of prejudice against the Gypsies, foreigners and other subjects that are commonly stigmatized. In particular, for some questions we have chosen to use the Bogardus scale (illustrated above), while in the next question we asked to give a dichotomous answer for each figure in the list, and according to three different degrees of closeness with the interviewee (working environment, friendship group and family group). This dimension seeks to measure any existing social distance towards disadvantaged groups.

Through successive markers, related to the social personality of the sample, we head the propensity to transgress social norms, both from the point of view of one's view of life and that of the society. We try, in other words, to establish the degree of acceptance by the individual, of certain practices, such as "travel on public transport without paying", "use contraceptive", or "assuming drugs", identifying also the idea that they have of the society and its degree of acceptance of behaviors that violate social norms. Other indicators identified are related to the use of television programs and reading newspapers and magazines; political orientation; the future vision of the world and of their own life; the religion and the degree of religiosity; finally, they detected information about the type of their own home and ownership of the same and the monthly economic resources available to the family.

The formulation of the questions in the questionnaire followed a two-pronged approach: there are open questions that require to enter a response formulated directly by the interviewee (place of birth, own level of education and employment, father's and mother's level of education, eventual reasons of the avoiding certain parts of the city, religious affiliation), and closed questions, which provide pre-coded answers. These provide semantically autonomous responses (e.g., sex: male or female, marital status: never married, married, separated, divorced, widowed, employment status: employed, unemployed, unoccupied, retired, student, housewife, etc.) and replies only partially autonomous, as in the case of using a scale (very, somewhat, a little, not at all agree with the statement included in the question or in the series of questions). The variables used are mostly dichotomous, while for others, since that the purpose is to measure an attitude, we used Likert scales. In only one case we used a semantic differential technique: the measure of the political orientation of the respondents (Savini 2017).

The questionnaire also included an open-ended question asking to justify the failure to attend certain parts of the city. This qualitative element, according to the initial hypothesis, should complete, or at least integrate, the statistical analysis of quantitative data, from the analysis of the content of those answers it is possible to further deepen the interpretation of the studied phenomenon.

Once administered the questionnaire and built a fundamental tool for the statistical analysis of the data, namely the matrix cases for variables, with the implementation of the information collected, we can proceed with the next and the last stage of research: data analysis and presentation of results. In fact, as we noted: "The organization of the information is the natural conclusion of the detection phase, and set the start of the data analysis, the time at which the information is arranged to be analyzed" (Frudà 2007: 361).

The data analysis of this research is univariate and bivariate, namely it is based on a variable at a time or on two variables considered simultaneously. As independent variables were determined to "age" and "district of residence": basically, these are the core from which unravels the research; sex is also an independent variable, but it is related only to some of the dependent variables because, as occurs in other research: "If the sample of respondents consider women separately from men, disparities guidelines and attitudes about the social distances are rather weak" (Tacchi 2010: 115).

23.3.3 Sample

The sample was calculated on the basis of the total population of the city of Pescara and has been defined to the extent of 247 individuals (109 males and 138 females, aged between 18 and 90 years, almost all Italians and with a higher education) according to a precise logic in relation to the universe, that is more than 100,000 inhabitants, assuming a 95% confidence level and an adhesion percentage to the research by subjects of 20% (assumed expected prevalence), with a maximum acceptable sampling error of ±5%, we get the definition of the sample size.

$$n = \frac{t^2 P(1 - P)}{D^2}$$

where n is the sample dimension, t is the confidence level, P is the expected prevalence and D is the absolute accuracy desired.

We initially hypothesized to build a simple proportional sample in which was present a cohort of at least 50 individuals for each geographical area (district). According to the consideration of additional variables related to the structure of population, on the basis of demographic surveys, we decided to integrate the process by sampling a reasoned choice, in order to intensify the resident portion in Castellammare as we believed that this is the most heterogeneous neighborhood. This area, unlike other cities in the last twenty or thirty years, has experienced a huge change: new layers of population increased. They were made of people of no EU origin above all, living and working permanently and taking over even in the tertiary sector, the main area of historic to the old residents of the middle and upper classes, Castellammare Adriatico economy. Therefore the "historical" living sedimentation and the highest proportion of foreign residents have made it necessary to its over-representation.

We decided not to refer to the segmentation currently made by the municipality for the electoral purpose because Pescara, due to the particular historical and urban development, presents a social, economic, cultural and urban homogeneity, and it is for this reason that the division made by this research lies in the five geographical

Table 23.1 Empirical sample of interviewees distinguished by age and residence. Absolute values

Age	Q1	Q2	Q3	Q4	Q5	Total
18–29	10	12	11	8	10	51
30–40	14	22	14	17	14	81
41–50	7	9	8	4	9	37
51–60	5	18	9	4	3	39
>60	8	11	1	12	7	39
Total	44	72	43	45	43	247

Table 23.2 Empirical sample of interviewees distinguished by age and sex. Percentage values

Age	Female	Male	Total
18–29	47.06	52.94	100
30–40	46.91	53.09	100
41–50	72.97	27.03	100
51–60	66.67	33.33	100
>60	58.97	41.03	100
Total	55.87	44.13	100

Base 138 Females + 109 Males = 247

areas identified the most representative. In order to seize these unique and homogeneous characteristics that define at least five souls of the city, the area was divided into sub-areas that would represent the most of its natural boundaries, contrary to those "artificial" like "electioneering", administrative or geographical demographic boundaries, created by Pescara. They are:

- Q1 "Centrale" (from the central station to the so-called Borgo Marino, an ancient and now disappeared fishing village located near the river);
- Q2 "Castellammare" (from the central station to the northern border with Montesilvano);
- Q3 "Colli" (it includes the entire hilly area, both north and south of the river);
- Q4 "Portanuova" (Via Conti di Ruvo up to Villaggio Alcyone);
- Q5 "San Donato e Fontanelle" (peripheral zone to the west of the city, which includes the neighborhoods at risk of Rancitelli and Villa del Fuoco).

Citizens were divided—as well as by age group—by the district of their residence because of the nature of the research project. This allows to make comparisons and considerations about citizens' opinion on the other districts (Tables 23.1 and 23.2).

23.4 Results

The results of this research are more than "some" but it is impossible to report the whole work in this context, which is primarily a Ph.D. thesis. Then, in this chapter we will introduce only the main achievements.

The perceived social distance refers to the distance or proximity that respondents express, or they feel to have, in respect of a number of subjects; the acted distance is the distance that we place between ourselves and others. To detect it there are two questions: the first attempts to capture the data by querying the subject of how he behaves towards that persons he does not want to have business with; and the second question asks for explanations. The same thing happens to the experienced distance that is the perception of being pushed away that the interviewees have to from certain people. Then we asked, also, to indicate the supposed causes of that.

The other indicators refer to the dimension of prejudice against the Gypsies, foreigners and other stigmatized subjects. This dimension tries to measure the possible social distance existing related to disadvantaged groups. The stereotype is a conceptual system *"that allows us to simplify our representations"*[4] (Arcuri 1995: 126); it is a "mental shortcut"—an inferential process—used to simplify the reality, because it allows us to collect some superficial information about someone and categorize him in a certain group. Precisely because of this excessive tendency to simplify, to superficialise and undue generalization (Allport 1954), stereotypes may lead to wrong interpretations and, consequently, erroneous judgments. In the interpretation of Lippman, in fact *"the consequences of stereotypes tend to be negative because of their rigidity, for the fact of being impermeable in front of the experience's disconfirmation and for their potential function of reality distortion"*[5] (Arcuri 1995: 127). Prejudice has a typically negative connotation because the pre-packed idea, referring to the object, that the individual shares, leads it to turn away and not accept a change of opinion. In reference to social distance, prejudice is meant as a distancing device between individuals and groups and that's why it is considered a dimension of social distance.

The social distance related to Gypsies and foreigners (mainly from Africa) was measured through the Bogardus scale. The answers, as shown by numerical values, are likely to present a result that we can briefly explain in these terms: in respect of the Gypsies, while in the central and hilly districts it has a greater rate of tolerance, a higher level of acceptance and therefore a potential proximity, in the peripheral areas with the greatest Gypsies density, in which the cultural level is lower, the socio-economic status are more compressed, the Community participation are more isolated but above all there is, in the perception of indigenous suffering the massive presence of deviance and negative consequences, and therefore there is a clear rejection of which is measured in terms of a higher social distance (Table 23.3).

The Bogardus scale highlights that the rejection of the Gypsies is revealed from its first "steps" since though the majority of the sample would not drive them from the country (59.11%) and would admit them at least as visitors (59.51%), however, would not concede them citizenship (53.85%). As for the other elements of the scale, we found that the gypsy is accepted as a fellow worker (61.94%), he is not

---

[4]Translation from Italian to English by the author.

[5]Translation from Italian to English by the author.

Table 23.3 Assessments of the respondents about Gypsies groups present in Pescara, distinguished according to their district of residence. Percentage values

Items	Mod	Q1	Q2	Q3	Q4	Q5
You would exclude them from the country	Yes	29.55	31.94	27.91	57.78	46.51
	No	68.18	66.67	67.44	37.78	51.16
	NA	2.27	1.39	4.65	4.44	2.33
You would admit them as visitors to the country	Yes	61.36	59.72	55.81	60.00	60.47
	No	36.36	38.89	37.21	33.33	34.88
	NA	2.27	1.39	6.98	6.67	4.65
You would give them citizenship	Yes	56.82	50.00	46.51	22.22	34.88
	No	40.91	48.61	48.84	73.33	60.47
	NA	2.27	1.39	4.65	4.44	4.65
You would accept them as working partners	Yes	59.09	69.44	69.77	44.44	62.79
	No	38.64	30.56	25.58	53.33	32.56
	NA	2.27	0.00	4.65	2.22	4.65
You would accept them as neighbors	Yes	43.18	52.78	55.81	37.78	32.56
	No	54.55	45.83	37.21	57.78	65.12
	NA	2.27	1.39	6.98	4.44	2.33
You would accept them as friends	Yes	47.73	61.11	58.14	33.33	46.51
	No	50.00	37.50	34.88	62.22	48.84
	NA	2.27	1.39	6.98	4.44	4.65
You would accept him as a husband for your daughter	Yes	22.73	33.33	32.56	8.89	23.26
	No	72.73	63.89	62.79	84.44	72.09
	NA	4.55	2.78	4.65	6.67	4.65

(*Base* 44, 72, 45, 45, 43 = 247)—*NA* Not Available

desired as a neighbor (51.42%), but he is accepted as a friend (50.61%). Finally, the 70.45% of the sample excluded that would accept a gypsy as husband of his daughter.

Distinguishing the answers of the respondents according to their original district, it can add more information about this kind of distance.

The trend is: greater openness of the residents of the central and hill areas, indicated by Q1, Q2 and Q3 compared to those of Rancitelli neighborhoods, San Donato and Fontanelle (Q5). Q4 indicates instead Portanuova and Alcjone Village, up to the border with Francavilla al Mare.

Another thing happens when the same questions were referred to other foreigners who are not Roms, that is asking questions about residents attitudes, opinions and stereotypes regarding African, Indian, Romanian and, in general, immigrants who, with varying percentage, are present in the city and its neighborhoods. Against this minority sample manifests a wider acceptance. The 78.95% of the sample would not drive them from the country, in large majorities admit them as visitors (81.38%), and would grant them citizenship (64.37%). This proximity is supported by the subsequent answers: respondents would accept them as fellow workers (82.59%), as neighbors (76.11%) and as friends (76.52%). The 46.56% of the sample responds yes to the question "would you accept a gypsy as the husband of your daughter?", proving that a foreign immigrant is more accepted than a gypsy (the same question receives a degree of acceptance of 23.10%) and the 50.61% answered no, highlighting that opening the doors of your place to a stranger is still a taboo for half of the respondents.

The analysis continues with the distinction of respondents' answers by place of residence, in order to identify some differences between neighborhoods, since the largest concentration of African, Indian and Oriental immigrants is in the area around the Central Station, that is an area highly frequented and the meeting place of walking and shopping.

In this case, we perceived that in central and hilly districts, both in the peripheral and marginal areas, the rate of potential social proximity, that is, tolerance, availability of contact and social inclusion is far higher. This is a sign that the specific context of the Gypsies in Pescara is what records much lower rates of environmental incompatibility (Table 23.4).

Table 23.4 Assessments of the respondents about foreigners' people present in Pescara, distinguished according to their district of residence. Percentage values

Items	Mod	Q1	Q2	Q3	Q4	Q5
You would exclude them from the country	Yes	20.45	15.28	23.26	28.89	16.28
	No	79.55	83.33	76.74	68.89	83.72
	NA	0.00	1.39	0.00	2.22	0.00
You would admit them as visitors to the country	Yes	84.09	76.39	74.42	93.33	81.40
	No	13.64	19.44	25.58	4.44	16.28
	NA	2.27	4.17	0.00	2.22	2.33
You would give them citizenship	Yes	70.45	65.28	65.12	51.11	69.77
	No	29.55	30.56	34.88	46.67	25.58
	NA	0.00	4.17	0.00	2.22	4.65
You would accept them as working partners	Yes	81.82	86.11	81.40	75.56	86.05
	No	18.18	12.50	18.60	22.22	11.63
	NA	0.00	1.39	0.00	2.22	2.33
You would accept them as neighbors	Yes	77.27	79.17	81.40	64.44	76.74
	No	22.73	19.44	18.60	28.89	23.26
	NA	0.00	1.39	0.00	6.67	0.00
You would accept them as friends	Yes	79.55	79.17	79.07	60.00	83.72
	No	20.45	19.44	20.93	33.33	16.28
	NA	0.00	1.39	0.00	6.67	0.00
You would accept him as a husband for your daughter	Yes	40.91	51.39	60.47	28.89	48.84
	No	59.09	44.44	39.53	66.67	46.51
	NA	0.00	4.17	0.00	4.44	4.65

(*Base* 44, 72, 45, 45, 43 = 247)—*NA* Not Available

Excluding the ethnic conditions but heading instead to personal conditions of categories of people who are stigmatized as mentally ill, drug addicts, prostitutes and homosexuals, prisoners, alcoholics and people with disabilities, in this case the social distance is not related to intercultural relations but to the distance between persons belonging to a standard of "normality" and people who for various reasons are "marked" and socially marginalized.

The acceptance of diversity is researched and studied in correlation with the social space in which the daily life of people is spent. The dimensions are considered the working environment, friends and family, placed in this same progression, in order to highlight the degree of acceptance with increasing proximity to the most intimate personal sphere, which is the family.

Thanks to the bi-variate analysis of this indicator with the variable "residence district" emerge more or less the same levels of general acceptance: they are always quite high for all figures except for schizophrenics. With regard to other subjects, in the former detainee, a former drug addict, recovering alcoholic, disabled, depressed, a former prostitute and homosexual, it is to believe that the trend is that of greater tolerance and acceptance as possible, which makes imagine an initial trend of Pescara being against these particular subjects, but also of a substantial coexistence.

Why schizophrenics record this particular distance? We should consider that the interpretation is a significant rejection of mental illness because it is unpredictable and because it can somehow give rise to the outbursts of aggressive behaviors that are perceived as potentially dangerous and uncontrollable and that can affect the subject.

We should add a further consideration: while most accepted figures do not allow identification processes with the "other from inner self" that is, with the dark side of the personality, as everyone has moments of dissociation, although physiological, warns that in condition of schizophrenia there is a kind of contamination in the statement of recognition of the deep self and then there is an aversion in consideration of how much the mental troubles can deconstruct their own and others' personalities. The "contagion", so to say, with an alcoholic or with a disabled person is considered of minor impact and not so susceptible of invasiveness as with the schizophrenic. The results of the survey in relation to the age of the respondents are of greater importance. The prisoners have a great acceptance of maximum by the young's and the olds, the same goes for the depressed, the former prostitutes and homosexuals. In the case of former prostitutes and homosexuals, it has been identified a growing tolerance with the growing of the age, so that older people seem to reject the possibility of a social acceptance of these figures although overall the acceptance percentages are wider than they are in the case of the schizophrenic subject, which records for all age groups the greater rejection among all the figures (Tables 23.5 and 23.6).

Table 23.5 Acceptance/rejection levels in the workplace towards the subjects on the list, by the respondents, divided by age. Percentage values

Workplace Items	Mod	Age 18–29	30–40	41–50	51–60	>60
Former detainee	Yes	72.55	69.14	72.97	66.67	71.79
	No	25.49	28.40	21.62	25.64	25.64
	NA	1.96	2.47	5.41	7.69	2.56
Former addict	Yes	68.63	66.67	72.97	56.41	51.28
	No	29.41	30.86	18.92	35.90	43.59
	NA	1.96	2.47	8.11	7.69	5.13
Former alcoholic	Yes	76.47	72.84	75.68	69.23	56.41
	No	21.57	24.69	16.22	23.08	38.46
	NA	1.96	2.47	8.11	7.69	5.13
Person with a disability	Yes	94.12	98.77	89.19	89.74	87.18
	No	5.88	1.23	2.70	2.56	2.56
	NA	0.00	0.00	8.11	7.69	10.26
Subject with depression problems	Yes	62.75	62.96	75.68	66.67	66.67
	No	35.29	35.80	16.22	25.64	28.21
	NA	1.96	1.23	8.11	7.69	5.13
Schizophrenic subject	Yes	29.41	22.22	24.32	20.51	20.51
	No	66.67	75.31	62.16	69.23	74.36
	NA	3.92	2.47	13.51	10.26	5.13
Former prostitute	Yes	90.20	87.65	83.78	76.92	66.67
	No	7.84	11.11	5.41	15.38	30.77
	NA	1.96	1.23	10.81	7.69	2.56
Homosexual	Yes	94.12	92.59	94.59	82.05	76.92
	No	3.92	3.70	0.00	7.69	20.51
	NA	1.96	3.70	5.41	10.26	2.56

(*Base* 51, 81, 37, 39, 39 = 247)—*NA* Not Available

Finally, it must add that in the case of the former alcoholic and former drug addict the age variable suggests a greater compatibility and understanding on the part of younger age than there is with those of older people. A separate case is the relationship with the disabled, that records the greater inclination of all the figures to a very wide social inclusion condition; very rare cases of rejection occur in younger age groups, but they are of little importance as all age groups show a high acceptance (Table 23.7).

Table 23.6 Acceptance/rejection levels into friends' environment towards the subjects on the list, by the respondents, divided by age. Percentage values

Friends environment

Items	Mod	Age 18–29	30–40	41–50	51–60	>60
Former detainee	Yes	74.51	74.07	62.16	58.97	48.72
	No	23.53	23.46	24.32	28.21	41.03
	NA	1.96	2.47	13.51	12.82	10.26
Former addict	Yes	64.71	76.54	67.57	56.41	41.03
	No	33.33	20.99	18.92	28.21	56.41
	NA	1.96	2.47	13.51	15.38	2.56
Former alcoholic	Yes	78.43	82.72	75.68	64.10	51.28
	No	19.61	16.05	10.81	25.64	46.15
	NA	1.96	1.23	13.51	10.26	2.56
Person with a disability	Yes	98.04	100.00	89.19	92.31	84.62
	No	1.96	0.00	0.00	2.56	7.69
	NA	0.00	0.00	10.81	5.13	7.69
Subject with depression problems	Yes	84.31	85.19	75.68	76.92	69.23
	No	13.73	13.58	13.51	15.38	25.64
	NA	1.96	1.23	10.81	7.69	5.13
Schizophrenic subject	Yes	37.25	46.91	37.84	28.21	20.51
	No	58.82	48.15	43.24	53.85	74.36
	NA	3.92	4.94	18.92	17.95	5.13
Former prostitute	Yes	74.51	80.25	75.68	69.23	46.15
	No	21.57	16.05	10.81	17.95	51.28
	NA	3.92	3.70	13.51	12.82	2.56
Homosexual	Yes	88.24	90.12	83.78	79.49	69.23
	No	9.80	6.17	5.41	10.26	28.21
	NA	1.96	3.70	10.81	10.26	2.56

(*Base* 51, 81, 37, 39, 39 = 247)—*NA* Not Available

We identified several indicators that, while not directly affecting the most intimate dimensions of people, who may be embarrassed to expose themselves about sensitive issues such as the ability to marginalize or the feeling of being excluded, can help to explain its causes. Besides the above mentioned, one concerns the avoided neighborhoods. We asked which area is not attended and to explain the answer. This qualitative element, according to the hypothesis, completed, or at least supplemented, the statistical analysis of quantitative data. The sample in fact was built taking into account the structure and the division of the city. We identified five significant areas, each of which showed a different reaction to the questions proposed.

Table 23.7 Acceptance/rejection levels into family environment towards the subjects on the list, by the respondents, divided by age. Percentage values

Family environment						
Items	Mod	Age 18–29	30–40	41–50	51–60	>60
Former detainee	Yes	60.78	66.67	45.95	46.15	46.15
	No	37.25	29.63	40.54	33.33	43.59
	NA	1.96	3.70	13.51	20.51	10.26
Former addict	Yes	62.75	64.20	59.46	35.90	41.03
	No	35.29	33.33	32.43	43.59	51.28
	NA	1.96	2.47	8.11	20.51	7.69
Former alcoholic	Yes	68.63	70.37	64.86	46.15	46.15
	No	29.41	27.16	27.03	35.90	48.72
	NA	1.96	2.47	8.11	17.95	5.13
Person with a disability	Yes	92.16	95.06	97.30	74.36	74.36
	No	7.84	2.47	0.00	10.26	15.38
	NA	0.00	2.47	2.70	15.38	10.26

Family environment						
Items	Mod.	Age 18–29	30–40	41–50	51–60	>60
Subject with depression problems	Yes	86.27	80.25	78.38	56.41	56.41
	No	11.76	16.05	13.51	28.21	33.33
	NA	1.96	3.70	8.11	15.38	10.26
Schizophrenic subject	Yes	54.90	53.09	48.65	30.77	25.64
	No	43.14	41.98	35.14	53.85	64.10
	NA	1.96	4.94	16.22	15.38	10.26
Former prostitute	Yes	70.59	60.49	78.38	41.03	35.90
	No	27.45	35.80	13.51	38.46	35.90
	NA	1.96	3.70	8.11	20.51	7.69
Homosexual	Yes	82.35	92.59	81.08	66.67	41.03
	No	15.69	6.17	13.51	12.82	48.72
	NA	1.96	1.23	5.41	20.51	10.26

(*Base* 51, 81, 37, 39, 39 = 247)—*NA* Not Available

23.5 Conclusions

The perceived social distance between the citizens of Pescara, measured in terms of socio-cultural, economic, social and professional differentiation, seems to have an economic base but by comparison with another indicator, namely the existence of economic barriers, cultural values, social and political barriers, there is a discrepancy. People in the sample feel close to those who, for example are poor, but they believed that those who have different incomes have no point of contact between them.

The results about social distance towards the Gypsies, foreigners and other excluded groups, show that social distance is more obvious to the ethnic level as in Pescara, despite the river divides the city in half, the urban and social order is continuous (Montesilvano, Pescara, Francavilla al Mare, Chieti and form a single metropolitan area) and it does not seem to be social distance between the inhabitants of both banks of the river. However Pescara citizens do not seem to avoid all foreigners without distinction: only a specific group, the Gypsies, especially those stationed in what was at first the area outside of the urban center (called Rancitelli). This area was used to build a kind of "ghetto", in which allocate fringes of population in difficulty or considered problematic. Over time, the growing urbanization of the city had broken the boundaries of the "distant" periphery until it was incorporated into the urban texture. Despite this, the "ghetto" still exists and has defined borders, sometimes impenetrable.

The distancing attitude towards the minority of this suburb is shown by the residents of Portanuova (which live, basically, in the border with the district of Rancitelli) but not by those of other districts such as the one called "Centrale" and Castellammare. Portanuova, the so-called "bearing" district, must deal with the supposed turbulences of the "lively" neighborhood, and as it arises from the judgments of the respondents, it is labeled as a dangerous and risky neighborhood to be avoided.

The conclusion that can be drawn from this research is that physical and social distance (subjective) seem to be, in the specific context of Pescara, inversely proportional. The decrease of physical distance (i.e. spatial) corresponds to the increase of social distance. Residents of the northern districts of the city seem unaware of the problem of this neighborhood and they express a greater tolerance about the sedentary nomads now firmly embedded. In other words, the prejudice against them by those who live far away seems not to exist.

Regarding the presence of Gypsies, our research has revealed that there is another neighborhood with a historical presence of Gypsy families and it is the hilly one, but contrary to the previous case, there is no distance related to Gypsies, as in the areas of "Centrale" and Castellammare. The difference between the two groups is that the latter become embedded in the social, cultural and even working context of the city (the latest generations are educated and working or looking for a job) while in the neighborhood of Rancitelli, Gypsies seem to be known for their collusion with the organized crime.

In addition to this, the evaluation of social distance related to certain categories of people commonly stigmatized, with regard to the spatial dimension, highlighted that people with a gloomy or messy past (i.e. former prisoners, former drug addicts, alcoholics and former prostitutes) and homosexuals are generally accepted in the workplace but not much in the family. When the physical distance decreases, the social distance towards these categories of people increases. In other words, it is pointed out that there's a display of greater closeness shown in public in front of a genuine and effective intimate social distance (measured from the data revealing the degree of their acceptance in the family). In terms of exclusion/inclusion, one can assume that, according to Goffman (1959), we are more inclusive and available when we are in public than we are in the private sphere of our own group and family, where we are subjected to the comparison with and the judgment of the others.

Finally, it was possible to construct indices. Among these, there is the index of social extraction: our data seems to underline that as the level of social extraction increases, the social distance expressed by respondents' decreases. The strategy preferred by those who have a generally low level of extraction is trying not to meet that unwelcome person which is, as respondents say, considered just a little; on the contrary, those who have a high average level of social origin emphasizes the use of the escaping strategy but do not believe that the avoided person is inferior. A low level based on social origin seems to be connected with a greater tendency to stay away from the Gypsies and foreigners, while those with higher social origin show a more tolerant and inclusive behavior.

References

Allport, G., (1954) *The nature of prejudice*, Cambridge, Mass, Addison-Wesley Publishing Company.

Arcuri L., (1995) *Manuale di psicologia sociale*, Bologna, Il Mulino.

Bauman Z., (2000) *Liquid Modernity*, Cambridge.

Bichi R. (a cura di), (2007) *La distanza sociale. Vecchie e nuove scale di misurazione*, Milano, Franco Angeli.

Bogardus E. S., (1925a) *Social Distance and Its Origins*, in Journal of Applied Sociology, 9.

Bogardus E. S., (1925b) *Measuring Social Distances*, in Journal of Applied Sociology, 9.

Bogardus E. S., (1926) *Social Distance in the City*, in Proceedings and Publications of the American Sociological Society, 20.

Bogardus E. S., (1933) *A Social Distance Scale*, in Sociology and Social Research, 17.

Bottero W., Prandy K., (2003) *Social interaction distance and stratification*, in British journal of sociology, 54, Issue 2, pp. 177–197.

Cesareo V., (2007) *La distanza sociale. Una ricerca nelle aree urbane italiane*, Milano, Franco Angeli.

Coco A., (2011) *La distanza sociale. Reggio Calabria: le condizioni sociali in una terra del Sud*, Milano, Franco Angeli.

D'Amato M., (2009) *La distanza sociale. Roma: vicini da lontano*, Franco Angeli.

Fantozzi P., La Spina A., (2010) *La distanza sociale. Distanti e disuguali nelle città del sud*, Milano, Franco Angeli.

Frudà L., (2007) *La distanza sociale. Le città italiane tra spazio fisico e spazio socio-culturale*, Milano, Franco Angeli.

Goffman, E., (1959) *The presentation of self in everyday life*, Garden City, NY Doubleday & Co.

Hošková-Mayerová, Š., Rosická, Z. (2015). E-Learning Pros and Cons: Active Learning Culture?. Procedia - Social and Behavioral Sciences, 191, no. June 2015, pp. 958–962.

Lazarsfeld, Paul F. & Barton, Allen H., (1951) *Qualitative Measurement in the Social Sciences. Classification, Typologies, and Indices* in Daniel Lerner & Harold D. Lasswell (Eds.), The Policy Sciences (pp. 155–192), Stanford University Press.

Marx K., (1921) *Das Kapital. Kritik der politischen Oekonomie*, Hamburg.

Park R. E., Burgess E. W., McKenzie R. D., (1925) *The city*, Chicago, The Chicago University Press.

Park R. E., (1950) Race and Culture, New York, The free press.

Pascuzzi E., (2010) *La distanza sociale. Politica e società a Messina*, Milano, Franco Angeli.

Savini V. (2017) Microcredit and Quality of Life: An Analysis Model, (Eds.), Recent Trends in Social Systems: Quantitative Theories and Quantitative Models, Decision and Control, Vol. 66, Maturo 43–52.

Simmel G., (1908) *Soziologie: Untersuchungenüber die Formender Vergesell-schaftung*, Leipzig.

Sorokin P., (1927) *Social mobility*, New York, Harper & Brothers.

Statera G., (1990) *Metodologia e tecniche della ricerca sociale. Una introduzione sistematica*, Palermo, Palumbo.

Tacchi E. M., (2010) *La distanza sociale. Milano e i ghetti virtuali*, Milano, Franco Angeli.

Weber, M., (1934) *Die protestantische Ethik und der Geist des Kapitalismus*, Mohr, Tubingen.

Chapter 24
Socio-Vital Areas Analysis a Qualitative Approach to Sociological Analysis of Urban Spaces and Social Life

Gabriele Di Francesco

Abstract We present the theoretical synthesis and the technical and methodological setting of a research, aimed to analyzing and evaluating the urban places and spaces of social life, where the human interactions take place and the city come to life.

Keywords Socio-vital areas · Urban spaces · Social interactions

24.1 Urban Spaces and Socio-Vital Areas

In the daily experience of contemporary cities the image of the streets and squares, spaces, the same empty spaces between buildings and between neighborhoods seem to have taken over the last few years is a symbolic dimension which appeared to be the legacy of the past and instead responds "to a growing social demand for a better habitat" (Amendola 2009, p. V).

The analysis of the space where social phenomena are born becomes more and more important, more central, both as part of the design, and in that of social life.

The reflection on urban design, usually only architectural, must be even and especially social, draw on the experiences of individual, personal and familiar life, on those of common life, of the relational dynamics that take place between people, of the community life, which will be played on those same urban contexts in terms of interaction with others.

Socio-vital area, moreover, could be considered any space, not necessarily purpose-built and designed to carry out some social function, but that could be a possible option for socializing, from the most basic (the chance encounter along a street, a square, a café) to the highest (the *agora* or however the places where political, economic and institutional decisions are made for the life and develop-

G. Di Francesco (✉)
Dipartimento di Economia Aziendale, Università degli studi G. d'Annunzio
Chieti-Pescara, Viale Pindaro 42, 65127 Pescara, Italy
e-mail: gabriele.difrancesco@unich.it

© Springer International Publishing AG 2017
Š. Hošková-Mayerová et al., *Mathematical-Statistical Models and Qualitative Theories for Economic and Social Sciences*, Studies in Systems, Decision and Control 104,
DOI 10.1007/978-3-319-54819-7_24

ment of a community). Or even the perception and representation that you have of a space, a street, a neighborhood, of an organization or an institution, where it spends the daily lives of people.

These socio-vital areas generally shall be considered only in part in the designing process, because they depend a lot from life and choices—which is not always rational or fully aware—of the inhabitants. Behaviors ranging from an almost absolute randomness to a strict ruling. Chosen or not by those who live in a specific urban context, seem to represent in each case the elements essential to the life of a community where the faces meet, where there is the relationship and knowledge, where we recognize part of the same habitat, where we organize and decide, acquire a large part of social identity and build the shared memory.

They seem to be precisely those elements—but the list cannot be exhausted the number and variability—to making symbolic the places and make significant the spaces for human interaction.

Identify the peculiarities of these spaces is not easy, because there are those who live there and make it the center of their business, but also those who are there occasionally. In many ways it seems to approach the representation of "Abode of the community" in which Walter Benjamin describes Paris and its streets. *"The roads are the abode of the community"*, observes the author of *Passages* and goes on to say: *"The community is a reality perpetually awake, perpetually in motion, which lives between the walls of the buildings, experiments, invents and knows like the individuals cover the four walls of their home. For this community the bright enamelled plates of companies represent an ornament of their walls like and, perhaps, more than an oil painting in a bourgeois living room and the walls with "défense d'afficher" are on their desk, newsstands their libraries, letterboxes their bronzes, benches furniture of the bedroom and the cafe terrace veranda, from which oversees the life of his house. There, where the road menders hang the jacket to the grille, is the vestibule, while the carriage gateway that, by the flight of the courtyards, leads to outdoors, is the long corridor that scares the bourgeoisie and represents his way to the rooms of the city, the* passage *is their living room. In it, more than elsewhere, the road makes himself known as* l'intérieur *furnished and lived by the masses"* (Benjamin 1982, p. 474).

The Socio-vital area finally looks like a magmatic reality, impossible to catch in its entirety if not for an imaginative effort that will provide access to the whole, the overall landscape, the simmelian landscape, which is not the single detail that we "perceive with very different degrees of attention, "but it is something bigger and deeper.

It is difficult to perceive this reality in its entirety. Like the landscape for Simmel, *"we perceive with our consciousness a unitary whole, which overcomes the elements, without being related to their particular meanings and being mechanically composed by them—solely this is the landscape. (...) A vision complete in itself, perceived as independent unit, but nevertheless intertwined with something infinitely more extended, floating, included into limits that do not exist for the feeling— just deeper—of the divine unity, of the natural totality. From this feeling autonomous boundaries of every landscape are constantly touched and loosened, and the scenery, though separate and independent, is continuously spiritualized by the obscure consciousness of this infinite connection"* (Simmel 2006, pp. 53–55).

In the literary field a step of Marcel Proust seems to echo, although with other connotations, the difficulty of perceiving the fragmentation and multiplicity of reality, as well as the boost to catch the unity of something that is hard to find out: *"tout d'un coup un toit, un reflet de soleil sur une pierre, l'odeur d'un chemin me faisaient arrêter par un plaisir particulier qu'ils me donnaient, et aussi parce qu'ils avaient l'air de cacher au delà de ce que je voyais, quelque chose qu'ils invitaient à venir prendre et que malgré mes efforts je n'arrivais pas à découvrir"* (Proust 1939, p. 256, it. tr. 1990).

It is the variability and mutability together with the multiplicity of forms to constitute the substrate of the socio-vital spaces, such as the landscape. In this regard, Simmel emphasize again how "the material of the landscape, which is provided by the mere nature, is so infinitely manifold and variable from time to time, that the points of view and forms, which in individual case with these elements produce the impression unit, will be very different" (Simmel 2006, p. 57). Multiple views and impressions, then, that seem to be repeated in another context, that of the *flâneur* by Walter Benjamin when he says that the city "can unfold to the passer from everywhere like a landscape without threshold" (Benjamin cit., p. 484), in that dialectic of the *flânerie* that sees "on one side the man who feels observed from everything and everyone, the suspect par excellence, on the other the unavailable and hidden individual" (Ibidem. Vol. I, p. 470).

The problem at this point is how reflexively catch the essence of the social and living spaces without losing sight of the unity of interpretation of a given urban context and trying indeed to perceive the subjective elements, emotional, that flow from this reflection, they can provide one unified vision.

"Him Who sees without understanding is much more... restless than he who understands without seeing" observes Simmel, almost underlining the centrality of observation in the study of human relations in large cities and pointing out that the view—especially—and hearing are the most important sensory activities for the analysis of social phenomena.

"Celui qui voit sans entendre est beaucoup plus... inquiet que celui qui entend sans voir. Il doit y avoir ici un facteur significatif pour la sociologie de la grande ville. Les rapports des hommes dans les grandes villes... sont caractérisés par une prépondérance marquée de l'activité de la vue sur celle de l'ouïe. Et cela... avant tout, à cause des moyens de communication publics. Avant le dé-veloppement qu'ont pris les omnibus, les chemins de fer, les tramways au dix-neuvième siècle, les gens n'avaient pas l'occasion de pouvoir ou de devoir se regarder réciproquement pendant des minutes ou des heures de suite sans se parler" (Simmel 2006, pp. 26–27).

The problem is how reflexively grasp the essence of the socio-vital areas without losing the unity of interpretation of a given urban context and looking rather to perceive the subjective and emotional elements, arising from the reflexion to provide one unified vision.

24.2 Socio-Vital Areas Analysis, a Methodological Approach

The social life moreover is "permeated by acts of observation that each actor puts into practice every day" (Bichi 2007, p. 207) that measure and test the "social behaviors" (Cipolla 1996) of individuals in places in which they occur.

The purpose of these acts of observation is to "be able to see the world from the point of view of those who inhabit it: understand their perspective; grasp their motivations and the meanings they attach to the daily practices "and is particularly useful" as many times as you want a complete and detailed picture of a behavior and the context in which it takes place in a verbal behavior (what is said, or not said) or non-verbal (what is done, or not done). (…) The information is collected through the act of "look" (…) term used to indicate not only the act of seeing with the eyes, but in a much broader sense. Means test (have experience), not only the view but also with the five senses (De Lillo 2010, pp. 35–36). "Watch" on the other hand is a way to understand and empathize (Dal Lago and De Biase 2002)

The observational approach in the social sciences has a long history and has many examples, from the studies of Malinowski's on the *Western Pacific islands* to the analyzes of Simmel, however departing from the observational elements of social reality. For sociology the basic reference is obviously to the experiences of so-called School of Chicago, founded in 1892, and studies of its exponents, from Anderson (The Hobo 1923), to Park-Burgess-McKenzie (The City 1925), from Robert and Helen Lynd (Middletown 1929) to William F. White (Street Corner Society 1943). The history of observational techniques in sociology, however, includes numerous other examples, from Banfield to Goffman, from the "grounded theory" to ethnomethodology.

The fact remains that element constant and significant of the observation carried out in "natural environment" is the attempt to approach "as much as possible to people, groups, relationships, practices" (De Lillo cit., p. 36; Ferri 2017; Maturo Hošková-Mayerová 2017).

In the survey about Levittowners, conducted by Gans since 1958, the scholar says that "By living in the community in the first two years of its existence I proposed to observe the development of neighborly relations and social life and to realize the formation of organizations and institutions. (…) I discovered the nature of the everyday life, especially in my way, where I could watch the neighbors and myself in our roles as hosts and inhabitants of the block" (Gans 1967).

The research of Gan on Levittowners is the classic example of the analysis of social life in urban environments with the use of the techniques of participant observation and follows somehow the open road at the time by William Foote White investigation of Street Corner Society.

It is known how the Whyte same to warn how the observation is not immune from risks and how, also in the participant observation, more often it is necessary not to be too intrusive to read the behaviors and penetrate the sense of observed attitudes.

"Sometimes I had a doubt," writes White, "that loiter on the corner of the street could not be an important enough activity to be classified with decent name of "search". *I told myself that maybe I should ask questions to these men. However it was also necessary to learn what was the right time to ask your questions and what you ask". And meditating on the affirmation of Doc, "When you make those questions the people are right away on you. If people accept you in the company, you can always be a spectator and you will learn the answers over time, without even need to ask questions, "he understands that this was very true, so much so that observes," When I was apart and I listened, I had often answers that I could never get if I searched for my information only on the basis of the interviews"* (White 1955, pp. 385–386).

The reference to the need to make observations, but in a more detached and less intrusive, seems to be present also in the words of Edgar Morin (1967), who, in the appendix to his research on the town of Plodémet and its metamorphosis, joking about the possibility of understanding of quantitative sociologists, says the indispensability to capture the "instant", of what can be grasped only by dazzling intuition.

The observer directly, as opposed to participating observer is freer and less intrusive; he can remain concealed without declaring its real goals and his true identity. Who looks tend not to be perceived by those who are observed. Adopt a sort of camouflage that is standardized which makes it completely confused with the social actors that act on the set observed.

"The complete observer", as he calls Earl Babbie (2008), "can study a social process without becoming part of it, maintaining a certain detachment. Probably, his discretion will ensure that the subjects of the investigation do not realize being studied".

Another element to take into due consideration is the level of structuring of the tools used for the collection data and informations. If direct observation, everything's free, it leaves a large margin of subjectivity, the use of highly structured instruments (a questionnaire, a survey form, rating scales, etc.) lose the "naturalness" of observation.

For the study of socio-vital areas is preferable to use the direct observation, that adopts semi-structured survey instruments, in order to conduct the observation for short periods (if necessary to the purposes of the survey, repeated several time throughout of the day) and with the subsequent immediate recording of what is observed.

Finally, it is to perform a qualitative research using the technique of direct observation in the "natural environment", i.e. tense "to detect behaviors that occur spontaneously in the context in which it usually takes the act of the subjects, namely in the field and in situations of "real" life "(De Lillo cit. p. 45), obviously without violating the rights of the people.

This research is conducted in the field by the researcher-observer, who must perceive all that is located and lives in the context observed, all that is observable— including their emotions and their feelings—and take a mental note, and then make, but elsewhere, the recording of what he observed, evaluated, also perceived emotionally.

A search of this kind may have ethical implications, especially because "hidden", "blanket" not apparent towards the subjects observed. This is a big problem that involves the right to privacy of the people observed, of which, in the course of direct observation, we see behaviors, listening to snatches of conversation, we perceive the opinions and attitudes especially with the nonverbal communication analysis.

Even in the case of the study of social and living spaces, however, although it involves observations to be conducted in public spaces (a street, a square, the bus stop or the inside of a moving bus, a cafe, a station, public building, etc.) the issue of respecting the confidentiality arises both in the observational stage, both in the time of analysis of data and dissemination of results.

It is therefore necessary to act in order not to violate the rights of people and to do it so as not to damage them either physically or psychologically. If it appears the absolute limit that cannot be exceeded is ethical duty of the researcher to protect in every way the identity of the observed subjects and their interactions, such as with the use of pseudonyms and trying that all data, not only those so-called sensitive, are not connected with individual subjects.

Risks and difficulties of the observational research conducted in public spaces also seem less strong in consideration of the results that can be achieved, and especially when compared to the results reached by the use of other techniques of qualitative research (focus groups, life stories, etc.) or quantitative (surveys by questionnaire).

The application of direct observation to socio-vital areas, public spaces in which human interactions take place, however, is particularly advantageous when you want to identify and study social action with its own rules, its formality, as well as reports (of use, livability, coercion, stability, movement, etc.) that bind people to the same observed spaces.

Exemplary in this regard, are the extraordinary studies of Erving Goffman (1971) on relations in the public (and, in particular, the rules that govern daily situations, such as taking place in a subway car; divide the elevator car with strangers; along a crowded sidewalk; choose where to settle down in a crowded beach). "*Goffman describes the gestures, body movements, words, rituals with which, in these and other situations, we establish and manage relations with our neighbors, and together, "defend" our personal space and our identity, trying to avoid and/or opposing others' violations. These aspects of "normal appearance" making up the reality: an order perhaps "minimum", but crucial, that we are all daily committed to maintain*" (De Lillo cit. p. 59).

In this regard it seems interesting report an E.A. Poe sentence in which the writer, the late nineteenth century, precisely describing the crowds of passers-cluttering the streets of Paris: "*Le plus grand nombre de ceux qui passaient avaient un maintien convaincu et propre aux affaires, et ne semblaient occupés qu'à se frayer un chemin à travers la foule. Ils fronçaient les sourcils et roulaient les yeux vivement; quand ils étaient bousculés par quelques passants voisins, ils ne montraient aucun symptôme d'impatience, mais rajustaient leurs vêtements et se dépêchaient, D'autres, une classe fort nombreuse encore, étaient inquiets dans leurs mouvements, avaient le sang à la figure, se parlaient à eux-mêmes et gesticulaient, comme s'ils se sentaient seuls par le fait même de la multitude innombrable qui les entourait. Quand ils étaient arrêtés dans leur marche, ces gens-là cessaient tout à coup de marmotter, mais redoublaient leurs gesticulations, et attendaient, avec un sourire distrait et exagéré, le passage des personnes qui leur faisaient obstacle. S'ils étaient poussés, ils saluaient abondamment les pousseurs, et paraissaient accablés de confusion"* (Poe, 1886, p. 89, in Benjamin cit., p. 497).

24.3   Variables of Experimental Research on the Socio-Vital Areas

The reality of social sites is also something fluid and changeable, a process sometimes chaotic that often builds in later times (usually no short) and which feeds on absolutely subjective elements: memory, feeling, affection, the strength, the attraction-repulsion dynamic, etc.

> *"Social groups"*, Bergamaschi points out in its reflection on Halbwachs *"shape and organize the space according to their needs but the resistance they encounter require them to comply with the constraints of the "mechanical and material things", intended as a reification of experience and practices of other groups who in the past have inhabited it. Lived space becomes a place of continuous reworking processes, the adjustments of appropriations, of past and present configurations inseparable"* (Bergamaschi 2008, pp. 85–86).

Analyze everything, try to grasp every aspect in this context is very difficult. "Observe everything is cognitively impossible," how Rita Bichi points out (cit., p. 222). It did not seem fanciful, however, try to experience the observational paths that allowed to read the social and living spaces of the city by focusing on some significant aspects.

The specific aspects of the observation—the thing to observe—have been reclassified within nine key variables, and modulated according to the indications of Spradley (1980, p. 78), which are summarized below:

(1) the space or field of observation;
(2) the objects and artifacts in the research situation;
(3) the actors (people involved in the observation);
(4) the actions, i.e. movements, modes of communication (verbal and nonverbal) of the single subjects observed in the field;
(5) the activities related to each other by people observed (actors) and the identification of groups and activities;
(6) the situations and social distance;
(7) the set of activities and related events that actors play in the socio-vital areas;
(8) the alleged objectives, hidden and overt of the actors observed;
(9) the feelings and emotions in others and experienced subjectively observed during the observation by the observer.

It is nine-dimensional areas each of which was in turn formed by a specific set of indicators.

The analysis of the observation field has also focused on some specific indicators, namely:

(1) freedom of stability, that is, the acceptance of the place, the ability to stop, to find answers to possible needs, verifying the presence in the observational places of businesses, service providers such as banks, newsagents, pharmacies and similar, etc.;

(2) the freedom movement (possibility to move within the spaces and in the routes of entry and exit, bus stops, parked cars, for work blocks, etc.);

(3) offered opportunities of socialization (informal stopping points where you stand still in group, but also areas equipped perhaps with benches, tables and bar chairs, coffee etc., and other areas such as bus shelters for waiting bus, etc..).

Additional considered variables were related to the interpretation of economic and socio-cultural level reference attributed to the observed area (e.g. Residential area, palace, housing, suburban, town center, public space, political space, commercial elite, local of services, etc.), as well as the eventual level of physical and human degradation.

For the analysis of socio-vital spaces is also important to identify "pedestrian paths", (e.g. functional to purchase products, access to buildings, offices, shops, stalls, etc.), paths of "*custom stroll*" and "stabling" where you wait, you come across, where the clusters are formed, the cliques and informal groups, drawing also one or more positional maps of places and spaces. Spaces not just physical, but also rational, designed to detect even the social distance of the subjects that affect the urban area.

The reference in this regard was, in addition to the maps used by Foote Whyte in the study about Street Corner Society, also the psychosocial analysis of Lidia De Rita on the sociometric control over the neighborhoods of the community of Matera's "Sassi" (1954) study which had as its focus the neighborhood understood not only in the topographical sense, but also in the social sense with a precise value determined by the local culture.

"The place welcomes the imprint of the group," observed Halbwachs, "and it is mutual. Then all group practices can result in spatial terms, and the place that occupies is the meeting of all of the terms" (Halbwacks, p. 218, reported by Bergamaschi 2008, p. 86).

The movements of social actors have moreover contents of communication—verbal and nonverbal—that give meaning to the same contexts, which seem to be aimed at the building in the space of recognizable community, in order to preserve the common memory, essential factor for the transformation of space in an anthropological place (Augé 1993).

In methodological terms the analysis of some factors is considered crucial to determine a clear framework of life and relationships of a social group in the spaces to socio-vital areas observed.

Specifically it is necessary:

(1) observe the verbal and nonverbal communicative behaviors;
(2) pay attention to the relationships of the actors observed (alone or in a group), to posture, attitudes and their behavior, trying to grasp the common spatial directions (e.g. a point or place in particular towards which all go together) or if you make explicit common interests (e.g. more sellers in the same market stall) or a group activity (e.g., talking excitedly, rest or play on a bench, chatting or working at an outdoor cafe, etc.);
(3) to review the situation, i.e. to try to interpret everything that happens in the physical and interactional context between social actors during the observation;
(4) to capture the explicit and implicit meanings of the activities and events taking place in the context of human observation;
(5) to analyze particularly the situations in order to become conscious of what is being place under his own eyes and to build the great mosaic of everyday representation of reality, positioning, like tiles in a mosaic, the elements of cultural, social, economic political, ideological, ethical, symbolical, etc.;
(6) to identify and interpret the personal and interpersonal spaces between subjects and to evaluate the physical and social distances, to grasp the possibility of interaction ritual, the possibility customary, the importance of subjective (private) and collective (public) of each action.

It seemed also interesting to consider the possibility of the physical proximity, to determine the acceptance of others, the lack of personal and social inhibitions, and conversely the natural confidentiality on the part of the subjects or actors, etc. Examples are the perception of body necessitated proximity (e.g. on a bus) and the sensations arising from this. In such contexts it is appropriate to check the distance also physical that must be maintained to others not to appear invasive of another's privacy.

Likewise it is appropriate to refer to the links between bonds and interpersonal activity and bodily, social, hierarchical, functional, professional prestige, distances etc. and related social and living spaces.

It is also believed that among the factors to be detected was particularly important to account for collective behavior that could present the elements of the event, given the possible meanings not only contingent and apparent, but also of values, symbols and rituals.

Specifically for social-life event is meant the totality of the significant interactions observed in an area at a given time, which is the measure, for example, of a very busy place or otherwise isolated and lonely, and that largely characterize also a space as industrial or residential, business, or intended for leisure, of rapid consumption (the stage for a concert) or more long lasting use (theater, hospital). It emphasizes the randomness and the different (multiple, infinite) interconnections and combination possibilities.

The event does not depend on subjective impressions or feelings, but on the quantity and quality of the activities observed between the actors involved. It is not, finally, to cite one simple example, the single aperitif drunk at a specific time, but the set of aperitifs drunk in a given place (bar or coffee), the ritual mode shared by the group, in some sense, the customary semblance (appearance) and conventional salience (meaning and importance) attributed to the same observed event.

Meant as "quid evenit", is the set of observable random everyday activity—individual or associated—made by all the different parties involved in the process of observation, in one context or socio-living space in a determined or determinable time or timeframe.

Together with the morphological study of the environment the analysis of the purposes and opportunities in a social-life context also seem relevant in a sociological path of research also lead to the establishment of possible collective representations of a human habitat, whether urban or rural, agricultural or industrial, technologically advanced or backward, and the needs arise from it.

These variables seem indeed to have the potential to specify the meanings of the actions, activities and observed situations, and help to clarify the connections with aspects, the connections and the emotional and affective reflections that can be traced there.

The goal remains, however, to be able to assemble in a comprehensive framework of variables, those mosaic tiles represented by socio-vital areas, which are too complex to read and interpret, those manifold *topoi* where come to life interpersonal dynamics and are intertwined human relationships, where the reality of social groups occurs, people construct the shared memory and can acquire common and defined social identities.

References

Amendola G. (a cura di) (2009), Il progettista riflessivo, Laterza, Roma-Bari.
Anderson N. (1923) The Hobo. The Sociology of the homeless man, The University of Chicago Press, Chicago.
Augé M. (1993), Non luoghi. Introduzione a una antropologia della submodernità, Elèuthera, Milano.
Babbie E. (2008), The Basics of Social Research, i.t. tr. (2010) La ricerca sociale, Apogeo, Milano.
Benjamin W. (1982), Das Passagenwerk, Suhrkamp Verlag, Frankfurt am Main, it. tr. a cura di E. Ganni (2010), I «passages» di Parigi, 2 voll., Einaudi, Torino.
Bergamaschi M. (2008), Città e spazio nel pensiero di Maurice Halbwachs, in Sociologia urbana e rurale, n. 68.
Bichi R. (2007), Le tecniche dell'osservazione, in Cannavò L., Frudà L. (a cura di), Ricerca sociale. Dal progetto dell'indagine alla costruzione degli indici, Carocci, Roma.
Cipolla C., De Lillo A. (a cura di) (1996), Il sociologo e le sirene: la sfida dei metodi qualitativi, FrancoAngeli, Milano.
Dal Lago A., De Biase R. (2002), Un certo sguardo: introduzione all'etnografia sociale, Laterza, Roma-Bari.
De Lillo A., Arosio L., De Luca S., Ruspini E., Saia E. (2010), L'osservazione, in De Lillo A. (a cura di) Il mondo della ricerca qualitativa, UTET, Novara.

De Rita L. (1954), Controllo sociometrico di vicinati in una comunità lucana, in *Bollettino di psicologia applicata*, agosto-ottobre 1954.

Ferri, B. (2017), Social Sustainability in Urban Regeneration: Indicators and Evaluation Methods in the EU 2020 Programming, *Studies in Systems, Decision and Control, Vol. 66, Maturo (Eds.), Recent Trends in Social Systems: Quantitative Theories and Quantitative Models,* 75–88.

Gans H. J., (1967), Levittowners, Random House-Pantheon Books Division, it.tr. Indagine su una città satellite USA, Il Saggiatore, Milano, 1971.

Goffman E. (1971), Relations in public: microstudies of the public order, Basic Books, New York.

Lynd R. S. and Merrell H. (1929), Middletown. A Study in Contemporary American Culture, Harcourt, Brace and Company, New York.

Maturo F., Hošková-Mayerová, Š. (2017), Fuzzy Regression Models and Alternative Operations for Economic and Social Sciences, *Studies in Systems, Decision and Control, Vol. 66, Maturo (Eds.), Recent Trends in Social Systems: Quantitative Theories and Quantitative Models,* 235–248.

Morin E. (1967), Commune en France, Fayard, Paris, tr. it. (1969) Indagine sulla metamorfosi di Plodémet, Il Saggiatore, Milano, pp. 318–319.

Park R. E., Burgess E. W., McKenzie R. D. (1925), The City, The University of Chicago Press, Chicago.

Poe E.A. (1886), Nouvelles histoires extraordinaires, it.tr. Ch. B., Paris, in Benjamin W., 1982.

Proust M. (1939), *Du coté de chez Swann*, Paris, p. 256, it. tr. (1990), *Un amore di Swann*, De Agostini, Novara.

Simmel G. (2006), Saggi sul paesaggio, a cura di M. Sassatelli, comprende i saggi Philosophie der Landschaft (1913) (it.tr. di L. Perucchi), Die Alpen (1911) (it.tr. di M. Sassatelli), Die Ruine (1907) (it.tr. di M. Sassatelli), Böcklins Landschaften (1907) (it.tr. di L. Perucchi), Armando, Roma.

Spradley J. (1980), Participant Observation, Harcourt Brace Jovanovich College, Fort Worth.

White W.F. (1955), Street Corner Society. The Social Structure of an Italian Slum, The University Chicago Press, it. tr. di M. Ciacci, (1968), Little Italy. Uno slum italo-americano, Laterza, Bari.

Chapter 25
Knowledge Creation Processes Between Open Source Intelligence and Knowledge Management

Stefania Fantinelli

Abstract Connections between intelligence studies and social sciences have been already underlined in different fields: anthropology and cultural intelligence; connections are found between social and cognitive psychology and the competences of the analyst of intelligence; social sciences methodologies and tools are used in intelligence as well. This work will focus on Open Source Intelligence (OSINT), defined as the activity of discovering, discriminating, gathering, validating, analyzing and distributing information derived from sources which are open, public, accessible and unclassified (Fleisher in Inteligencia y Seguridad, 2008). Eight interviews on a sample of Italian OSINT analysts and experts revealed what is the common use of OSINT methods, how they are linked to the knowledge creation and knowledge management processes. Aim of the study is to explore the OSINT methods in a social psychology view and evaluating its relation with knowledge management in organizations.

Keywords Open source intelligence · Knowledge management · Knowledge creation

25.1 Introduction

The definition of intelligence as a social science is the theoretical premise to this work, and the statement proposed argues that intelligence studies can be relevant also for matters other than national security or military issues; in particular Open Source Intelligence (OSINT) is implementable in almost every human activity related to information. It is at first necessary to define the main concepts that will be treated: Intelligence is defined as the activity of discovering, discriminating,

S. Fantinelli (✉)
Department of Languages, Literatures and Cultures, University G. d'Annunzio,
Chieti – Pescara, Via dei Vestini, 31, 66100 Chieti, Italy
e-mail: stefania.fantinelli@unich.it

© Springer International Publishing AG 2017
Š. Hošková-Mayerová et al., *Mathematical-Statistical Models and Qualitative Theories for Economic and Social Sciences*, Studies in Systems, Decision and Control 104,
DOI 10.1007/978-3-319-54819-7_25

gathering, validating, analyzing and distributing information in order to get a sort of strategic advantage on a competitor (Nacci 2014).

An important element to consider when talking about intelligence—meant as national policy activity—is the difference between this informative activity and espionage, as there is a very rich cinematographic and literature history around this kind of topic that contributes to create a secret and mysterious halo surrounding the matter.

Intelligence is the collection and validation of available information concerning national security; while espionage concerns the discovery and trade of confidential information using illicit techniques (Sidoti 1998), such as hidden devices for interceptions, manipulations or blackmail. Furthermore intelligence—meant as a formal discipline—is not only applied in military issues, there are many different areas of interest, such as business, marketing, private or public organizations and in accordance with Sidoti (1998) the common denominator is the security purpose.

If we refer to Open Source Intelligence (OSINT) it implies that the sources are open, public, accessible and unclassified (Fleisher 2008); these clear availabilities and the natural openness concept are in contrast with that traditional idea of secret services. Open is an information born to be spread or communicated, but it is necessary to highlight the concept of the relativity: the openness is not only regarded as a feature of the information, it is more precisely a characteristic of a specific situation. This means for example that an information which has been classified, can be open for a person in a specific context, but the same information cannot be seen as open for another person who can have a different role in the same context. A very good figurative explanation can be the Escher's piece *Relativity*, which clearly describes different points of view about the accessibility, different perspectives and dimensions. Source is intended as every entity with an informative content and a narrative attitude (Nacci 2010) such as an institution, single person or groups, associations, databases, data from the Internet, forum, blog, e-mail, websites, etc.

Intelligence studies have a long history in the field of strategic, military and political activities[1], along centuries the intelligence context has been slowly expanded covering many others areas such as those concerning commercial and social issues; already in 1947 Dulles (who was a former director CIA) began a sort of debunking of intelligence activities, related to the secret aspects. Recently Caligiuri (2016) argued that as Intelligence is relevant for individuals and institutions and it is in relation with knowledge, it can represent a link or a common point among human sciences. Simultaneously it is inevitable to point out that Intelligence has often been related to social sciences as well (Neuman 2006; Best 2008). Even if human sciences and social sciences cannot be exactly equated, there are some matters embedded in both areas; in order to make clearer the affinity between intelligence and those relevant fields in human and social sciences and to highlight

---

[1]Sun Tzu in the V century B.C. elaborated a strategic military thinking becoming famous in the Oriental literature.

the relevance of a multidisciplinary approach to the intelligence field (Nacci 2014), some examples of connections will be described.

It was in the 1957 that Platt, among others, explained how Intelligence specialists used to borrow suggestions and methodologies from social sciences; they also adopted some social science references in order to deepen their own studies about both disciplines (Platt 1957).

An intersection point with social psychology and cognitive psychology regards the analyst's abilities; it was Heuer one of the first and more prolific scholar who dealt with the psychology of intelligence analysis. He described how heuristics are extremely relevant in the analysis work, referring for example to cognitive models that affect everyone's way of perceiving the reality: an analyst's work rely on his personal values, past experiences, rules and information related to the context (Heuer 1999).

Another and more recent trend of intersection with social sciences can be identified in Social Media Intelligence (SOCMINT), it is defined as the intelligence analysis on social media (Omand et al. 2012). But what is exactly meant by social media and how they are relevant in Open Source Intelligence? There have been many attempts in defining the social media concept: Kaplan and Haenlein (2010) included in the definition the foundation elements of web 2.0 because of its collaborative and participatory strengths, and User Generated Content, that can be created and exchanged through social media. Social media can provide informative content, originally created to be spread and communicated to whom can have access to it, representing a possible material for an intelligence analysis, as well as a group of people, a community or an organization (Hošková-Mayerová and Rosická 2015; Hošková-Mayerová 2017).

The acronyms SOCMINT has been coined in 2012 by Omand and colleagues, who leveraging the spreading use of social media, claimed that social media science should be developed as a new academic discipline and became officially part of intelligence family.

Along with the common points regarding applied methods in social sciences and intelligence, it could be useful to mention a social research definition: Bailey defined it as the collection and interpretation of data in order to understand the society (1982). In this case the main objective is the comprehension necessary for a subsequent intervention on the context. Several authors state that Intelligence is a social discipline in the application field, we may wonder if it is considered the same by social scientists. Could it be useful for both research fields to create an enriching dialogue?

25.2 Theoretical Framework

Having defined OSINT as the collection, validation and analysis grounded on open information and sources with the aim of a strategic superiority, a consistent matter to deepen concerns the type of information processing implemented and the potential connection with the creation of knowledge. Can be OSINT intended as a knowledge management method?

Information seem to have a leading role in our lives: we live in the so called information era; we can get information every time, in every part of the world, from every kind of device, there is also a large amount of literature defining the concept of information overload (Rosenberg, 2003), informative overdose (Da Empoli, 2002, cited in Caligiuri 2016), information war (Zhang and Xu 2013). Those are all conditions that make even more harder to individuate, select and utilize the right information for our purposes. On the other hand we can suppose that the large quantity of available information has affected the way in which we gather information and has also changed the human awareness about the important role played by information in taking decisions. Furthermore technology can often support our information selection and subsequent choices: choosing the daily route to work, buying a new product, deciding where to go out for dinner, even in managing diplomatic relations. As a consequence the need for an effective way of information gathering, management and knowledge creation is very compelling, both in private and public contexts. My personal attention is oriented to the information management process, to the knowledge management (KM) and creation in organizations, meant as «social systems who coordinate people through roles, rules and values» (Katz & Kahn, 1966, cited in Haslam 2004).

Along with this information revolution someone has argued that the definition of knowledge has changed: it would be now connected to and described even as the capacity of predicting the future, instead of the knowledge of the past (Mayer-Schönberger & Cukier, 2013, cited in Caligiuri 2016); in fact in intelligence analysis one of the possible output is the creation of alternate scenarios. The intelligence analysis is able to convey a sort of logical interpretation of information (Agostini & Galmonte, cited in Caligiuri 2016) and this is one of the reasons why OSINT methodology can fit well for the purpose of a knowledge management process.

During the information selection process carried out by humans, one of most common mistakes is to search for those information that tend to confirm the initial theory, instead of deny it. This is due to a psychological bias: the common human tendency to look for confirmations and proves, which can cause a false hypothesis confirmation (Snyder 1981); this informative distortion seem to be the most frequent in organizational communication dynamics (Stohl & Redding, 1987, cited in Haslam 2004), (Hošková-Mayerová 2017).

The idea that knowledge has a central role in organizations it is not a new issue and according to a recent summary about the KM theories and definitions (Tardivo 2008) we can state that it took its first steps in the late '80s, through the works of Itami (1987) and Drucker (1998). The Nonaka and Takeuchi's work (1995) provides one the first and most quoted definition of *organizational knowledge creation*: «the capability of a company as a whole to create new knowledge, disseminate it throughout the organization, and embody it in products, services and systems» (Nonaka and Takeuchi 1995). It was Stewart 1997 that represented an additional innovative explanation about KM: the intellectual capital is not something that an organization can own, it is rather knowledge that transforms raw materials in something more valuable (Stewart 1997). So it seems that the constant themes are dynamicity, transformation and change.

The definition provided by Davenport and Prusak (1998) adds the relevant potentiality of the unknown: «knowledge is the identification, management and promotion of what an organization knows or could know»; there is a continuous balance between the elaboration and the exploration of new knowledge (Crossan et al. 1999). It is also important to distinguish knowledge from data and information: the difference between knowledge and information is the crucial factor in framing KM because the real gap between them is due to the human factor (Parraguez Ruiz 2010), individuals and social interactions add dynamicity and complexity to information. Indeed one of the most common disapproval to the first IT application for KM was exactly about the impossibility for machines to grasp the implicit meaning shared by humans (Parraguez Ruiz 2010).

According to the categorization elaborated by Argote and colleagues in 2003, KM can be analyzed by four different points of view which embody four possible approaches: organizational learning, relational, technological and innovation. As far as I am concerned the relational approach would be the perfect frame in explaining OSINT as a method for KM. The systemic theory is the leading rationale in the relational approach: every individual in an organization is interdependent to others, the resulting network has value both in terms of people's ties and resources' accessibility (Argote et al. 2003). Of course this kind of interpretation does not fit very well with a traditional hierarchical organization structure, the relational approach implies a productive and multidirectional communicative exchange that could make easier to overcome some psychological limits related to social identity categorization in communication. According to several scholars (McGarty et al. 1994) the positive persuasive effect in a communication is strictly affected by the current social categorization: people involved in the communicative process have to perceive themselves as belonging to the same social category. Furthermore the receiver of a communicative process evaluate the message as useful and informative according to the interlocutor social categorization (McGarty et al. 1993).

25.3 Research Methodology

E-mail interviews were the selected method for this research, being aware of their advantages and disadvantages which have already been pointed out by many authors (Burton 1994; Young et al. 1998; Curasi 2001). The most relevant benefit in our case was the convenience about time and space, as the specialists involved had different time and location needs. Together with these positive aspects it is necessary to go along with other drawbacks, such as the absence of face to face interaction and the loss of the non-verbal communication and spontaneity; nevertheless e-mail asynchronous communication offers participants more time to reflect, so that they can provide thoughtful answers (Bowden 2015), usually well complete in terms of spelling, grammar and ideas (Kazmer and Xie 2008).

Interviews focus was on open source definition, applied methodology, knowledge, skills and abilities and on the description of knowledge management practices. The sample was constituted of eight expert OSINT Italian analysts, for what concerns the exiguous number of participants it has to be noted the really hard availability of OSINT analysts, as it is often seen as a classified and confidential activity and it is not so frequent an exchange between academic research and this professional area.

A bottom-up qualitative analysis has been conducted, following a grounded theory's perspective (Glaser 1992), to broaden and enrich the leading systemic thoughts; data have been coded in an inductive way (Braun and Clarke 2006), trying to identify those themes strictly linked to the data (Patton 1990). Some key patterns have been detected, according to Braun's theorization about thematic analysis, the *keyness* (Braun and Clarke 2006) of themes has been defined in terms of relevance to the main research interest.

Frequent participants' comments and statements represent meaningful information able to identify the common definition of open source, biases and potential limits of the discipline and the link with knowledge management.

25.4 Analysis and Interpretation

The definition of open source, for almost every participant, is only related to the internet (social network, website, blog, etc.), this is an unexpected opinion since open sources existed even before the internet, maybe it could be a bias due to the information high availability and easiness of access from each part of the world. Each expert addressed the amount of available information as both a positive and a negative matter, some have also mentioned the Big Data issue, showing a shared perception of an information overload.

Ex. 1: «the documental volume is always growing, in an almost exponential way».
Ex. 2: «the collection is the fundamental aspect in our over-informative system».

The information overload is a controversial issue: many different authors claim that there should be a big concern about the increasing amount of data available affecting negatively the cognitive process (Malhotra 1982; Schick et al. 1990) and the performance accuracy (Muller 1984), but on the other hand there are also few scholars who define the information overload and more specifically the big data matter as a fictitious problem. For example in the field of philosophy of information Floridi describes the issue as a mere problem of small patterns (Floridi 2012) meaning that first of all the problem is epistemological rather than technological and moreover that it is relevant to identify specific data features. According to Floridi it is not worrying neither important the increasing amount of data, but it is crucial learn how to detect those little aggregations of substantial data that really add a new value (Floridi 2012).

The majority of participants described OSINT activity as related to political and military contexts (political stability, counter terrorism, reputational risk), someone defined it in more general terms as strategic activities, relevant also for a business organization. In this sense it could be useful briefly explain what competitive intelligence is: «it involves the use of public sources to develop data on competition, competitors, and the market environment» (McGonagle and Vella 2002), as a consequence having information concerning other firms or organizations became a competitive advantage.

The need for comparison can be explained in the same terms as for social comparison; this theory was formulated by the social psychologist Festinger (1954) in order to explain how people evaluate their opinions and abilities. The central assumption is that a personal comparison can be done with other persons and the most realistic evaluation is the one derived from the comparison with others perceived as equal (Festinger 1954).

It could be possible to talk of an *organization comparison* (Fantinelli and Sivilli 2015) that could be useful during crisis periods or as a competitive strategy, in order to strengthen the organization integrated communication or to make a self-evaluation.

In the participants' explanation of the methods there are few descriptions of the human factor, with IT tools being the most quoted; this preference reflects the above mentioned difference with KM that praises its relation with human factor.

Ex. 3: «an OSINT analyst has to know information technology tools, such as web applications and software».

It is worth to mention online text mining as one of the most quoted and utilized techniques in competitive intelligence, marketing and business contexts (Zanasi 2001); text mining is a specific type of linguistic and semantic analysis applied to a large amount of data in order to extract effective knowledge.

Some participants express critical concern about the analysis accuracy, information credibility and sources reliability that analysts tend to solve adopting a systemic analysis approach, as stated in the relational approach for KM, there is however a common attention in the application of a tidy and logic methodology.

Ex. 4: «a significant aspect of methodology is the *information crossing*, to create integrated narrative elements».
Ex. 5: «the customer is already a source».

There are few statements that make clear another key difference between OSINT methodology and KM practice: the first is oriented to a specific objective and matter to solve, the latter has a mainly collaborative purpose, the knowledge creation is oriented to the organization development. Some examples will clarify the concept:

Ex. 6: «we deal with specific information that have to lead to a solution for the expressed question».

Ex. 7: «our efficacy is measured through the solution of the specific problem».

The observed sample is maybe affected by the fact that their job efficacy and productivity is measured through the client satisfaction about the specific question, as the most of participants have stated.

From a more specific psychological point of view it is possible to describe the analyst's job according to Holland's professional codes; in 1959 Holland elaborated a career theory about professional vocational choices, describing six different profiles linked to specific professional environments, activities, specific requirements and skills. This model is used in the field of vocational and counseling psychology in order to estimate job satisfaction derived from the correspondence between job characteristics and worker's personal predisposition or abilities. The model is also used to define a specific work environment according to the categorization in Holland's codes, that are: realistic, investigative, artistic, social, enterprising, conventional.

Several participants highlight the need for dynamic cognitive competencies, it is a very active work given its unpredictability and the variable time schedule. The match of this competence with the Holland enterprising code (1959) is explained by the following quotation:

Ex. 8: «the open source analyst has to be liquid in terms of competencies, flexible, open minded».

The adjective liquid suggests that the job is dynamic, never the same and the analyst needs to possess adaptability.

Many others participants underline the importance of applying critical thinking, analytical and forecasting capacity, meant as the use of alternate scenarios in the analysis (Oppenheimer 2012).

Ex. 9: «the current emergency about terroristic threat imposes predictive skills».

Ex. 10: «the activity of knowledge management and information analysis produces advantages about the prediction of what can happen in the observed area».

It is very frequent the mention to the skill described as open attitude or open mind, that can be explained as the ability in gathering different information and knowing different facts before having an opinion in order to make the best possible decision.

Ex. 11: «fundamental skills are mental openness, initiative attitude and curiosity».

For these reasons it has been detected also a minor congruence with Holland's artistic code, it is a kind of job basically unconventional. Several participants pointed out some general knowledge requested for this kind of job: there is a constant reference to a multidisciplinary approach, political sciences and international relations knowledge. According to the transcription we can identify Holland' investigative type for both the prevalent analysts' attitude and the job environment, even though in the Holland model the coexistence of enterprising and investigative code represents an inconsistency in the job description. Furthermore someone affirms:

Ex. 12: «the research extends over a long period of time, it becomes almost investigative».

According to several participants' opinion OSINT is being applied frequently to social network and this is a recent thread also in social sciences research, defined as network analysis or Netnography. It could be worth to deepen this research topic since it opens a new link between Intelligence, Technology and Social Sciences.

25.5 Conclusion and Future Perspectives

The Italian academic interest for intelligence studies is quite recent, it is in fact around '80s that some researches focused on the intelligence's failures; it is instead very rich and constant the interest for the discipline in terms of military history (Neri 2014).

This gap in the Italian context and the literature review preceding this research highlight that there are no previous similar studies; the hope is that this result could become the basis for a deeper and larger future exploration. Some research limitations are certainly evident: the small sample and the impossibility to generalize our results; having considered only the OSINT analysts' point of view makes possible to further investigate the complementary perspective, private organizations and public institutions could represent the sample for future studies.

Intelligence and more in particular Open Source Intelligence can be seen as daily activities in our lives, as information and knowledge have both determinant roles in decision making; but it is necessary a more comprehensive understanding of which contact points there are with other scientific fields.

It is an American common way of saying that Intelligence and decisionmakers should stay one arm distant that is almost funny since American people are used to metaphorically measure the right social distance with others through an elbow. But we may add that a longer distance between each other does not interrupt the communication flow neither the influence process; so it could be worth to better investigate the analyst/decisionmaker relationship under multiple points of view: how is this distance perceived from both sides? How is accountability distributed? On one hand it is surely right to keep this distance for security and objectivity reasons, but on the other hand it is inevitable that the analysis communication would affect the decisionmaker's opinion. As past studies have demonstrated in social psychology (McGarty et al. 1993) the different social categorization between interlocutors is detrimental for the communicative process; so it could be helpful to further work on the communication process between analysts and decisionmakers.

Some scholars have demonstrated as the time perspective (work together for a limited period of time or having a longer collaborative future perspective) correlated with way of interacting (face to face or mediated communication) can represent a determinant variable in pursue an effective communication and quality decisions in groups (Alge et al. 2003); since computer mediated collaborations are even more frequent, this is another relevant issue to expand in knowledge creation processes.

For what concerns the intelligence analysis applied to internet sources it has to be noted that SOCMINT is already being used for public health monitoring, that is the activity of health experts whose are learning to scan tweets and Google searches/trends to identify pandemics earlier than traditional methods (it has already happened for H1N1 virus). This is another research thread relevant for social scientists as well, we may wander of what other social use of this technique can be implemented outside the corporate area?

References

Alge, B. J., Wiethoff, C., Klein, H. J., (2003) *When does the medium matter? Knowledge-building experiences and opportunities in decision-making teams,* Organizational Behavior and Human Decision Processes, 91, pp. 26–37.

Argote L., MC Evily, B., Reagans, R., (2003) *Managing Knowledge in Organisations. Creating, Retaining and Transferring Knowledge*, Management Sciences, 49, VVII.

Best, J., (2008) *Social Problems*, W.W. Norton & Company, New York.

Bowden, C., Galindo-Gonzalez, S., (2015) *Interviewing when you're not face-to-face: The use of email interviews in a phenomenological study*, International Journal of Doctoral Studies, 10, pp. 79–92.

Braun, V., Clarke, V., (2006) *Using thematic analysis in psychology*, Qualitative Research in Psychology, 3, 2, pp. 77–101.

Burton, P. F., (1994) *Electronic mail as an academic discussion forum*, Journal of Documentation, 50, 2, pp. 99–110.

Caligiuri, M., (2016) *Intelligence e Scienze Umane*, Rubbettino, Soveria Mannelli.

Crossan, M. M., Lane, H. W., White, R. E., (1999) *An organizational learning framework: from intuition to institution*, Academy of Management Review, 24, 3, pp. 522–537.

Curasi, C. F., (2001) *A critical exploration of face-to-face interviewing vs. computer-mediated interviewing*, International Journal of Market Research, 43, 4, pp. 361–375.

Davenport T., Prusak L., (1998) *Working Knowledge*, Harvard Business School Press.

Drucker, P. F., (1998) The coming of the new organization, *Harvard Business Review on Knowledge Management*, pp. 1–19.

Fantinelli, S., Sivilli, D. F., (2015) *Open source intelligence's methodology applied to organizational communication*, Mediterranean Journal of Social Science, 6, 2, pp. 233–239.

Festinger, L., (1954) *A theory of social comparison processes*, Human Relations, 7, pp. 117–140.

Fleisher, C., (2008) *OSINT: Its Implications for Business/Competitive Intelligence Analysis and Analysts,* Inteligencia y Seguridad, 4, pp. 115–141.

Floridi, L. (2012) *Big Data and Their Epistemological Challenge*, Philosophy & Technology, 25, pp. 435–437.

Glaser, B., (1992) *Basics of grounded theory analysis,* Sociology Press.

Haslam, S. A., (2004) Psychology in organizations. The Social Identity Approach, 2nd edition. In Cortini, M., Pagliaro, S., (Eds.) *Psicologia delle organizzazioni*, Maggioli, Santarcangelo di Romagna.

Heuer, R. J., (1999) *Psychology of Intelligence Analisys*, Center for the Study of Intelligence, CIA.

Holland, J. L., (1959) *A theory of Vocational Choice*, Journal of Counseling Psychology,6, 1, pp. 35–45.

Hošková-Mayerová, Š. (2017) Education and Training in Crisis Management. In: *The European Proceedings of Social & Behavioural Sciences EpSBS, Volume XVI*. Future Academy, 2017, p. 849–856. ISBN 2357-1330.

Hošková-Mayerová, Š., Rosická, Z. (2015) E-Learning Pros and Cons: Active Learning Culture? *Procedia - Social and Behavioral Sciences*, 191, no. June 2015, pp. 958–962.

Itami H., (1987) *Mobilizing Invisible Assets*, Harvard University Press, Cambridge.

Kaplan, A. M., Haenlein, M. (2010) Users of the world, unite! The challenges and opportunities of Social Media, *Business Horizons*, n. 53, pp. 59–68.

Kazmer, M. M., Bo Xie, (2008) *Qualitative interviewing in internet studies: playing with the media, playing with the method,* Information, Communication & Society, 11, 2, pp. 257–278.

Malhotra, N. K., (1982) *Information load and consumer decision making,* Journal of Consumer Research, 8, pp. 419–431.

McGarty, C., Haslam, S. A., Hutchinson, K. J., Turner, J. C., (1994) *The effects of salient group memberships on persuasion,* Small Group Research, 25, pp. 267–293.

McGarty, C., Turner, J. C., Oakes, P. J., Haslam, S. A., (1993) *The creation of uncertainty in the influence process: the roles of stimulus information and disagreement with similar others,* European Journal of Social Psychology, 23, pp. 17–38.

McGonagle, J. J, Vella, C. M., (2002) *Bottom line competitive intelligence,* Quorum Books, Westport.

Muller, T. E., (1984) *Buyer response to variations in product information load,*. Psychological Review, 63, pp. 81–97.

Nacci, G., (2010). *Intelligence da Fonti Aperte: per una ontologia Ingenua*, Intelligence & Storia, n. 3.

Nacci, G., (2014) *Open Source Intelligence Abstraction Layer*, Edizioni Epoké, Novi Ligure.

Neri, C., Pasquazzi, S., (2014) Intelligence failures. Teorie, casi empirici e fattori correttivi, *XXVIII Convegno della Società Italiana di Scienza Politica*, Università di Perugia.

Neuman, W. L., (2006) *Social Research Methods: Qualitative and Quantitative Approaches,* Sixth Edition, Pearson Education Inc., Boston.

Nonaka, I., Takeuchi, H., (1995) *The Knowledge-Creating Company: How Japanese Companies Create the Dynamics of Innovation*, Oxford University Press, New York.

Omand, D., Bartlett, J., Miller, C. (2012) Introducing Social Media Intelligence (SOCMINT), *Intelligence and National Security*, pp. 1–23.

Oppenheimer, M. F., (2012) *From Prediction to Recognition: Using Alternate Scenarios to Improve Foreign Policy Decisions,* SAIS Review, 32, 1, pp. 19–31.

Patton, M.Q., (1990) *Qualitative evaluation and research methods, second edition*, Sage, New York.

Parraguez Ruiz, P., (2010) *Knowledge Management Oxymoron's*, MSC in Innovation and Technology Management, University of Bath. Retrieved from http://www.openinnovate.co.uk/papers/KMandHumanFactors.pdf.

Platt, W., (1957) *Strategic Intelligence Production,* Frederick A. Praeger Publishers, New York.

Rosenberg, D., (2003) Early modern information overload. *Journal of the History of Ideas,* 64(1), pp. 1–9.

Schick, A. G., Gorden, L. A., Haka, S., (1990) *Information overload: A temporal approach*, Accounting Organizations and Society, 15, pp. 199–220.

Sidoti, F., (1998) *Morale e metodo nell'intelligence,* Cacucci, Bari.

Snyder, M., (1981) On the self-perpetuating nature of social stereotypes, in D.L. Hamilton (ed.), *Cognitive Processes in Stereotyping and Intergroup Behaviour,* Erlbaum, Hillsdale.

Stewart T. A., (1997) Intellectual Capital. The New Wealth of Organizations, Nicolas Brealey, London.

Tardivo, G. (2008) L'evoluzione degli studi su Knowledge Management, Sinergie, 76, pp. 21–42.

Young, S., Persichitte, K. A. & Tharp, D. D., (1998) *Electronic mail interviews: guidelines for conducting research*, International Journal of Educational Telecommunications, 4, 4, pp. 291–299.

Zanasi, A., (2001) Web and Text Mining for Open Sources Analysis and Competitive Intelligence, presented at *Intelligence in XXI Century*, feb 14–16, 2001, Priverno.

Zhang, K. J., & Xu, W., (2013) "Big fire control" in the condition of information war doi:10.4028/www.scientific.net/AMR.760-762.1269.

Chapter 26
Cultural and Natural Heritage Challenges

Zdena Rosická

Abstract Every cultural heritage object and its content has its unique character and calls for an individual approach considering safety, protection, security, risk-preparedness and further viable use. The public usually know about high-value losses of cultural property caused by burglary, fire of flood when the mass-media report them; however, physical care, including environmental and conservation control, property transfer and transport, personal access, thefts from exhibits during the day, incidents of smash bring about higher-cost internal losses, which are sometimes not reported at all. In case any disaster strikes, harm to cultural treasure is sometimes serious and losses irreplaceable unless relevant measures are taken in time.

Keywords Heritage challenges · Disaster · Protection · Safety

26.1 Introduction

Cultural and natural heritage is unique treasure of every nation; it indicates and demonstrates the level and development of community, philosophy, religion, science, technology, arts, and culture of every country. This asset has to be protected against its damage, vandalism, robbery and other devastating and deteriorating affects.

Disasters or technological accidents usually focus the eyes of the world on the ever-present risk surrounding significant cultural and natural heritage. The power of the latest media is able to draw citizens in all parts of the world into the human drama being played out on site. We are immediately ready to offer and give the time, money, and energy, our fullest support to measures to repair damage and to improve prevention strategies to avert future loss. Nevertheless, once the event is past, the media review when and what happened and tragic examples of visible losses that grip our attention start slowly fading out.

Z. Rosická (✉)
Mendel University in Brno, Zemedelska 1, 613 00 Brno, Czech Republic
e-mail: zdena.rosicka@mendelu.cz

© Springer International Publishing AG 2017
Š. Hošková-Mayerová et al., *Mathematical-Statistical Models and Qualitative Theories for Economic and Social Sciences*, Studies in Systems, Decision and Control 104,
DOI 10.1007/978-3-319-54819-7_26

26.2   View About Disasters

A rather fatalistic view about disasters prevailed in prehistory and much of early historical time up to about two centuries ago. They were primarily seen as the result of astrological or supernatural forces. This is illustrated by the fact that the word *disaster* etymologically entered the English language from a word in French— *desastre*—which in turn is a derivation from Latin words *dis, astro*, which combined roughly meant *formed on a star*. In its early usage, the word had reference to unfavourable or negative effects, resulting from a star or a planet.

In time, the word disaster was applied more to a major physical disturbance such as earthquakes and floods or what came to be traditionally known as actions attributable to be supernatural. In time, disasters were formally labelled in the legal system of many countries as God will, with the implication that nothing could be done about their occurrence. Such a fatalistic attitude of cultural value does not encourage the development of new social groups or approaches to deal with or manage disasters. With the development of secularism, particularly in Western Europe and accompanying development of science as a new way of getting knowledge, a different perception of the source of disasters appeared. That time they were seen as Acts of Nature, and the responsibility was shifted from the scare to a secular view of phenomena.

The shift to a focus on Acts of Nature latently set the stage for an even more drastic shift in perception. As Voltaire said about the large casualties and losses in the 1755 Lisbon earthquake, it should not be perceived as God will but as resulting from building without heed in a highly seismic zone in Portugal (Dynes 1994). However, another and different view of the source of disasters appeared. The God will was displaced by Act of Nature, and the stage was set for the displacement by another view, i.e. that disasters resulted from Act of Men and Women. There were two trends which affected the development of this new perception. Of the secondary importance was the slow appearance of disasters resulting from technological accidents and mishaps. These disasters were seen as resulting from inappropriate actions of human beings. The assumption was that these kinds of disasters could be prevented and their negative effects mitigated or reduced. Since this view spread, it spilled over as a possibility for all kinds of disasters.

The view developed among scholars and researchers on the topic that disasters result directly and indirectly from the actions, intended or otherwise of human beings. If people are living in unprotected flood zones, in non-earthquake protected proof buildings in known seismic zones, or close chemical plants, they are creating the necessary conditions for a hazard to generate a disaster.

The earliest systematic but limited human efforts to try and to adjust to cope with some kinds of disasters were generated by recurrent fires and floods. Fires led eventually to the development of fire departments, floods evoked certain kinds of specific engineering efforts. Neither of these two kinds of agent-specific social reactions constituted any kind of social invention to develop protection system generally, although they represented specific attempts to deal with particular kinds of disasters.

26.2.1 Fires and Floods in History

The Romans were probably the first to establish organized groups to fight fires. These bands were known as Familia Publica and were composed of slaves. They were very inefficient and slow to respond. When a fire in 6 A.D. burned almost a quarter of Rome, the Emperor Augustus abolished the bands and created the Corps of Vigiles which had full time and trained personnel and specialized equipment. They were first professional fire services in the world. They expanded from Rome into the rest of Europe, for example to Britain by at least the 5th century A.D. Such services slowly disintegrated with the decline of the Roman Empire. It was only in 13th century England that building regulations started to appear aimed at reducing the threat of fires along with the later appearance of fire insurance for adjustment to suffered losses. Fire engines, privately run by insurance companies appeared in England. Nevertheless, the Great Fire in 1666 London which left 200,000 homeless and burned out the heart of the city, led to a massive reorganization of the fire services in the city. The new arrangement became the model for the structures and functions that fire departments have in most places of the world today.

In the history, fire departments were overwhelmingly concerned only with fires and not with disasters in general. Recently some organizations have become a little more involved with the crisis periods or disasters. This is because in western type societies they have taken over general function of providing ambulance services and to an extent the providing of emergency medical services (Hošková-Mayerová 2017; Švarcová et al. 2017).

The development of police departments while leading to local specialized groups did not constitute any kind of social invention to cope with disasters generally. The police everywhere have not had any kind of disaster agent as part of their focus or responsibility. However, as a usually present community group with resources they often responded in whatever way they could at the crisis times of local disasters although primarily with the idea of maintaining social order.

As to floods, there have been human efforts to try and prevent or reduce their effects that go deep into the prehistory of the human mankind. There is archaeological evidence that the ancient Egyptians and Chinese made major attempts to control recurrent floods. In Egypt in the 20th century B.C., Amenemher II completed southwest of Cairo what is history's first substantial river control project, the irrigation canal and a dam with sluice gates. There are stories that a Chinese emperor 23 centuries B.C. deepened the ever flooding Yellow River by massive dredging. Nevertheless, historical accounts report that dams for flood control purposes were built as far back as 2600 B.C. in Egypt and in 1260 B.C. in Greece. These and other preventive and mitigative efforts in many other societies were seldom continuously attempted, probably because most were not too much successful. There is still little indication that specialized protection system for disasters was ever developed. Although engineering efforts to cope with floods have been a function of many societies in the ages none directly led to the evolution of any long lasting organization dealing with disasters in general.

Almost any report about a major disaster will note the presence of the military in the situation. The role in disasters or civilian emergencies by the armed forces in any society has been and can be rather diverse. In almost every society, at the crisis period of disasters the military will provide relevant personnel, equipment and facilities. In Japan, for example, where as early as the 1880s a central government disaster organization had been established; the very extensive disaster planning is totally independent of any possible wartime situation. Military units played major roles in response to major disasters in contemporary Japan, as we could observe from the 1964 Niigata earthquake and later after the earthquake in Kobe.

26.3 Benefit and Economy

Based on the Hague Convention, Geneva Convention and UNESCO initiatives, in view of the national legal frameworks, one may conclude that monuments are under special protection. However, is that protection sufficient?

In recent years, there is a tendency in hazard prevention to withdraw from rigidly defined security requirements and focus on solutions based on the results of risk evaluation conducted for a given object or area. Such an approach allows for optimum effects, e.g. ensures fire security at reasonable costs. This also applies to the protection of cultural heritage as many legal acts require cultural property administrators to ensure adequate fire security systems, without naming specific actions that will guarantee sufficient fire security level, Under procedures, instructions and plans prepared for this circumstance, it is necessary to conduct hazard analysis to apply appropriate technical and organizational security measures. General economics deals with relationship among benefit, costs and limited, rare factors or means. Microeconomics understand benefit B as a positive for a consumer, and most authors consider only positive yield which grows depending on quantity up to zero point, i.e. marginal value or zero growth.

Risk economy is considered, in fact, reverse microeconomics, risk is characterized as negative yield. Different approach is considered for enterprising risks. The result or benefit consists in prevention, i.e. decrease of risk marked as B and

$$B = R_0 - R_p \qquad (1)$$

where R_0 is risk without prevention and R_p is risk after prevention. Therefore benefit or yield B or its decrease is always given by subjective assessment or evaluation aggregation. There can be used

- model calculation,
- expert evaluation,
- assessment by respondents' aggregation.

In case the model calculation is possible, it is used preferably. However, a model always involves a subjective approach to the risk; therefore expert evaluation is used applied in most cases. Having got the result or assessment B we can obtain data related to:

- loss specified in money,
- probability,
- empirically specified phenomena,
- mutual risks comparison,
- risk priorities.

Nevertheless, B determination is always a problem, particularly if applied to knowledge management risk.

26.4 Fire Protection Expert Teams

Expert teams are characterized by high level of knowledge, abilities, creativity and information. It is one type of co-called knowledge management. Talking about the public administration transparency, the public should know particular decision-making procedures related to emergency planning, disaster prevention policy, threats to life, property and environment, etc., and the problem of information uncertainty components become more urgent. Sometimes, regardless the number or experts in teams, the system of decision-making inputs and processes cannot be fully accomplished, and this phenomenon should be considered seriously. Experts can generate ideas, suggestions or variants which are solved afterwards, assess and select given or existing alternatives. Figure 26.1 presents a simplified chart with two-way links between the segments. The system is framed in initial information and ideas at the input section and decisions at the output. It is a time-varying system with interactive and iteration procedures. Pair hierarchy of individual steps is evident; the cycling can be arranged at creating solutions themselves. General conclusion cannot be accomplished; every step is of high importance, at decision making they are sometimes "filled" subconsciously.

26.4.1 Fires in Cultural Heritage Objects

In terms of the reasons for fires occurring in historical objects, human negligence such as carelessness in fire handling, improper maintenance of technical devices, installations, etc. can be classified among crucial ones; therefore it becomes highly urgent to take measures to improve the 'culture of safety' among administrators, owners, operators and occupants. Only systematic training and appropriate informational campaign might be useful to improve the current undesirable safety situation (Hošková-Mayerová and Rosická 2015). As a source of particular concern there are high numbers of fires caused by deliberate action, often to cover up prior

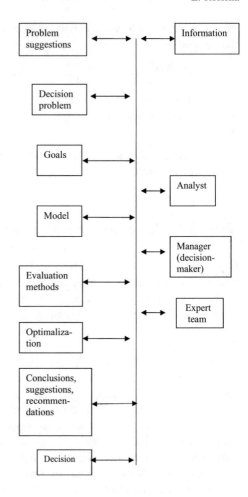

Fig. 26.1 Segments of decision-making system

robbery. In addition, it is very difficult to effectively protect historic objects where wooden structures are used.

Apart from irregularities related to improper evacuation conditions, often connected with the nature of the historic object's structure, there are numerous cases of inadequate protection that can be commonly referred to by one word—negligence. These are particularly as follows:

- lack of bad operation of hand extinguishing equipment,
- no periodic tests of technical devices and installations,
- lack of personnel's scarce knowledge of fire protection instructions.

A typical group of irregularities are deficiencies in signalling and alarm installations. These are mostly related to the fact that the owners or administrators of particular historic objects fail to provide mandatory equipment for the so-called fire monitoring.

26.4.2 Damage to Buildings and Their Contents

The problem can be divided into categories as follows:

- full or partial destruction of objects and building element by burning,
- damage from heat, smoke and combustion by-products (mostly soot) to structures, interior finishes and objects. Particularly at risk are organic elements such as wood, although high temperatures associated with fire can reduce the structural capacity of non-organic materials without visible signs of deterioration,
- water damage resulting from the effects of fire-fighting efforts to arrest the spread of fire.

Some measures for fire protection are usually in place in most buildings, given the high risk of fire in all human settlements. These measures often fail to address the full range of both passive and active measures by which protection can be improved. An effective fire prevention and mitigation strategy for properties should include measures in a full range of complementary organizational, technical and physical areas. Particular objectives to be met in protecting the property's historic values should also be classified (Czech regulation 1998).

Everybody can still remember consequences of Pernštejn castle fire 10 years ago (see Fig. 26.2) when a medieval jewel of the sixteenth century caught fire due to

Fig. 26.2 Pernštejn castle on fire

Fig. 26.3 Krásna Horka

chemical self-ignition of material at repair and reconstruction activities. Human negligence caused in 2012 irreparable damage to a large gothic castle Krásna Horka in eastern Slovakia: two boys aged 11 and 12 were trying to light up a cigarette, set grass at the castle hill on fire and the castle subsequently caught fire (see Fig. 26.3).

26.5 Heritage Crimes

Historical buildings are frequently at risk from vandalism, theft and robbery. At a time of high demand for raw materials, metal theft is becoming a significant problem of many historical places because they are targeted by thieves in search of copper, lead and zinc: roofs, lighting conductors, pipes or gutters are highly attractive because of the raw material price. Artworks and sculptures made of metals are usually exported abroad since it is the theft to order, however, sometimes, huge pieces are also sold for scrap.

Therefore, efficient protection should be applied to prevent building perimeter as well as its valuable content. The "attacker" intending to commit a crime can move to the "target" using various methods. He/she approaches from outside to reach the protected object and is usually familiar with the physical protection of the object. He/she could even practise before how to overcome measures installed on the object perimeter, within the perimeter area and how to overcome future barriers on the predetermined route. He/she would benefit from the object vulnerability, e.g., low detection probability, low mechanical resistance of barrier devices. His/her route does not have to be the shortest or fastest: the goal is to avoid the detection. The unauthorized entry can also be accomplished using different methods, e.g., with help of cleaning services, service organizations or visitors. These people can enter the building legally and do not have to use force to overcome the physical protection measures.

26.5.1 Methods to Overcome Physical Protection Measures

How can intruders enter buildings? What methods are used? Crawling through windows, attacking masonry with a hammer or industrial tools, using glass cutting tool, driving a vehicle through the wall or applying minor explosive or welding equipment? What about digging elaborate tunnels or removing roof tiles or using just brute force?

Barriers have different resistance and require various times to be overcome. Some object can be entered within 1 min; therefore their resistance is very low. The casual intruder can use just small simple tools and physical violence, e.g. kicking, shoulder charging, lifting up or tearing out. Burglars, who need longer time up to 3 min, usually use screwdrivers, pliers, wedges; in case of grilles and hinges, use of small handsaws is very efficient. Within a 5-minute period necessary to enter the object, the attacker uses a crow bar, other screwdrivers and hammers, pin punches and a mechanical drilling tool: thus, he/she is able to attack vulnerable locking devices. 10–15 min are sufficient to gain the entry if more mechanical, battery-powered and electrical tools are available: an axe, chisels, electric drill, jig saw, sabre saw, an angle grinder: the effort brings a reasonable reward, however he/she has to take a greater risk because of noise he/she generates. A highly trained and experienced intruder is prepared to take a high level of risk, is well organized and resolute in his/her effort, and does not usually concern about the level of noise he/she produces. Within twenty minutes, he/she, in addition to above mentioned efficient tools uses a spalling hammer, and an angle grinder with a disc up to 230 mm diameter. All the tools can be operated by a single person and have high performance.

Having entered the object, there are further barriers to be overcome: however, no tools are needed; those are construction barriers such as stairs, lifts, ventilation shafts, pipeline interconnections and doors that are permanently open in one direction. The barriers enable the intruder to pass construction boundaries and to move between different zones within an object. Next, detection and monitoring devices and systems are used to stop or record the unauthorized entry. There can be applied a variety of alarm elements, glass break-up protection systems, magnetic-mechanical contacts, movement detectors, microwave and ultrasonic sensors, passive infrared sensors, CCTV cameras, thermo-cameras or radars. Access control systems began to be used as a substitute for physical inspection of the site in places where identification cards could be easily misused or are considered insufficient.

26.6 Intervention, Prevention and Viable Use

Many heritage buildings need upgrading to comply with current standards of safety and security. It is beneficial and essential for their long-term protection or it helps to secure a viable use. In most cases, heritage buildings can be modified to meet the intention of current building regulations and suitable security measures can be put in place. The challenge is to do it without losing the qualities that can give the building their heritage significance (Rosická 2011).

Speaking about safety standards, modification-oriented proposals consider a building code, personal and public safety and the place insurance.

To find a good solution to improve safety and security without harming the cultural heritage value of a place, it means to follow a logical and systematic approach: to establish the significance of the place, undertake the audit of current safety and security conditions, specify diagnostic techniques, prepare a safety and security policy, and to prepare a schedule.

Considering prevention, there are still some barriers, which are being solved and removed; unfortunately, very slowly. Many of these barriers are still rooted in perception prevalent among professionals at municipalities dealing with heritage; low-level communication and 'passive' resistance, which usually changes in the moment of critical situation (flood calamity, etc.). The focus of the 1996–99 ICOMOS triennium—The wise use of heritage—offers an admirable framework for increasing the attention given to cultural heritage at risk and communication related.

Many built-heritage professionals are still more accustomed to planning for intervention than for prevention. Interventions are visible and dramatic, and permit explicit exploration of various conservation philosophies; approaches focused on maintaining the existing state of the resource rarely carry the same professional appeal of interest. Preventive approaches extend the life of cultural heritage at a lower long-term cost; authenticity is maintained at higher levels if refashioning episodes can be avoided. However, improving preventive measure-focused communication offers many benefits: the extension of the life of cultural heritage properties brings a tangible benefit upon these properties; adopting a cultural-heritage-at-risk framework refocuses conservation attention from the curative to the preventive, from the short-term to the long-term, and consequently offers property owners significant opportunities to perform long-term savings.

The effect of environment on humans has both temporary and long-term nature and long-term exposure can cause lifelong fixation relationship or feeling. People perceive cultural dimensions of space intentionally; these are cultural symbols, architectural objects, symbols, shapes, colours typical for the territory and landscape (Rosická 2010). Another factor that affecting perception results from education and overall outlook, technical focus, age, type of personality, experience, length or intensity of perception. The paradox of perception is the fact that for residents of certain areas of the city, the historic centre becomes invisible; however, it still exists in their subconscious. Residents perceive architectural specifics, details, shapes, colours, unlike tourists, for whom it becomes this amazing environment and deep experience.

26.7 Conclusion

Cultural and natural heritage preservation is an important component of sensitive equilibrium between social, economic and cultural development. The built heritage is often exposed to the threats of demolition, alteration and burglary; however, it is an invaluable asset when properly maintained, preserved and protected: it brings benefits to local communities, employment, improved living conditions and unforgettable genius loci. Unique architectural and natural assemblages, parks and open space can offer new preservation, protection and prevention approach based on reasonable cooperation between public and private sectors to keep unrepeatable sites phenomena, functional diversity and variety. Residents´ attitudes towards the cultural and natural heritage are very important as well as considering their opinion (Rosická 2015). There is always a risk that development and change may threaten the cultural continuity and traditional qualities will be lost at the expense of innovation and uncontrolled development. New projects should contribute to the enhancement of critical areas and adaptations of historical buildings for residential, working or community use and make them viably habitable. The public and residents always highly appreciate if aims of municipality programs emphasize improving living conditions and physical conditions of buildings. The development plans should be in accordance with the ICOMOS Washington Charter, i.e. the introduction of contemporary elements in harmony with the surroundings can contribute to the enrichment of an area (ICOMOS 1987). Any change should be handled carefully in order to encourage new activities such as currently popular adventure tourism, sustainable development, services and infrastructure networks.

References

Czech regulation 137/1998: *General technical requirements for building construction.*
DYNES, R.R. (1994). *Community emergency planning: False assumptions and inappropriate analogies.* International Journal of Mass Emergencies and Disasters No. 12/1994, pp. 141–158, ISSN 0280-7270.
Hošková-Mayerová, Š., (2017) Education and Training in Crisis Management. In: *The European Proceedings of Social & Behavioural Sciences EpSBS, Volume XVI.* Future Academy, 2017, p. 849–856. ISBN 2357-1330.
Hošková-Mayerová, Š., Rosická, Z. (2015) E-Learning Pros and Cons: Active Learning Culture? *Procedia - Social and Behavioral Sciences, 191,* no. June 2015, pp. 958–962.
http://www.firebrno.cz/pred-10-lety-horel-hrad-pernstejn.
http://www.slovakia.com/castles/krasna-horka/.
http://www.hradkrasnahorka.sk/index.php?page&lng=en.
ICOMOS Charter for the Conservation of Historic Towns and Urban Areas (1987). Washington DC, USA.
Rosická, Z (2008) *Emergency preparedness related to cultural heritage,* Institut Jana Pernera, o.p.s., Pardubice, Czech Republic.

Rosická, Z. (2010). *Human factor resulting from protection of landscape, culture and history considering ethics.* Conference Proceedings, Žilinská univerzita v Žiline, Fakulta špecialneho inžinierstva, pp. 607 – 611. ISBN 978-80-554-0201-7.

Rosická, Z. (2011). *Deteriorated Objects should be Converted to Prevent Risk of Decay.* Deterioration, Dependability, Diagnostics. Monograph. 1 ed. Brno: Hansdesign Brno, pp. 125–128. ISBN 978-80-260-0633-6.

Rosická, Z. (2015). *Cultural heritage interests position as a part of municipality planning process.* Housing Policies and Urban Economics, Italy, 2(1), pp. 31–40, ISSN 2385-1031.

Švarcová, I. Hošková-Mayerová, Š., Navrátil, J. Crisis Management and Education in Health. In: *The European Proceedings of Social & Behavioural Sciences EpSBS, Volume XVI.* Future Academy, 2017, p. 255–261. ISSN 23571330.

Chapter 27
Cultivating Storytelling Practice: *"Every Uses of Words to Everybody"*

Fiorella Paone

Abstract The chapter aims to contribute to reflection about new form of emerging literacy starting from communicational changes which characterize new social and cultural paradigm due to more and more use of electronic media. Therefore, we will show some of most recent and critical studies; besides, it will show new characteristics of the linguistic competence and hypothetical educational programme able to receive and value the above-mentioned characteristics. According to this, the *narrative paradigm* thought as linguistic practice of construction of liquid identities will be considered as a possible operative strategy able to promote personal development in a holistic perspective, enforcing both personal and social self.

Keywords Competence · Communication · Story · Literacy · Media · Identity · Linguistic education

27.1 Introduction

Communicational changes pervade every field of our *everyday life* and they have such an importance to characterize society we live in—from time to time named *information society* (Bell 1979), *knowledge society* (EC 2000)[1] or *network society* (Castells 1996)—leading to a new organization both of ways in which social interactions work and ways of knowledge production and sharing (Paone 2016). In fact, media influence socialization process of individuals because they influence

---

Rodari G., *Grammatica della fantasia* (1973), Einaudi, Torino, p. 6.

---

[1]We can consider such a definition universally recognized starting by 2000 when European Council of Lisbon underlighted european need to accelerate the transition to *knowledge society*, investing on growth of mass informatization and, above all, on current educational system.

F. Paone (✉)
d'Annunzio University DEA, Via dei Vestini 31, 66015 Chieti, Italy
e-mail: f.paone@unich.it

© Springer International Publishing AG 2017   401
Š. Hošková-Mayerová et al., *Mathematical-Statistical Models and Qualitative Theories for Economic and Social Sciences*, Studies in Systems, Decision and Control 104,
DOI 10.1007/978-3-319-54819-7_27

dominant standard of knowledge construction, organization and transmission starting by which each person manages self-processes of thought and action (Mc Luhan 1962). In effect, media are thought as: *exchanging and sharing spaces of values that influence processes of identification* (Morcellini and Cortoni 2007: 8–9). Therefore, media need to be thought both by a technological and, above all, cultural perspective considering their effect both on society and single person: *words transformations from oral to written to printing and electronic stage produced and keep producing changes on ways of thought and changes perception that are transmitted by means a mental and speech organization independent from explicit contents and, so, from awareness by which individuals express* (Ong 1982: 8).

To this propose, they speak about new emerging languages, new alphabets, and new competences. *Literacy* by the nature is deeply-rooted to socio-cultural context in which grow and is affected by changes that new-media produce on knowledge construction processes (Street 2003). *Literacy* essential modalities to use a medium and transform a technology (writing, printing or electronic) to instrument of communication conform and modify according to communicational context in which they work becoming more and more well-structured, as we will show in the following paragraphs.

27.2   Literacy and Communicational Changes

In the last 30 years concept of *literacy* faced a deep interpretative revision, moving from a monoconceptual approach to a pluriconceptual one which involve more current and new semantic aspects. Traditionally, the word *literacy* refers both to possess and mastery of code and its own rules and process by means social agents become familiar with practices of making literate of society (Vertecchi 2000). According to Smith, *literacy* allows to give meaning and using own linguistic opportunities of specific culture we live in Goleman et al. (1984). *Literacy, so, should* be considered as a complex system of resources that individuals use when deal with texts production and comprehension (regardless of different media platforms by which above-mentioned texts are produced). This implies that such a concept connotes itself as intrinsically evolutionary concept: linguistic mastery moves from first form of competence developed in natural contexts (*emergent literacy*) to formalized scholastic learning (formal literacy teaching) and to an aware use in complex sociocultural practices. Therefore, *literacy* is an important component of socialization process since it influences development of specific modalities of knowing, organizing, sharing experiences and, as a consequence, acting in an environment of specific sociocultural interactions. *Literacy* is not a set of "neutral" techniques and instruments; it is built in the interaction itself, it's authentic social practice (Banzato 2011) which can influence specific way of thinking, social behavioural rules, values and expressive modalities.

By this point of view, we can use PISA project definition of *literacy*, considered as: *mastery of a determined cultural domain of an individual on such a level able to allow active participation to social life* (Castaldi 2009: 37). This meaning of *literacy* activates an interactive process that makes possible to reach individual and social aims. Nowadays, in fact, concept of *literacy* has not to be considered a list of technical skills or basic competences to be acquired once for all, but it has to be meant as a life-long learning construction of own knowledge as the learning is a constant attempt to adapt to a social environment that changes over and over again (Banzato 2011).

The emerging challenge takes us to meaning matter of understanding how individual thinks, produces and shares knowledge at this communicational revolution stage, *"where the whole is not ascribable to only one episteme, to only one truth or only one literacy"* (Banzato 2011: 27–28), but spreads to pluralities of visions, interpretations and *literacies*, born of which is related to emergence of problems of managing a great amount of information decoded by different media.

During '80s and '90s international scientific debate, promoted above all by American Librarian Associations, takes to the introduction of the concept of *information literacy,* considered as a cognitive activity based on management and analysis of information coming from different technological channels and it requires a critical thought approach of comprehension and use of information (Bawden 2011). Such a competence stands as *integral educational agent, key for an educational complex curriculum* (Lenox and Walker 1992: 16) able to promote the learning of meta-cognitive processes which lead to implement methodologies and basics of research, selection and management of contents. *Information literacy* becomes a landmark on which building new teaching/learning models. According to this perspective, a citizen can be defined *information literate* when *has learned to learn.* Finally, from 2000s *"'information literacy' is not a specific part of literacy but an aspect that changes literacy itself"* (Banzato 2011: 46) and it has becomes a focus of international educational policies. In other words, this competence becomes extension to traditional *literacy* concept that progressively has included possibilities for individuals to comprehend information, in each way they are offered, together with *critical awareness that analyses information by philosophical and ethics aspects* (Johston and Webber 2003).

Another landmark is represented my *media literacy*, a word that was born in United States around '70s (Calvani 2010). According to Buckingham, it implies development of a meta-language able to describe shapes and structures of different modalities of communication and able to comprehend how these work in different context of life (Buckingham 2003). In this sense, *media literacy* implies ability to use and analyse different media as well as a deep analytical comprehension enriched by multiplicities of social, cultural and economic aspects, becoming as *ability of access and analysis of the whole expertise of symbols, words and sounds which we face in our everyday life* (EC 2003: 15).

Moreover, some *literacy* theorists suggest to use the plural of this word due to coexistence of multiple channels of communication: they suggest to speak about *literacies, multiple literacies, digital literacy, and new media literacies,* introducing a pluralities of definitions not yet defined by scientific community.

The *New London Group*,[2] has coined the word *multiliteracies* that is meaningful by a socio-pedagogical point of view; *multiliteracies* deals with modalities of multiple representation of reality and different languages of current society, more than an alphabet. This issue suggests to think alphabetic *literacy* with a holistic and epistemological vision of *multiliteracies* based on awareness of pluralities and modalities of production languages of significance related to current cultural paradigm change.

According to this point of view, we can speak about *digital literacy*, a word that works *like an umbrella which has inside other literacies* (Banzato 2011: 82) time by time involved and enriched by ethic, political and philosophical dimensions depending on the focus. Gilster was the first to introduce *digital literacy* defined as *ability to understand and use information in multiple ways starting from a wide variety of sources* (Glister 1997: 1) referring to a cognitive act that means *handling ideas and not a mere combinations of buttons* (Glister 1997: 1). By this focus, such a competence is applied both to '90s world wide web Glister refers to and each complex informative source not only digital. So, the *digitally literate* will have to be able to understand, assimilate and above all manage these new ways of multi-modal and multidimensional communication. Rheingold systematizes 5 key cores defined *5 literacy* indispensable to become a *digitally literate*: attention, participation, cooperation, critical use of information and web-oriented intelligence. Rheingold highlights how *free digital fluxes of information can enrich people if correctly used as well as become unproductive, insane and toxic for the society if we don't know how receive them (or selectively exclude them), analyse, assimilate and enrich them with our participation and cooperation* (Rheingold 2012: 11). In order not to fall in negative consequences we have to acquire and experience specific skills that make us able to use new media in *intelligent, human and aware* way (Rheingold 2012: 5). Being good digital citizen is a learning competence. As well as we learn to read and write, we can also learn to manage our attention, to look for information on the web and verify them, to effectively participate to virtual network and offer initiatives which could positively affect society. However, Rheingold highlights *the importance to reflect on own communicative practices* (Rheingold 2012: 14): recognizing what media and what social activities we use to avoid, which we are attracted or bother from, which one represent a guide for or mislead us and reflecting on why we act in a certain way.

Jenkins introduces concept of *new media literacies* focusing on some key concepts (Jenkins 2009). First of all, this concept includes both traditional competences that evolved through print culture and more current cultures related to mass media and digital media. So, it's essential to emphasize how process of competences extension happens in an inclusive way and not in a substituting way as expressed in the "Gutenberg revolution". In this sense, ability to comprehend and

---

[2]Team composed of 10 experts coming from different fields (pedagogists, linguist, sociologists, media experts, curriculum and social policies experts) that met up on 1994 to analyze impact new technologies of knowledge production and transmission (in particular Internet) have on teaching and learning styles.

produce texts (*textual literacy*) is essential: *students must be able to read and write time before they will be involved in participating culture* (Jenkins 2009: 92). Besides, *literacy-related competence of XXI century* should be defined as socio-communicative skills, modalities of interaction within a wider community and not thought as individualized skills to use only for individual expression. In effect, *social production of significance is much more than mere multiplication of individual interpretations* (Jenkins 2009: 95): it implies qualitative difference in ways by which we give sense to cultural experiences and, as consequence, implies a deep change of perspective on the way we reflect about competences. *New media literacies are given most relevant assignment: guaranteeing development of that empowerment process which allows young people to experience active and aware citizenship in the information society they have to live in* Ferri and Marinelli (2010): 27–28.

27.3 Linguistic Education and Communicative Competence

New shapes of emerging *literacies* should be combined with a meta-reflection about their constitutive principles and it should rest on grounded cultural basis able to define set of competences development-oriented. Owning a *literacy* suitable to current context is a precondition both to reach scholastic success and actively participate to current multimedia and multi-modal communicational practices in order to access each cultural context with a critical sense. So, it is precondition of individual growth and active experience of right to citizenship. More and more, young persons (and other) don't have ability to discern between different messages they receive and so they are impeded to build grounded basis for expression and action (Turkle 2011). For these reasons it's essential considering media context where they live first communicative interactions in order to understand nature of social and cultural processes by means they build knowledge and criteria of action.

Linguistic education helps us to start solving this matter. Traditionally, linguistic education aim is implementing knowledge system of individuals for the adequate and effective use of a language in relation to linguistic dimension (knowing language), extra-linguistic (integrating language), socio-practical (doing with language) and meta-linguistic (making language) (Balboni 2008). As De Mauro and GISCEL[3] expressed in '70s with their *Ten Thesis for a democratic linguistic education*, linguistic education is essential to act principle of equal social dignity of all citizens as ratified by the 3rd article of Italian Constitution. Republic must remove economic and social barriers which impede the whole develop of person without discrimination of sex, race, language, religion, political opinions, personal and social conditions (Ferreri and Guerrieri 1998). This reflection expresses an

---

[3]www.giscel.org.

essential democratic right to support by means policies realization and practices of linguistic education based on acknowledgement of different cultural background of individuals took as a resource to value and enrich (Milani 1967).

Nowadays, we have to admit that 40 years after *Ten Thesis* publication the linguistic education aim and practices should be enlarged, getting deep in and specifying the concept of communicative competence too. This because cultural background of students has changed a lot and communicative competence take a new role from the past since it has to compare itself with newmedia messages decoding and production (cfr. Jenkins 2009).

Debating about this matter, De Mauro expresses that linguistic education should stimulate and cultivate a critical and aware attitude towards the whole system of information (De Mauro 2004). Likewise, *new need of communication needs old needs which old ones contribute to glorify: self-defining and self-introducing, self-inner-reading and analysing, talking and recognizing the Other, socializing and interacting, arguing, understanding different points of view and expressing the own one, giving words for pain as well as love, expressing needs and comprehending requests, defining diseases, looking for solutions by means of words* (Ferreri 1998: 5) and cooperation, immediacy, recombination, flexibility could be added to the list (Jenkins 2009).

Concepts above-mentioned lead us to wondering about essential elements that allows to build a communicative competence functional to personal and collective communicational well-being, trying to catch that set of cognitive and social strategies which lead to an aware use both of old and new media.

As a rule, being a competent subject in current society implies to be able *to mobilize and integrate cognitive and emotional resources both internal and external in order to face unknown problems which don't have a known solution and show up in significant real context giving effective and efficient performance with both ethically and socially share modalities* (Calvani 2010: 36). Pelleray enriches debate about competence defining it as *ability to face a task, or a set of tasks, succeeding to mobilize and organize own inner, cognitive, emotional, willing resources and using external ones available in a coherent and fruitful way* (Pellerey 1998: 12). Finally, knowledge dimension is involved and, above all, motivational, socio-emotional and meta-cognitive ones.

About communicational matter, concept of competence is strictly related to *literacy*. This should mean that relation between traditional communicative competence and social and digital one have to be recognized and empowered. The actor of *information society* has to be given conditions to get such "active citizenship" that also comes from ability to read, produce and transmit old and new languages[4] in order to be able to experience participation.

By an international perspective, they debated on a redefinition of key competences needed to live and work in current society, with particular attention to role

---

[4]UNESCO highlited this aspect during first World Summit on the Information Society (WSIS), held in 2003 in Givevra and in 2005 in Tunisi.

and function of socialization agencies. According to this approach, *literacy* stands as a new goal that educational systems should reach to in order to enhance development of essential abilities for a lifelong learning process.[5] In December of 2006 a new European Recommendation (EC 2006) included knowledge, skills and attitudes essential for an individual in a society digital competence-based in "common European framework of reference". According to this, digital competence is defined as ability to *use with familiarity and critical sense technologies of information society for the job, free-time and communication* (EC 2006: 40). This issue is supported by basic ability owned in ICT; it supposes a solid awareness of role, opportunities and potential risks we face on the Web and communicating by means of electronic platforms and all this implies a critical and reflexive attitude towards available information and a responsible use of interactive means of communication. Besides, we refer to communicative competence with mother-tongue language and foreign/second one which importance has been already recognized and formalized in the common European framework of reference for languages (EC 2001).

Therefore, new communicative competence goes on with a trans-medial and plurilanguage perspective able to compare with codes, experiences and values different from the belonging ones, it's *ability a person (considered as a social actor) has to use a wide and expanded expertise of linguistic and cultural resources to communicate and take part in inter-cultural interactions* (Luperto 2013: 43).

By this perspective, linguistic education has a new challenge more to face (in addiction to traditional ones): promoting acquisition of plurality of codes and languages which draw structure of contemporary culture with a horizon open to encountering cultures which we live with and an aware use of new media (Rackova and Hoskova-Mayerova 2013).

27.4 Storytelling as Practice of Identity Construction

The following words by De Mauro are more than ever able both to express the deeper and basic meaning of educative goals related to language and open to possible practical strategies of principles concerned: *linguistic education means educating to what lives in the language: our other people's stories (...) it's operative attitude and practicality* (De Mauro 2001). De Mauro shows a definition of linguistic education considered not as speculative process end in itself but dynamics with open nature to life and action, related to concreteness, thus, flexible, evolving, open to change and able to face sociocultural transformations. This assertion is a stimulus to reflect on goals of current linguist education and social and educational

---

[5]This issue should lead to a revision both of traditional curricula and educational practices and competences of teachers. *E-learning programme* (2004–2006) promoted by European Parliament and Council aims to digital competence development.

practices more functional to their realization. Identification of *our and other peo-ple's stories* as something living in the language and goal and possible instrument of linguistic education seems to foster need of inter-cultural and inter-medial meeting which represent new challenge above-mentioned.

Each *story* (myth, legend, biography, tale, scientific or historiographical inter-pretation ecc...) is always an orderly set of facts, personal and not personal events, true or imaginary. The *stories* are a well-organized system of information, logically coherent and able to recognize themselves through change, but opened and flexible, suitable to context, to its actors and related narrative shapes. In short, each *story* is narrative fact, both communicative practice able to show a certain experience in a certain context of communication and system of facts, concatenation of events, particular modality of organizing them (Blezza 2015).

Narrative fact considered as communicative practice the shared experience is highlighted; as system of facts the attention goes to organization of what told, to structures what it's made of Rivoltella (2016). In both of the conceptions above-mentioned, each *story* is composed of representations and structures that allows the *Voice-over* to build complex systems of knowledge; these systems of knowledge help to understand, analyse, orient and build new meanings and, so, activate a hermeneutic process (Demetrio 1996).

Person thinks in terms of *stories* and in this way shapes him/herself, his/her experiences and environment in a constant dynamics of re-interpretation of his/her internal and external environment (Bateson 1984). As Elliade highlights, telling means reveals as fact happened and still happen: *modern person loves to hear telling stories and tell, because it's a way to join again an articulated and sig-nificant world* (Elliade 1963: 205). We can add that *stories, as manual skill, have shaped our thinking organ as well as (we could say) great aggregation of indi-viduals. Hundreds of ancient myths of different and far populations have told, their way, this truth, describing creation of the world as narrative act of a poet-god who give universe a start by means of storytelling* (Nandropausa 2002). *Storytelling* could be the main social act, universal and omnipresent, that leads individuals and communities to explore symbolic dimension both individual and collective in order to understand its operating principle.

Such a social act risks to be braked by media flow that gives only one model of thought, only one conforming *story* in order to maintain *status quo,* to reify and construct rigid identities (Barthes 1982). This kind of *story absorbs critical power and contradictory nature of statements* (Berardi 1997: 125). As a solutions to these kind of risks, Naussbaum (1997) speaks about *narrative imagination* that is ability to understand other people's point of view, their *stories,* their desires and needs through empathic intuition intended as driving force of a migrant and open process of identity construction, able to resist to conformation and reification risks.

Socio-educative responsibility means activating a meta-cognitive process able to activate *narrative imagination*, giving voice to pluralities of *stories* of individuals and communities and its different way of telling, making them available for all communicative competence and hermeneutic and civil instruments of storytelling language. Storytelling perspective, in fact, gives a new point of view about such competences, highlighting aspects of creativity and awareness. Narration is not only putting in common, giving information (lat.: *communicare*) but deals with knowledge as basis for aware and interpretative action. In effect, etymologically word *narration* comes from *gnarigare*, which root is *gna* (to know) plus verb *ager* (to make, to do, to act). Telling and telling about him/herself a person knows him/herself and his/her environment by means a dialectic process that starts from a self-awareness to awareness of being part of a community. Storytelling practice, so, is both educative and social since answer to the need of construction and transmission of one's self and self-culture (Ricoeur 1987). Storytelling practice, besides, answers to a social question of reconstructing imagine and sensibility of a fragmented social body (Hošková-Mayerová and Maturo 2016) compressed by a *sad passions age* (Benasayag and Schmit 2003). through formal and informal educative programme storytelling can become mythopoeia and aware practice able to create a *never-ending flux of living story* (Nandropausa 2002), to put in contact with possible modalities of the existent, to find again *our and other people's stories* in the everyday life, making them emerging, giving them voice, body, colour, making them sounding inside and outside people, giving an open and nomadic way, building a network of "singing streets", as alternative to conforming *story* and only one global thought.

In a socio-educative perspective, concepts above-mentioned give space to different development programs due to build new competences *literacy-related* of XXI century. In fact, a solid *literacy* allows to use masterly and functionally every communicative languages to storytelling. This allows every citizen to own and put in circle information in newest and powerful modalities, distancing from those media contents produced only by a part of people and consumed only by another part but going toward a world where each citizen as a more active role in culture production and in different *stories* which it is made of. As Jenkins explains, such a possibility is not automatically practicable; that is to say that we need education and institutional mediation able to involve citizens to *critical debates that help them to empower their intuitive comprehension of such experiences* (Jenkins 2009: 76).

Education and experience empower storytelling competence making it intentional, enhancing processes of identification and both cultural and inter-cultural cohesion able to use and integrate codes and processes of each media contexts because community doesn't exist without shared storytelling, to be intended as identity space able to integrate the following dimensions: relationship (present), memory (past) and hope (future).

27.5 Conclusion

Storytelling allows to systematize intra- and inter-personal experience making in relation internal with external modality that is matching personal elaboration of self experience with social context that stands as landmark and background for each *story* (Blezza 2011). Our culture offers pluralities of resources and symbolic instruments; comprehending and learning to use them in creative way make *story* recognizable by us and others. We can find two levels of possible implementation and development of reflections. The first is related to construction of sociocultural conditions able to build bridges between different cognitive paradigms and symbolic systems that characterize each communicative device, networking both formal and informal realities of socialization of territory. The second one is acting on the education, offering educational programs which allows everybody to access to each tool of communications, starting from the construction of a solid *literacy* considered both as prerequisite for personal well-being and active experience of citizenship.

References

Balboni P. E. (2008). Fare educazione linguistica, TORINO, Utet.
Banzato M. (2011). *Digital Literacy. Cultura ed educazione per la società della conoscenza.* Milano: Mondadori.
Barthes R. (1982). Introduzione all'analisi strutturale dei racconti, in AA.VV., *L'analisi strutturale del racconto.* Milano: Bompiani.
Bateson G. (1984). *Mente e natura.* Milano: Adelphi.
Buckingham D. (2003). *Media Education. Literacy, learning and Contemporary Culture*, Polity Press, Cambridge.
Bawden D. (2001). Information and digital literacies: a review of concepts. *Journal of Documentation*, 57(2), pp. 218–259.
Bell D. (1979). *The social Framework of the Information Society*, in Dertouzos M e Moses J., *The Computer Age: A Twenty Year View*, Cambridge (MA): Mit Press.
Benasayag M., Schmit G. (2003). *Les passions tristes. Souffrance psychique et crise sociale.* Paris: La Decouverte.
Berardi F. (Bifo) (1997). *Exit. Il nostro contributo all'estinzione della società.* Genova: Costa&Nolan.
Blezza F. (2011). *Pedagogia della vita quotidiana.* Cosenza: Pellegrini.
Blezza F. (2015). *L'arte della parola che aiuta,* Roma: ilmiolibro Gr.Ed. L'Espresso.
Calvani A. (a cura di) (2010). Tecnologia Scuola, Processi Cognitivi. Per un'ecologia dell'apprendere, Franco Angeli, Milano.
Castaldi M. (2009). *Valutare le competenze. Pecorsi e strumenti.* Roma: Carrocci Ed.
Castells M. (1996). The Information Age. Economy, Society and Culture. vol. I: *The Rise of Network Society* Blackwell Publishers, Malden (Mass).
Council of the European Union (2001). Common european framework of reference for languages: learning, teaching, assessment.
De Mauro T. (2001). Apprendere nella società complessa. *Minima Scholaria.* Bari: Laterza.
De Mauro T. (2004). Nuove tesi per un'educazione linguistica democratica, *Insegnare* n. 4: 39–43.
Demetrio D. (1996). *Raccontarsi. L'autobiografia come cura di sé,* Milano: Cortina.
Elliade M. (1963). *Myth and Reality.* W. Trask, New York: Harper and Row.
European Commicion (2003). *eLearning: better eLearning for Europe,* Brussels: Directorate-General for Education and Culture.
European Commision (2006). *Recommendation of the European Parliament and of the Council of 18 December 2006 on key competences for lifelong learning,* (2006/962/EC). http://eur-lex.europa.eu/LexUriServ/LexUriServ.douri=OJ:L:2006:394:0010:0018:en:PDF.
Ferreri S. (1998). Spunti in tema di educazione linguistica. *Quaderni del Giscel* n. 2: 5. Firenze: La Nuova Italia.

Ferreri S., Guerrieri A.R. (1998). *Educazione linguistica vent'anni dopo e oltre.* Firenze: La Nuova Italia.

Ferri P. Marinelli A. (2010). *New media literacy e processi di apprendimento,* in H. Jenkins, *Culture participative e competenze digitali. Media education per il XXI secolo.* Milano: Guerini e associati.

Gilster P. (1997). *Digital literacy,* New York: Wiley.

Goleman H, Oberg A., Smith F. (1984). *Awakening to literacy,* Exeter, NH, Heinemann.

Hošková-Mayerová Š., Maturo A. (2016). Fuzzy Sets and Algebraic Hyperoperations to Model Interpersonal Relations, *Studies in Systems, Decision and Control,* Vol. 66, Antonio Maturo et al. (Eds): Recent Trends in Social Systems: Quantitative Theories and Quantitative Models: 221–223.

Jenkins H. (2009). *Confronting the challengers of participatory culture: Media Education for the 21st century.* Cambridge: MIT Press.

Johston B., Webber S. (2003). Information literacy in Higher Education: a review and case study. *Studies in Higher Education,* 28: 335–52.

Lenox M.F., Walker M.L. (1992). Information literacy: challenge for the future. *International Journal of Information and Library Research,* 4(1): 1–18.

Lisbon European Council (2000, 23/24 March). *Conclusions of the Presidency.* Retrieved from http://www.europarl.europa.eu/summits/lis1_en.htm.

Luperto A. (2013). *Educazione linguistica nella scuola primaria.* Roma: Anicia.

Mc Luhan M. (1962). *The Gutenberg Galaxy: The Making of Typographic Man.* Toronto: University Press.

Milani L. (1967). *Lettera ad una professoressa.* Firenze: LIBRERIA ed. fiorentine.

Morcellini M., Cortoni I. (2007). Provaci ancora, scuola. Idee e proposte contro la svalutazione della scuola nel Tecnoevo, Erickson, Trento.

Nandropausa (2002). Retrieved from http://www.wumingfoundation.com/italiano/Giap/nandropausa3.html#garrett.

Naussbaum M. C. (1997). *Cultivating Humanity: A Classical Defense of Reform in Liberal Education.* Cambridge: Harvard University Press.

Ong, W. J. (1982). *Orality and literacy. The Technologizing of the word,* London and New York: Methuen.

Paone F. (2016). Strategies for a Sociological Diagnosis of Communicational Environment of Students, *Studies in Systems, Decision and Control,* Vol. 66, Antonio Maturo et al. (Eds): Recent Trends in Social Systems: Quantitative Theories and Quantitative Models: 267–281.

Pellerey M. (1998). *L'agire educativo.* Roma: LAS.

Rackova P., Hošková-Mayerová Š. (2013) Current Approaches to Teaching Specialized Subjects in a Foreign Language. *ICERI2013 Proceedings.* Sevilla, Spain: Iated Digital Library: 4775–4783.

Rheingold H. (2012). *Net Smart. How to Thrive Online.* Cambridge: MIT Press.

Ricoeur P. (1987). *Tempo e racconto.* Milano: Jaka Book.

Rivoltella P. C. (2016). *Narrazione.* in Levere, Rivoltella P. C., ZANACCHI A. *La comunicazione. Dizionario di scienze e tecniche.* http://www.lacomunicazione.it.

Street B. (2003). What's new in the new literacy studies? Critical approaches to literacy in theory and practice. *Current Issues in Comparative education,* 5(2), 77–91.

Turkle S. (2011). *Alone Together. Why we expect more from technology and less from each other.* New York: Basic Books.

Vertecchi B. (2000). *Lettaratismo e democrazia:* 15–28, in Gallina V. *La competenza alfabetica in Italia. Una ricerca sulla cultura della popolazione,* Milano: Franco Angeli.

Chapter 28
Sociological Methods and Construction of Local Welfare in Italy

Vincenzo Corsi

Abstract In applied sociology, concepts, theories, models and patterns of explanation are used for the study of social phenomena; they provide a starting point to
plan social policies. The methods of social research are designed to identify, describe
and explain the social needs of individuals, groups and communities. Giving an
"objective" definition of the social needs of people is difficult but necessary. Frequently, the planning of welfare services is not orientated towards the actual needs
expressed by the citizens but follows a different course. The role of pre-evaluation is
strongly emphasised in the planning and construction of the welfare system. In this
chapter are described the methods of analysis of social needs and the models of local
welfare in Italy. In the first section I describe the importance of the study of the social
care needs of the population for the construction of the local welfare system. In the
second section I describe some aspects of local welfare system in Italy.

Keywords Sociology · Applied sociology · Social planning · Social policy ·
Methods of social research · Social care needs

28.1 Applied Sociology and Social Planning

Applied sociology uses concepts, theories and models with the aim to overcome
practical issues. It is involved in the study of social needs and in the planning of
social services and interventions. Therefore, applied sociology is a necessary tool in
the organisation of the welfare systems: through its social research methods it
allows to identify people's needs and to recognise the primary targets for social
policies. Sociological research describes and explains social phenomena, analyses
needs of people, groups, and communities; identifies cause relations between
variables; produces models useful for planning concrete interventions aimed to the

V. Corsi (✉)
Department of Business Administration, University of G. d'Annunzio Chieti-Pescara,
Viale Pindaro, 65100 Pescara, Italy
e-mail: vincenzo.corsi@unich.it

© Springer International Publishing AG 2017   413
Š. Hošková-Mayerová et al., *Mathematical-Statistical Models and Qualitative Theories
for Economic and Social Sciences*, Studies in Systems, Decision and Control 104,
DOI 10.1007/978-3-319-54819-7_28

social change. Applied sociology pursues social knowledge and allows to identify priorities in social intervention; through social research, it takes a leading role in the devising of the local welfare systems. In other words, sociology has an important role in planning and realising social interventions, aimed to improve the wellness of people and local communities.

Sociology is involved in two different areas of social intervention: (a) it describes and explains social phenomena in order to increase the knowledge of the society; (b) it makes this knowledge available to whoever is interested in realising models and practical interventions.

These different areas of social intervention don't always proceed simultaneously. Planning, management and evaluation of policies are aspects of a science applied to the solution of problems. Usually, sociology doesn't produce a knowledge directly involved in interventions; sociological knowledge aims to understand the society and to elaborate theories and models. Applied sociology translates the knowledge produced by the sociological research into actions in order to overcome social issues. What is normally identified as applied sociology is the scientific activity which brings together empirical research and intervention. Applied sociology is involved in all the situation where ideas, theories and models are employed in the solution of social issues.

In these situations, sociology is useful to analyse social problems, conflicts, economic or health disparities, in order to find solutions and create societies capable of responding to people's needs.

Sociological research produces a knowledge which is applied both in planning and social intervention; more generically, in the construction of the welfare systems. Sociology produces awareness of social phenomena, which is used with scientific implications in the planning and evaluation of social, sociocultural and socioeconomic policies, like social services or, for instance, public policies against unemployment and for socio-economic development.

The definition and the planning of social interventions, problem-solving, evaluation of policies and interventions, analysis of organisations, optimisation of the processes and management of changes, are all achieved through sociological knowledge.

Sociology is an empirical discipline that aims to collect data and to measure phenomena. Its goal is to describe and interpret the society in order to build theories capable of explaining social phenomena in their reality. The scientific dimension of the process is given by the ability to combine theory and empirical research.

Sociological knowledge is obtained through empirical research driven by quantitative and qualitative methods. Sociological theories are limits and opportunities that lead sociological research to find methods for the collection and the analysis of empirical data.

Sociological theories are sources useful to evaluate hypotheses and to interpret data collected concerning a defined social issue. Sociological knowledge is the product of a research aimed to describe and explain social phenomena. The study of social phenomena stays within a specific theoretical and methodological approach. The collection and the analysis of data require specific instruments in order to observe the social reality.

Sociology allows to analyse social needs and to identify the main problems which require an intervention, in order to foster the improvement of life conditions for the people, the general wellness and the social, health, economical equality.

Within this conception, in the context of the social policies, the study of the problems and the planning of interventions are important moments in the construction of the welfare. The attention for the society is focused on the welfare policies, when the intervention aims to change people's life conditions. In this cases, social policies are linked with the welfare system, with policies aimed to support the income of people and families, with policies directed to childhood, families, education, elderly, health, immigration, special needs, work, and housing. The concept of social welfare qualifies the systems that provide assistance to the community and is linked with the wider context of the strategies that the society uses in order to resolve people's problems.

Social policies are models and methods to identify and resolve the problems of the every-day life for people, groups and communities. The expression 'social planning' is referred to sectors of the society which are involved in the improvement of the life-conditions for the people in all their aspects: assistance, health, education, and housing. All these aspects of the social welfare are related to the public policies that constitute the welfare system, the intervention of which range from economic support to families in condition of need to the creation of social and health services.

As examples, we can point out either the initiatives aimed at preventing a situation of temporary and permanent inability to work, in cases such as illness or involuntary unemployment, or other forms of assistance for elderly and motherhood. Fields of intervention are many and range from social security to work policies, from health to social assistance, with different models of planning and intervention which can include a partial or total coverage of people's needs, balancing economic assistance and services on the territory (Ranci and Pavolini 2015).

28.2 Sociological Methods and Empirical Analysis

Sociology produces knowledge through the empirical analysis of facts and phenomena, through verified hypotheses and systematic observations of social behaviour. The aim of this discipline is to increase knowledge useful also for the intervention, like the planning, the implementation and the evaluation of social policies. The ambits where sociology is involved are different and diversified, they are in a functional relation with both specific sociologies and general sociological theories, concerning both micro and macro social. The sociological study of society is founded on sociological theories, conceptual categories, models of explanation, research methods for the collection and the analysis of data. The identification of the methods for the empirical sociological research and their correct use in each phase of the analysis produces scientific knowledge reliable and methodologically controlled (Bailey 1982; Corbetta 1999; Guala 2000; Ricolfi 1995). The theory leads the researcher to choose the object of the study, to identify methodological principles which are at the foundation of the research plan, to interpret results.

In the planning of scientific research it is necessary to understand the hypothesis, the goals to reach, the characteristics of the studied object, the methods for the collection and the analysis of data. All these preliminary conditions allow other researchers to repeat the study entirely or partially, even for a critical revision. It is important to explain clearly the plan of the research, addressing the object of the study, the hypothesis, the aims, the theory or the theories of reference, the methods employed and whichever other aspect of the study able to give useful information in favour of the scientific reliability of the results.

The method has a very important role because implies a link between sociological theory and the empirical research. Sociological tradition allows to differentiate and classify the methods of research in formal and informal. Each group incorporate a different research plan and different techniques to collect an analyse data. Each method has its own value in order to study specific aspects of the realty in relation with the different scientific aims.

In sociological research we find qualitative methods, defined also as descriptive methods: the ethnographic method and the participative observation belongs to this category. The ethnographic method uses a limited number of informers within a specific community, whilst the participative observation requires that the social researcher becomes a member of the community or of a specific social group in order to obtain direct information. Quantitative methods, on the other hand, are founded on research techniques which require mainly the use of statistics. In some circumstances, the combination of these two approaches can be considered as a consequence of the necessity of an in-depth study that requires both methodologies.

The difference between informal and formal methods in social research is to be referred mainly to the different aspects of the qualitative and quantitative sociology (Guba and Lincoln 1994). Qualitative sociology usually uses informal methods of research, whilst quantitative sociology uses mainly formal methods (Schwartz and Jacobs 1979). Qualitative methods are more useful than quantitative methods to understand the reasons behind individual behaviours and people's social actions. These methods are mainly employed to understand people's point of view in relation to the social situation in which they are involved. Conversely, quantitative methods are used to verify hypothesis of cause and effect. Sociological research led by qualitative methods aims to identify the point of view of the actors, the single motivations and the aspects of individual and collective life (Memoli 2004, 40).

There are situations in which qualitative methods are used together with quantitative methods. In these cases the researchers tend to obtain general empirical results applicable to similar situations. Generally, in social research quantitative methods may be combined with qualitative methods. The sociological research develops in relation to more sociological theories, as methodological approaches and methods for the detection and data analysis. The choice of methods in social research is defined in relation to the object of social research, the empirical situation in which the object is placed, the cognitive objectives which the research aims, the degree of generalizability expected, the hypothesis that the research aims to verify.

The combination of these variables determines the different methodological approach and the use of different methods, quantitative and qualitative, which are considered more suitable for the purpose. We must naturally consider the importance of quantitative and qualitative methods of social research for the analysis of social care needs of the population. We must not forget the importance of the stakeholder in the planning process of welfare services.

Planning, management and evaluation of social policies are aspects of science applied to the solution of problems; they are a practical use of sociological knowledge in the context of social intervention.

28.3 The Analysis of Social Care Needs

The study of people's social needs or of the needs of a local community is very complex. The problems are related primarily to the epistemological aspects and to the methods of data collection. Giving an objective definition of the social needs of the people is difficult, but necessary.

In social planning the study of needs is both a social research on the areas of intervention and a pre-evaluation of the life-conditions of the people who are the targets of the social policies (Bruni 2009). The study of social needs and health needs is complex because of the epistemological, political, methodological issues that are involved, together with the issues related with the choice of the research techniques to be employed (Palumbo 2001, 118–119). The aspect of the politically driven choice of the needs on which an intervention is planned implies that, sometimes, intervention projects are not oriented to address a real need of the person. For this reason, it is important, in social planning, to have a precursory study and a pre-evaluation of the needs of the population targeted by a project of local welfare. The social determinants of health like the socioeconomic status, the social structure and cultural factors are important to consider (Rebeleanu and Soitu 2017, 159).

The main methods of study, functional to the social planning, are the context analysis and the SWOT analysis. Context analysis is the study of the needs of a population, to be carried out in the moment when it is decided to enact interventions of planning of the local welfare services; it concerns the socio-economical and territorial background of reference. The SWOT analysis is an instrument for the strategic planning that allows to evaluate strengths and weaknesses, opportunities and threats for a project of social intervention. The above methodologies must be integrated with the participation of the stakeholders in the process of planning; the participative method is introduced in Italy by the law n. 328/2000 which establish the 'area plan' (*piano di zona*) as a strategic instrument for an integrated and collaborative approach to the planning of the local welfare (Battistella 2004).

Frequently, the planning of welfare services in Italy is not orientated towards the actual needs expressed by citizens but follows a different course. The role of pre-evaluation is strongly emphasised in the planning and construction of the welfare system. We must naturally consider the importance of quantitative and qualitative methods of social research for the analysis of social care needs of the population. We must not forget the importance of stakeholder in the planning process of welfare services (Corsi 2017, 68).

The participation of the stakeholders in the construction of the local welfare is very important, particularly in the constitution of the 'area plan', which is the planning instrument of the local welfare established by the law 328/2000. The 'area plan' operates in a planned and propelling way aiming to create a system of protection and promotion of the rights of the citizen; it guarantees a gradual development of social policies and services of the territory; for this reason the 'area plan' is a social planning instrument to be continuously redefined according to the analysis of people's social needs, considering the negotiations with the different stakeholders (Venditti 2004, 144).

The 'area plan' is the fist element of the construction of the local welfare in Italy; it includes a separate section called 'Local Social Profile' (*Profilo Sociale Locale*) which shows the analysis of territorial needs. In the 'Local Social Profile' social, demographic, economical and cultural aspects of the territory are analysed through the collection and interpretation of socio-demographic data; the secondary analysis the data; interviews with representatives of public and private agencies; interviews and informal conversations, or thematic meetings with representatives of certain categories of people. The aim of this document is to gain an understanding of social resources and social needs of the area in order to build a local welfare system.

The study of social needs can be divided in the following phases: the elaboration of need indices, the study of social epidemiology. In the first case, the needs and the social resources of the territory are highlighted, in the second case the distribution of social needs through the different stage of life, from childhood to the elderly, is identified. These needs can concern issues related to childhood, families, youth, people with special needs, mental health, poverty, unemployment etc. For the analysis of the social need is important to collect information from sources of a different nature, including the attendance data of the services already present on the territory and specific sociological studies.

The detection of social needs, necessary to the planning of the local welfare and of the social services, is based on an approach founded on macro and micro-data. The latter is related to participative methods with the aim to plan social interventions involving the people directly concerned (Venditti 2004, 101).

The methods to collect data can be surveys over opinion, focus groups, interviews; the methods can be quantitative, i.e. aimed to collect numerical information, or qualitative, aimed to collect information that can be expressed by words. Therefore, the methods to collect data are many and complex. Quantitative methods are useful to collect much information on feelings, opinions and behaviours; they allowed to generalize the results. Surveys, checklists, monitoring forms, forms to reveal the needs, forms to analyse already available data, are normally included into the quantitative methods. On the other hand, qualitative methods are limited to few examples, they don't allow to generalize but consent an in-depth analysis. Interviews and focus groups are considered qualitative methods. The latter uses, to analyze deeply an issue in short time, a group discussion involving the representatives of single categories of people.

The action of planning social services and interventions cannot be detached from an effective recognition of the needs on the territory, and, as a consequence, from the involvement of the different actors. Through the years, the role played by the participation in both the collection and analysis of the data and in the subsequent planning of the local welfare has increased. In Italy, participation is today the most common method to build the local welfare, through quantitative and qualitative research techniques, aimed to achieve a correct analysis of the social needs of the population.

The analysis of the social care needs of the people is carried out through qualitative and quantitative methods of social research, and through the participation of stakeholders. Participation is a method of social planning; we must take into account the demands of social services of the people and to incorporate these services into the local welfare system. The planning of welfare services cannot be separated from the effective recognition of the social care needs of the citizen people. The quantitative methods of social research are important but not sufficient, people must be involved, by qualitative methods, in the analysis and planning of local welfare system.

28.4 From the Synoptic Model to the Incremental Model

Participation is important from the early stage of the analysis of social needs and it remains important during the construction of local welfare and the enactment of social interventions. Every time that a project is enacted, it changes in relation with the social context where it is involved. The concerns expressed by the different stakeholders, who play a role in the construction of the local welfare (Bifulco 2015), are different, therefore it is important that the analysis of the needs is realised through a tangible contact with the reality, using a variety of methods where all the welfare actors are involved. These methods are: case studies, in-depth interviews, the instruments of quantitative analysis capable of identify the aspects which a considered key-points by the actors. The role of the beneficiaries of the social services has changed in time, from simple 'consumers' of the given services to being 'co-planners'.

In Italy, during the 70s, the synoptic model was employed by the social planning (Siza 2002, 45). It is a centralised model of planning, defined by a series of phases and procedures with a single decision maker who identifies the problems the priorities of the intervention and the goals to pursue. The logic behind this system priorities the efficiency in the employment of the economic and human resources, following a model of decision making centred on the rationality of the choices and the rational equilibrium between aims and means (Gherardi 1985, 74–75). It is founded on the predominance of some technical aspects like the analysis of needs, the definition of the aims, the selection of the aims according to the chosen priorities, the analysis of all the alternative of intervention and their comparison, the choice of the alternative which allows to maximise the results (Merlo 2014, 100).

The models allows to identify, from the beginning, the priority of the problems and the goals to reach, that must direct the choice of the solutions and of the interventions to be planned and enacted. This happens through the comparative analysis of the alternatives of intervention and of the means to reach the goals; afterwards the consequences of the enactment of each mean are analysed, and the mean which allows to reach the defined goals in the most efficient way is chosen (Leone and Prezza 2003, 240).

This model is driven by a rational decision-making process where the problem on which to intervene is clearly identified; the possible strategies of intervention are identified and it is possible to establish a list of priorities and foresee the consequences of the decisions taken. Behind the application of this model there is the belief in a decision-making process capable of a complete rationality in the analysis of the possible alternatives and in the choice of the most efficient one. The model implies a common poll of unchallenged values and a perfect faculty of analysis of the decision maker as well as the existence of a technically correct solution of the problems on which it is decided to intervene.

The criticism against the decisional model started with Simon (1947) who, in 1947, underlined the structural difficulties encountered by a decision-maker who uses a model of rational decision. The reason behind this position lay in the evidence of the impossibility of the decision maker to evaluate all the alternatives and the consequences of any of them. It is more realistic to think that the decision-making process aims to identify not the best solution but the most satisfying in a defined social context of problems. The reason of these behaviour is to be found in the limits of human rationality and in the complexity of the environmental context where decisions are taken (Siza 2002, 46).

The criticism against the model starts from the observation that a perfect knowledge of the situation, where the intervention is planned, is impossible, given the complexity of the problems and the quantity of the variables involved; as a consequence, it is impossible that a single decision maker can consider all the possible alternatives for the solution of a problem, therefore the best solution cannot be identified (Lindblom 1979). From this point of view, the construction of a local welfare made through a synoptic planning model, which involve a complete rationality, becomes pointless. The model of 'absolute' rationality shows different limits because the amount of information necessary for the complete understanding of a phenomenon is infinite, the availability of information is normally limited and the cognitive rationality is conditioned by the context where the policy makers are acting. The decision is not the result of a decisional process of complete rationality, but the result of an interaction between different actors of the social planning, involving many stakeholders (Siza 2002, 47).

In this model decision-making processes are de-centralised, there is no longer a single decision maker, but many subjects involved; through their interaction and negotiation the local welfare system is built, using the instrument of the participative planning where public and private institution take part. Participation is involved in all the phases of the planning of the local welfare, which demonstrates itself as a project of continuous construction and negotiation of services and social interventions, in response to the needs of the people who live in a defined territory.

The model of local welfare planning in Italy is participative. The planning uses a 'limited rationality' system and is based on the recognition of many goals and many actors. With the law 328, enacted in 2000, the planning assumes in Italy a systematic quality, allowing autonomy in the decision-making process to the intermediate and peripheral levels of the local government, with a strong recognition of the role played by the third section even in the decisions related to social planning. The law n. 328/2000 contains many important dispositions concerning the social planning for the construction of the local welfare and method indications.

The law identifies four levels of administration:

- The central government has the duty to identify the principles and the aims of the social policy through the National Plan of intervention and social services (*Piano nazionale degli interventi e dei servizi sociali*).
- The provinces help in the planning of the integrated system of intervention and social services.
- The regions have functions of planning, coordination and lead of the social interventions and of verification of their effective enactment on the territory; moreover they control the interactions between the different interventions.
- The municipalities provide the administration concerning the local social interventions and take part to the regional planning. To the municipalities is given the task of planning and enacting the local system of social services.

The constitutional law n. 3/2001, the reformation of the Title V of the Italian Constitution has redefined the responsibilities of the different levels of administration in the construction of the local welfare. The change in the law implies a new relationship between the different levels of administration: between the central government, the regions, the provinces and the municipalities. The central government is now responsible for the definition of the essential levels of assistance, the regions have the faculty of legislate on their own on the local social policies, the municipalities are responsible for the planning and the management of the local system of welfare though the 'Area Plan'.

This multiplicity in the levels of administration of the local welfare in addition to the duties given to the local health agencies (*Aziende Sanitarie Locali*) and to the third sector imply that social planning is by necessity participative, with the involvement of all the subjected interested in the different phases of the project. 'Area plans' are the most innovative element of the reform, they are at the base of the construction of the local welfare, they are the instrument to answer to the social needs of the population on a defined territory; they require an attitude towards negotiation and participation during the planning in order to be an useful instrument to build the local welfare system (Bifulco and Facchini 2015, 3).

Social planning in Italy today tends to involve many institutional actors and of the third sector. The municipality is at the centre of the system but the decision about the social services to be implemented are taken by many institution, public and private, with the participative method. The directing role of the local governments remains central in the system; it collaborates with the competences and decisions belonging to public and private organisations with the intention to combine social and health policies.

28.5 Conclusion

The social planning in Italy in the first half of the 70s plays a marginal role in the context of the public intervention and it is mainly linked to the socio-economic development. Only on the second half of the 70s planning starts to have a role in the reorganisation of the social services on a regional level; it is a prescriptive and centralised planning, applied in a limited way. Between the 80s and the 90s it assumes a local connotation, based on the municipalities, for specific sectors of interventions (i.e. elderly, childhood and adolescence etc.). Only with the law 328 in 2000 the social planning has a systematic order which gives autonomy of decision to the medium and peripheral levels of local government; moreover the important role of the third sector is recognised.

Social planning is an activity with a high level of rationality, uses different approaches and it is situated between model of 'total' rationality and model of 'limited' rationality. In the models of 'total' rationality the main idea is that social planning must be centralised; it is assumed that there is a possibility to have, from the beginning, a complete understanding of the phenomena and of the social problems (Burgalassi 2013, 37–38). On the other hands, the of 'limited' rationality are based on the idea that it is impossible to analyse social phenomena in their complexity; in the social planning the analysis of the information and of the decision taken by the different levels of administration are partial and limited; social planning must have a disposition orientated to evolution and enhancement. Decisions are taken by a multiplicity of actors and different levels of administration which are involved in a collaborative way in the definition of the plan and in the construction of the local welfare.

Social planning is a political project which allows to identify the solutions to the social problems of a territory. Participation allows to take into account, in the construction of the local welfare, different values and suggestions from different social actors and from many levels of the welfare management, it allows to take into account different interests and aims of the public institution, of the third sector and of the people to whom the social services are addressed.

Participation is the main model of social intervention through which, today, the local welfare system is built in Italy. To participate, in the construction of the local welfare, means to involve public institution, in the different levels of administration, and the different actors on the territory whenever it is necessary to choose intervention and strategies of action. The Italian model of local welfare planning shows a strong connection between the services and the territory and potentially a better efficiency in responding to people's social needs, because many actors are involved in the decision making process, including the citizens and the local communities.

References

Bailey, K. D. (1982). *Methods of Social Research*, The Free Press, New York.
Battistella A. (2004). *Costruire e ricostruire i Piani di Zona*, in Battistella A., De Ambrogio U., Ranci Ortigosa E. (2004), Il piano di zona. Costruzione, gestione, valutazione, Carocci Faber, Roma.
Bifulco L., Facchini C. (2015). L'anello mancante. Competenze e partecipazione sociale nei Piani di zona. *Autonomie locali e servizi sociali*, 1/2015, pp. 3–19.
Bifulco L. (2015). *Il welfare locale. Processi e prospettive*, Carocci Editore, Roma, 2015.
Bruni C. (2009). *La Pianificazione Sociale nel quadro della teoria sociologica*, in Ferraro U., Bruni C., Pianificazione e gestione dei servizi sociali. L'approccio sociologico e la prassi operativa, Franco Angeli, Milano.
Burgalassi M. (2013). *Politica sociale e welfare locale*, Carocci, Roma 2013.
Corbetta P. (1999). *Metodologia e tecniche della ricerca sociale*, Il Mulino, Bologna.
Corsi V. (2017). *Methods and Models of Social Planning in Italy*, in Antonio Maturo, Šárka Hošková-Mayerová, Daniela-Tatiana Soitu, JanuszKacprzyk Editors, Recent Trends in Social Systems: Quantitative Theories and Quantitative Models, Studies in Systems, Decision and Control, Vol 66, Springer International Publishing Switzerland.
Gherardi S. (1985). *Sociologia delle decisioni organizzative*, Il Mulino, Bologna.
Guala C. (2000). *Metodi di ricerca sociale. La storia, le tecniche, gli indicatori*, Carocci, Roma.
Guba E.G., Lincoln Y.S. (1994). *Competing Paradigms in Qualitative Research*, in Denzin N. K., Lincoln Y. S., Handbook of Qualitative Research, Sage, Newbury Park.
Leone L., Prezza M. (2003). *Costruire e valutare i progetti nel sociale*, FrancoAngeli, Milano.
Lindblom C. (1979). *Politica e mercato*, Etas, Milano.
Memoli R. (2004). *Strategie e strumenti della ricerca sociale*, FrancoAngeli, Milano.
Merlo G. (2014). *La programmazione sociale. Principi, metodi e strumenti*, Carocci Faber, Roma.
Palumbo M. (2001). *Il processo di valutazione. Decidere, programmare, valutare*, FrancoAngeli, Milano.
Ranci C., Pavolini E. (2015). *Le politiche di welfare*, Il Mulino, Bologna.
A. Rebeleanu, D.-T. Soitu (2017). *Perceptions of the Family Receiving Social Benefits Regarding Access to Healthcare*, in Antonio Maturo, Šárka Hošková-Mayerová, Daniela-Tatiana Soitu, JanuszKacprzyk Editors, Recent Trends in Social Systems: Quantitative Theories and Quantitative Models, Studies in Systems, Decision and Control, Vol 66, Springer International Publishing Switzerland.
Ricolfi L. (1995), La ricerca empirica nelle scienze sociali. *Rassegna Italiana di Sociologia*, XXXVI, 3, pp. 389–418.
Schwartz H., Jacobs J. (1979). *Qualitative Sociology. A Method to the Madness*, New York, The Free Press.
Simon H.A. (1947). *Administrative Behavior. A Study of Decision-Making Process*. Administrative Organization, Mc Millan, NewYork.
Siza R. (2002). *Progettare nel sociale. Regole, metodi e strumenti per una progettazione sostenibile*, FrancoAngeli, Milano.
Venditti M. (2004). *Il sistema sociale locale nelle sue dimensioni valoriale strutturale e funzionale*, Giappichelli Editore, Torino.

Chapter 29
Some Aspects of Social Life in Romanian Villages in the Interwar Period

Daniel Flaut and Enache Tuşa

Abstract In this study we will present some aspects of social life in Romanian villages in the interwar period. At the same time, we will describe the standard of living in rural areas, which varied depending on the ethnicity of the residents, using examples taken from the Dobrudja region.

Keywords Rural population · Learning · Health · Health care · Standard of living

29.1 Introduction

In today's society a large number of facilities available to modern citizens are taken for granted. Health care, access to education, electricity, ready supplies of quality food, paved roads, fast communication and transport represent elements of normality for the majority of us. However, if we cast our eyes almost a century ago the situation was completely different. Life in interwar Romania was not particularly leisurely either in urban or in rural areas, life in the countryside being considerably harder than town life.

The social realities of interwar Romania required actions and initiatives directed towards a better understanding of the territory, the individuals, the culture and traditions of the Romanian space. During the interwar period the Sociological (monographic) School of Bucharest initiated ample and thorough research on the Romanian village. The founder of this school was Dimitrie Gusti (1880–1955), a Romanian sociologist, ethician, Minister of Education, academy member and professor at the Universities of Iaşi and Bucharest (Popa 1996, 526). Organiser and leader of the monographic research sessions, of the "Prince Carol" ("Principele Carol") Royal

D. Flaut (✉) · E. Tuşa
Faculty of History and Political Sciences, Ovidius University, Constanţa, Romania
e-mail: daniel_flaut@yahoo.com

E. Tuşa
e-mail: tusaenache@yahoo.com

© Springer International Publishing AG 2017
Š. Hošková-Mayerová et al., *Mathematical-Statistical Models and Qualitative Theories for Economic and Social Sciences*, Studies in Systems, Decision and Control 104,
DOI 10.1007/978-3-319-54819-7_29

Foundation student teams, of "The Science and Social Reform Archive" ("Arhiva pentru ştiinţă şi reformă socială") and "Romanian Sociology" ("Sociologie româ-nească") journals, Dimitrie Gusti "militated for real knowledge of Romanian realities and in particular for guiding intellectuals towards an understanding of village life" (Vulcănescu and Diaconu 1998, 108). Dimitrie Gusti's research focused on the vil-lage, since he considered that this was "the most widespread and important reality of our social life" (Filipescu 2000, 311). According to Dimitrie Gusti, the village also represented "a sanctuary harbouring and maintaining the way of life of the Romanian people" and "an embodiment of Romanian life in a tiny corner of humanity" (Radu 2007, 162–163). Monographic research highlighted the diversity of interwar rural Romania. Not all villages were alike. They differed from one region to another, from one historical province to another, depending on the soil, climate, ethnic structure, social past, cultural and economic level (Gusti 1941, X).

This article aims to make use of statistical data to provide a brief insight into interwar Romanian villages, exemplifying some important aspects of Dobrudjan villages. Particularly useful sources in writing the article included studies by Dimitrie Gusti and his collaborators, such as Dumitru C. Georgescu, Sabin Manuilă or I. Measnicov, published in "The Science and Social Reform Archive" ("Arhiva pentru ştiinţă şi reformă socială") and "Romanian Sociology" ("Sociologie româ-nească") journals, and the Tulcea County monograph, an yet unpublished work written in 1938 by Const. A. Cristofor.

29.2 Some Aspects of Social Life in Romanian Villages in the Interwar Period

29.2.1 Population

After the end of the First World War, following the Great Union of 1918, the population of Romania doubled, increasing from 7,160,000 inhabitants (in 1912), to approximately 15,541,000 (in 1920) (Gojinescu 2009, 26).

According to the General Population Census of 29 December 1930, Romania had 18,052,896 inhabitants (Georgescu 1937, 68). The country's population den-sity, of 61.2 inhabitants per km^2, was higher than the European average of 44.3 inhabitants per km^2 (Manuilă and Georgescu 1938, 134).

At the beginning of the Second World War, in September 1939, Romania had almost 20,000,000 inhabitants (Banu 1944, 332).

It follows that over the course of two decades, the total population growth in Romania was of 28.3%, one of the highest in Europe (Axenciuc 1999, 366).

Statistics show that during the interwar period Romania was a mainly rural, agrarian country. Rural population represented 79.9% in 1930 (Georgescu 1937, 68) and 81.8% in 1939 (Axenciuc 1999, 370). In 1930 villages contained 79.2% of the households, 78.2% of the buildings and 55.2% of the total number of industrial and commercial enterprises, and had a population density of 48.9 inhabitants per km^2 (Georgescu 1937, 68–69). In the course of the same year, 75.3% of the rural population was Romanian and 75.7% of the village population spoke Romanian (Georgescu 1937, 70; Banu 1944, 336).

Population growth in Romania was almost exclusively due to rural surplus population, as urban natality was much lower than that of villages. Between 1921 and 1925 natality reached 41.8‰ in the countryside and only 21.7‰ in urban areas (Axenciuc 1999, 367).

Between 1931 and 1935, 95.4°% of the country's surplus population came from the countryside. Although rural natality followed a downward curve to 35.5‰, it remained higher than urban population, at 21.4‰. Mortality had largely similar values in the two environments, slightly higher in the villages, more specifically 21.1‰ in the countryside compared to 18.4‰ in urban areas (Georgescu 1937, 76). It is important to note that population growth in towns was not the result of urban population surplus, which was generally very low and sometimes inadequate. Inter-war Romania was no exception to the rule, urban population growth being almost exclusively the result of migration, with rural population surplus moving towards urban areas. A birth place analysis of the Romanian autochthonous population confirms this idea. In 1930, 76.3% of the inhabitants were autochthonous (that is born where their domicile was), with much higher values in villages than in towns, more specifically 82.6% in rural areas and 51.1% in urban areas (Georgescu 1937, 78).

Between 1935 and 1939, rural natality continued to follow a downward curve, reaching 32.2‰, while urban natality remained within the same limits (21–22‰) (Banu 1944, 309–310). On the other hand, mortality decreased in the countryside (19.6‰) and increased in towns (19.4‰) (Banu 1944, 313).

It can be noted that in 1930 men only accounted for 49% of the rural population due to their migration towards urban areas (Georgescu 1937, 69). However, unlike the situation visible in urban areas, village population was characterised by a high percentage of children and young people. In 1930, 48.2% of the rural population consisted of children and teenagers, with a maximum age of 19 (Axenciuc 1999, 368). A similar situation can be observed in 1937, when 48.23% of the village population had ages ranging from 0 to 20 years (Banu 1944, 332), Romania being in this respect one of the countries with the youngest population in Europe (Axenciuc 1999, 368).

29.2.2   The Inhabitants' Marital Status

A lot of marriages had fallen apart during the First World War. The two quiet decades after the war contributed to an increase in the number of marriages, which gave peasants the opportunity to start a family and settle down. Having one's own house, land and family was a source of pride to a peasant, which partly explains why the percentage of single people was lower in rural than in urban areas. As regards the marital status of the inhabitants, it can be noted that in 1930 the percentage of married people was higher in the case of the rural population (61.1%) than among town dwellers (50.6%). Statistics reveal that in 1930 28.3% of the people in rural areas were single, compared to 38.5% in urban areas, which suggests that there was greater family stability in the countryside. Thus, the percentage of divorcees was lower in villages than in towns. In 1930, 0.5% of the rural population had been through a divorce, compared to 1.1% of the urban population. The percentage of widows and widowers was approximately the same in villages and towns, that is 10% in rural areas and 9.5% in urban areas in 1930 (Georgescu 1937, 69–70).

29.2.3   The Inhabitants' Living Conditions

Dwellings in rural areas were family homes. Peasants did not marry their children off without building a house for them. In their opinion, a new house did not only represent a new family following marriage, but also a new dwelling for the newly-weds. This explains why the average ratio among village dwellers was of only 4.5 people per building, compared to 6.5 people per building in towns (in 1930) (Manuilă and Georgescu 1938, 136).

In 1930 almost half of the houses in rural areas were made of rammed earth, mudbrick and wicker, and 27% of these were covered with reeds, straw and corn cobs. In 1929 only 29% of rural dwellings had wooden floors, the rest having earthen floors. Moreover, approximately 20% of homes were only ventilated through the door because the windows were not mobile (Axenciuc 1999, 385–386).

Investigations carried out in 1938 by student teams organised by the "Prince Carol" ("Principele Carol") Royal Foundation into the living conditions of the rural population also reveal family lifestyle. The number of rooms per building increased from 1.6 to 2.3 rooms (Axenciuc 1999, 386). Nevertheless, as a general rule, members of peasant families, even wealthy ones, tended to squeeze in a single room, which was also used for cooking and which, moreover, "had an earthen floor and was consequently unsanitary". According to Dimitrie Gusti, the entire family "tended to squeeze in the few beds in the living room, even when they had a bed per family member in the rest of the house. Almost everywhere, there were at least two people sleeping in each bed" (Gusti 1938, 435).

There were almost no infrastructure changes in interwar villages. Most villages still had dirt roads. There were no ditches and pavements and during the rainy season all villagers had to wade through the mud. Only the national or county road going through the village was paved. Rural dwellings continued to be lit by means of kerosene lamps, but electric light also gained entrance to the countryside. In 1938, 66 villages situated in the vicinity of towns (out of a total of 15,201 rural settlements) had electric light. Only 1.25% of rural households had access to electricity. There was no sewage and no running water because villages had limited budgets (Axenciuc 1999, 389).

29.2.4   *Learning*

In 1930 Romanian literacy rates reached 57% among those over the age of 7 (Manuilă 1936, 933). Although it was lower than before, illiteracy still represented a major problem in interwar rural areas. The highest literacy rate was among 7–12 year-old children (72.4%) (Measnicov 1937a, 113) and the lowest among people over 65 (27.7%) (Manuilă 1936, 955). In rural areas the number of illiterate people was higher than in urban areas, with 48.7% in villages compared to 22.7% in towns (Georgescu 1937, 70). In urban and rural areas alike the literacy rate was higher among males than females. In towns, 84.5% of the men and 70.3% of the women could read and write. The situation was different in villages, where 64.9% of the men and only 38.7% of the women could read and write. Women from rural areas represented more than 40% of the Romanian population over the age of 7. In fact, rural female illiteracy accounted for the high illiteracy rates in Romania (Manuilă 1936, 938–939).

In 1930, half of the villagers over the age of 7 had never attended school (Georgescu 1937, 70). Some of the problems preventing access to education in rural areas included: the lack of cooperation between primary school teachers and parents, with the latter not being always aware of the benefits of learning; the fact that village children were assigned household chores, which limited the time available for study; the location of the school, which could also result in a higher number of illiterate people if it was far away from the children's homes; the inadequate resources of village families, etc. (Measnicov 1937a, 112).

The small number of primary school teachers in the interwar period represented another cause of the low literacy rates in rural areas. According to the 1930 census, Romania had 21.2 primary school teachers per 10,000 inhabitants, irrespective of the (rural or urban) area (Measnicov 1937a, 113). In the 1936/1937 academic year their number increased by 1.4%, reaching 22.5 primary school teachers per 10,000 inhabitants, irrespective of the (rural or urban) area (Measnicov 1937a, 116).

In 1930 most of the literate people in the countryside (93%) had only attended primary school. Very few village people attended secondary (4%) and vocational (1.4%) schools (Georgescu 1937, 70; Manuilă 1936, 933). In Romania 2.2% of the men and 0.7% of the women received higher education, with a considerable difference between rural and urban areas. In the countryside only 0.6% of the men and 0.1% of the women received higher education compared to towns, where the percentages were of 6.4% for men and 2.1% for women (Manuilă 1936, 952). The high cost of accommodation and tuition in town and the fact that young people had to do their share of the farm work made it impossible for over 99% of the parents to send their children to secondary school and especially university. There were only around 4–8 well-educated people per village (Axenciuc 1999, 387).

29.2.5 Health and Health Care

"The dire state of hygiene and health care" represented "the most alarming problem" in Romanian villages (Axenciuc 1999, 389). In general, rural sanitary and
medical assistance was a challenging problem in mainly agrarian countries, which
was also the case in interwar Romania. The life of country doctors was by no means
easy. They lived in isolation and experienced the typical hardships of rural areas: no
accessible roads for most of the year, no home comforts, considerable difficulties
getting food and supplies, no means of transport, harsh meteorological conditions,
uncertainty. The latter was also due to the fact that peasants were not used to paying
for health care, did not dispose of adequate means of payment or even lacked faith
in doctors. The peasants' lack of faith was occasionally fuelled by the so-called
'village elite' consisting of priests and primary school teachers who feared that
doctors might challenge the moral influence they already exerted in the village.
There were cases of doctors who had to leave their respective villages because they
could not deal with all these suspicions (Gheorghiu 1937, 80).

On 31 December 1921 the Ministry of Labour, Health and Social Welfare
budget provided for 1057 doctors, 237 in urban circumscriptions and 820 in rural
circumscriptions. Over the following years the number of doctors stationed in rural
areas increased, "yet the situation was still far from satisfactory". In 1925 there
were 4950 doctors, with only 964 working in rural areas. There were only 866 rural
circumscription doctors (Şandru 1980, 261). In 1934, Romania had approximately
one thousand country doctors who had to provide health care to a village population
of approximately 15.5 million inhabitants, which meant one doctor per 15,500
villagers. At the beginning of the following year, Romanian villages had 1360
doctors but would have needed at least 6500 (Şandru 1980, 266–267). Statistics
show that in 1937 there were 7162 doctors in Romania, with 1935 assigned to rural
settlements, that is 12.3 doctors per 100,000 inhabitants. 1371 of these were
employed by the Ministry of Health and 564 by other public or private institutions.
It is obvious that health care represented a rather serious problem in Romania,
compared to statistics published in other countries. For example, Mexico had, on
average, 1 doctor per 4237 inhabitants, Poland had 1 doctor per 3289 inhabitants,
Bulgaria had 1 doctor per 3059 inhabitants, England had 1 doctor per 1490
inhabitants, etc. (Gheorghiu 1937, 81–82).

It can be noted that in 1937 there was a significant percentage (14.6%) of female
doctors in rural areas. The ethnic structure of rural medical personnel represents
another extremely important aspect. In that year, 49.9% of country doctors were
Romanian, 31.5% Jewish and 18.6% of other nationalities. A doctor who did not
have the same nationality and religion as the superstitious and insufficiently educated
village population could rarely gain his patients' trust (Gheorghiu 1937, 83–85).

Rural areas had clinics and general hospitals. In 1937 the Ministry of Health had
in its subordination 247 general hospitals, approximately 150 of which were
exclusively assigned to rural areas. There were also 715 rural pharmacies, which
meant peasants did not have to travel to get the necessary medication. However, the

lack of medical laboratories, medical personnel and specialised sanitary equipment meant rural health care facilities could not provide adequate medical services to their patients (Gheorghiu 1937, 86). The 10–20 km distance to rural medical circumscriptions led patients to make do without seeing a doctor (Axenciuc 1999, 389). In 1933, 306,930 of the 527,391 births recorded in Romanian villages were assisted by unqualified midwives and 27,277 births had no kind of medical assistance (Banu 1944, 379). In 1936, 68.3% of the total number of deceased people had never been seen by a doctor or other medical personnel, and 78.8% of the deceased infants had died "without any kind of health care". Since calling in a town doctor was very costly (300–500 lei), the majority of sick people had to forego health care. The 'medical' treatments peasants mostly relied on were "traditional remedies and words of wisdom, spells and, in the happiest cases, the help of village midwives" (Axenciuc 1999, 390). As was also the case in towns, infant mortality was extremely high in villages. In 1930 it had almost identical values in rural (17.6 per hundred live births) and urban (17.3 per hundred live births) areas (Banu 1944, 316).

Hygiene conditions in villages were considerably worse than in towns. Peasants had no notions of hygiene, drinking water came from rivers and wells, there were no health care and hygiene facilities in the countryside and no public baths. In 1929, 22% of the 3.1 million rural dwellings had only one room and 53.55% had no water closets, "that is toilet rooms set up for the purpose" (Axenciuc 1999, 390).

The high mortality and morbidity of the rural population were also due to the villagers' poor nutrition. Surveys carried out in 1938 by "Prince Carol" ("Principele Carol") Royal Foundation student teams on people's living conditions highlighted the fact that peasants' diets were "inadequate as a result of the excessive consumption of corn, the low consumption of animal protein and fresh food" and "the lack of skill in food preparation". Their main flaw however resided in "their lack of balance, since the typical peasant family experienced stages of quantitative overeating in winter and long periods of characteristic malnourishment in summer" (Gusti 1938, 435).

School age children also received inadequate nutrition. A survey carried out over a period of 8 days in the course of the 1936/1937 academic year by a Căianul-Mic-Someş primary school teacher regarding the diets of 50 of his second grade students revealed the following: in 56 (14%) of the 400 cases analysed, students had come to school on an empty stomach in the morning, in 240 cases (60%) they had eaten plain polenta and in 104 cases (26%) they had eaten broth or meat with polenta. At noon, out of the 400 cases 78% of the students had eaten polenta with milk or sausages with broth, 18% had eaten plain polenta or bread, and 4% had come to school on an empty stomach. In the evening, out of the 400 cases included in the survey, 40% of the students had eaten plain polenta, 18% plain bread and 42% polenta with cream. Only one student out of the 50 had sufficient food. Store bought items used by the mothers of these students to prepare food were regarded as special day treats. For example, students drank tea in 10 instances and consumed sugar in 9 instances out of the 1200 meals over the 8 days (Vidican 1938, 381–382).

29.3 Some Aspects of Social Life in Villages from the Dobrudja Region in the Interwar Period

Dobrudja, the land between the Danube and the Black Sea, became once more part of the modern Romanian state at the end of the Russo-Turkish War of 1877–1878 (recorded in Romanian history as the War of Independence). The June-July 1878 Congress of Berlin acknowledged Romania's historical rights over Dobrudja. On 14 November 1878, at the time of its union with Romania, Dobrudja, including the Danube Delta and Snake Island, had a surface area of 15.776 km^2 and a population ranging from 80,000 to 100,000 inhabitants (Pătraşcu and Manea 2011, 241). In 1880, Dobrudja was divided into two counties: Tulcea and Constanţa (Nistor 2003, 178). Southern Dobrudja (or Cadrilater) was integrated in the Romanian state in accordance with the provisions of the Treaty of Bucharest (28 July–10 August 1913) at the end of the Second Balkan War (Anderson and Hershey 1918, 439) and ceded to Bulgaria under the Romanian-Bulgarian Treaty of Craiova (7 September 1940) (Preda-Mătăsaru 2004).

29.3.1 Population

In 1918 Dobrudja had 668.645 inhabitants, comprising 33.5% Romanians, 25.9% Bulgarians and 26.5% Turks and Tartars. The population increased in subsequent years due to rural surplus population resulting from colonization and natality (Limona 2009, 36).

According to the 1930 census, Dobrudja had a surface area of 23,262 km^2 and was divided into four counties: Constanţa, Tulcea, Durostor and Caliacra. Its 725 villages had a total population of 617,952 inhabitants (Georgescu 1937, 68), comprising 42.2% Romanians, 1.6% Germans, 5.3% Russians, 25.4% Bulgarians, 24.3% Turks (Banu1944, 337). It had the lowest population density in the country, that is 26.6 inhabitants per km^2. The average population was 852 inhabitants per village and the region had 129,916 households (Georgescu 1937, 68).

As was the case in the other Romanian provinces, Dobrudja experienced village to town migration caused by economic factors. However, the highest percentage of Dobrudjan rural population remained in the villages. The population shift from rural to urban areas did not have a noticeable impact on the age distribution of village population. At the same time, Dobrudja also experienced inter-village population shifts as a result of the marriages of young people who chose a spouse from a nearby or distant village. This was the main reason behind the relatively high percentage of people who lived in a certain village but were born in another village of the same county. Inter-village population shifts were also caused by economic factors, with richer but less populated villages welcoming a higher number of new inhabitants. A birth place analysis of rural Dobrudjan population based on the data collected in the 1930 census reveals that 73% of the men were born in the same village where their domicile was, compared to 68.1% of the women. Correspondingly, 15.4% of the women had changed domiciles within the same county, compared to 9% of the men. This was due to the fact that in most cases it was the woman who changed her domicile following marriage, moving into the husband's home. Dobrudja had the highest percentage of rural inhabitants coming from other provinces (6.1% of the men and 4.9% of the women) due to colonization from other regions. The percentage of inhabitants born abroad who settled down in various Dobrudjan villages (3.6% of the men and 3.3% of the women) was also higher than in other Romanian provinces, due to colonists from Greece or Yugoslavia (Measnicov 1942, 394–395).

Between 1931 and 1935, Dobrudja had a total natality of 40‰, whereas mortality reached 22.7‰. Rural areas had higher natality (43.7‰) and mortality (23.4‰) rates than urban areas, where natality reached 27.8‰ and mortality reached 20.2‰ (Georgescu 1937, 76).

Research carried out in 1934 in Tulcea County, where natality reached approximately 60‰, reveals that in Daeni village 275 women had given birth to 2173 babies, 1136 of whom were still living at the time of the study. It follows that a woman had an average of 8 births and 4.1 living children. 77 of the 275 women under observation had given birth to over 10 children each, with 7 of these giving birth to over 15 children each (Manuilă 1934, 79).

Unlike the rest of the country, Dobrudja had a higher percentage of males than females, both in the countryside, where the male population reached 50.3%, and especially in urban areas, where it reached 52.3% in 1930. The explanation resides in the fact that Dobrudja was a province largely populated through colonization and immigration (Manuilă and Georgescu 1938, 139).

29.3.2   Learning

In 1930, 47.5% of the people in rural Dobrudja were literate. The percentage was much higher among men (60.7%) than among women (34.1%) (Manuilă 1936, 940), with the highest literacy rate among 7–12 year-old children (67.3%) and the lowest among people over 65 (12.7%). In Dobrudjan villages over three quarters (75.9%) of women between the ages of 20 and 64 were illiterate (Manuilă 1936, 956). Most of the literate people in rural Dobrudja (94,1%) had only attended primary school. Very few of the villagers attended secondary (3.8%) and vocational (1%) schools. 1.9% of the men and 1.1% of the women in urban areas received higher education and it is interesting to observe that in rural Dobrudja the percentage of women who had received higher education was equal to that of men (0.1%) (Manuilă 1936, 947).

The statistics of the time show that in 1930 Dobrudja had 21.4 primary school teachers per 10,000 inhabitants, and in the case of the 7–12 age group there was 1 teacher per 56 children. It can be noted that Dobrudja had a relatively high percentage of literate people, notwithstanding the relatively low number of primary school teachers. The significant percentage of people who could read and write in

Dobrudja was due in part to the fact that a considerable percentage of the population of this province lived in urban areas (Measnicov 1937a, 113–115). A comparison between the situation in the 1936/1937 academic year and 1930 reveals a 3.8% increase in the number of primary school teachers relative to the population of Dobrudja (25.2 primary school teachers per 10,000 inhabitants) (Measnicov 1937a, 116).

In 1937, in the villages from Tulcea County there were 52 female school teachers in nursery schools and 412 school teachers (258 male and 154 female) in primary schools. In villages with a German population, in addition to the respective schools, there was "a position for German language and Catholic religion". In these villages from Tulcea County there was a total number of 3 school teachers for these positions (Cristofor 1938, 153). It must be noted that the Germans of Ciucurova, Malcoci and Atmagea paid considerable attention to the upbringing and education of their children and instilled in them the principles of order and good house-keeping. Households belonging to the inhabitants of these villages contained newspapers, specialised journals or general knowledge books (Cristofor 1938, 160).

29.3.3   Health, Health Care and the Standard of Living

Rural health care in Dobrudja was in dire straits. Between 1932 and 1934 the yearly average rate of infant mortality in Dobrudjan villages reached 20.4‰, Dobrudja coming second only to Bucovina in this respect. 58.8% of the births in Dobrudjan villages were assisted by unqualified midwives or had no kind of assistance. The lack of medical personnel was a problem that went beyond child birth. The majority of sick people in rural areas received no health care. Between 1934 and 1936, 85% of the deceased Dobrudjan villagers had never received any kind of medical treatment. This situation was caused by the peasants' poverty and ignorance, as well as by the shortage of doctors. Doctors settled down in places where they could practise medicine, that is where the population was used to visiting doctors and had adequate financial resources (Measnicov 1937b, 161–164). Dobrudja had the smallest number of country doctors per 100,000 inhabitants. Between 1934 and 1936, according to statistics, the total number of doctors in Dobrudja was 234, comprising 125 Romanians, 51 Jews, 1 Hungarian, 5 Germans and 52 doctors of other nationalities. Only 48 of these doctors practised exclusively in rural areas, which translated into 7 doctors per 100,000 inhabitants, with one doctor treating 14,247 inhabitants. The ratio of country doctors per 100,000 inhabitants differed from one county to the next, as follows: Durostor −4.1; Caliacra −5.8; Constanţa −8.8 and Tulcea −9.1 (Gheorghiu 1937, 82–84).

The standards of living of villagers differed from those of city and small town dwellers. There were also differences in terms of living standards amongst members of the peasant class. Moreover, since Dobrudja represented a conglomerate of nationalities, standards of living in the province's rural settlements also differed depending on the inhabitants' ethnicity.

In Caraibil (the old name of Colina, Tulcea County) (Cizer 2012, 164), a village with Tartar population situated north of Lake Razelm, dwellings were small, unsanitary, generally built out of clay and covered with rammed earth or reeds. Houses were heated with reeds or thistles gathered from the fields in autumn. The inhabitants regularly consumed bread and meat, even more so than in villages with Romanian population. Village natality was low (Cristofor 1938, 157).

In Uzlina, Litcov and Caraorman (Tulcea County) (Cizer 2012, 197), "villages inhabited by Russians and Lipovans" situated in the Danube Delta, on the Sfântul Gheorghe branch of the Danube, houses were small but more spacious than those belonging to the Tartars of Caraibil, covered with reeds and provided with bigger windows. Homes were heated only with reeds. The inhabitants' diets consisted of fish, polenta and bread, which was generally made at home. They consumed almost no beef and very little poultry. These villages were characterised by high natality but similarly high mortality due to the lack of health care and the excessive consumption of alcohol (Cristofor 1938, 157–158).

In Topolog and Dăeni (Tulcea County), villages inhabited by Romanians, houses were spacious, with bright and well distributed rooms. Homes were heated with wood and reeds. The inhabitants had adequate diets, frequently consuming meat and bread. The population had access to health care and mortality was low (Cristofor 1938, 158–159).

In Beidaud, a village situated in the southwest of Tulcea County (Cizer 2012, 193), whose population was mainly Bulgarian, houses were comfortable but below the inhabitants' means. The population did not abide by even the most elementary rules of hygiene, but plentiful food and living in the open air protected them against illness. Natality was high, mortality low and life expectancy higher than in the case of any other population (Cristofor 1938, 159–160).

In Ciucurova, Malcoci and Atmagea (Tulcea County), villages with German population, the houses were large, with bright and well distributed rooms. Houses were heated with reeds or wood. The villagers' diets were rich and varied, similar to those of many town dwellers. The locals consumed home-made bread every day (Cristofor 1938, 160). In all the villages inhabited by Dobrudjan Germans children's diets were rich and diverse, consisting of "milk, yoghurt, butter, poultry, pork, beef and mutton" (Stinghe and Toma 2007, 182).

29.4   Conclusions

This article has aimed to make use of statistical data to provide a brief insight into interwar Romanian villages, exemplifying some important aspects of Dobrudjan villages. It gives us great pleasure to observe the thoroughness, diversity and complexity of the data presented in these statistics, the conclusions drawn and in particular the predictions made. In this respect, one of these analyses foresaw for example that by the year 1960 the rural population of Romania was likely to reach 23–24 million inhabitants, provided that a constant population growth rhythm was maintained (Madgearu 1940, 51). This might have happened. Nobody anticipated the fact that the First World War was going to be shortly followed by a second one, bringing with it unprecedented atrocities. Likewise, nobody anticipated the tragic year 1940, when Romania lost Bessarabia, Northern Bukovina, Hertza, Cadrilater, and Northern Transylvania (the latter to be recovered in 1944). In 1947 Romania had a 238,391 km^2 territory, the same as its present day surface area. Romania lost a 56,658 km^2 territory and almost 3.8 million inhabitants, half of whom were Romanians (Trebici 1995, 113). The predictions made by the sociologist and economist Virgil Madgearu regarding the increase of the rural population of Romania did not come true. Nowadays, Romania has a total resident population of less than 20 million inhabitants, approximately the same figure as in 1966.

References

Anderson, F.M., Hershey, A.S.: Handbook for the Diplomatic History of Europe, Asia, and Africa, 1870–1914, 482 pp. Government Printing Office, Washington (1918)

Axenciuc, V.: Introducere în istoria economică a României. Epoca modernă şi contemporană, partea I., 448 pp. Editura Fundaţiei, România de mâine", Bucureşti (1999)

Banu, G.: Tratat de asistenţă medicală, volumul I. Medicina socială ca ştiinţă. Eugenia. Demografia, 430 pp. Casa Şcoalelor, Bucureşti (1944)

Cizer, L.-D.: Toponimia judeţului Tulcea: consideraţii sincronice şi diacronice, 164 pp. Editura Lumen, Iaşi (2012)

Cristofor, C.A.: Monografia judeţului Tulcea (manuscris), 164 pp. (1938)

Filipescu, I.: Dimitrie Gusti despre reforma socială şi temeiul ei ştiinţific (monografia sociologică). Revista Română de Sociologie, serie nouă, anul XI, nr. 3–4, 305–312 (2000)

Georgescu, D.C.: Populaţia satelor româneşti. Sociologie românească, Anul II, nr. 2–3, februarie-martie, 68–79 (1937)

Gheorghiu, C.C.: Asistenţa medicală rurală în România. Sociologie românească, Anul II, nr. 2–3, februarie-martie, 80–87 (1937)

Gojinescu, C.: Situaţia demografică a cultelor după 1919. Etnosfera, nr. 2, 26–32 (2009)

Gusti, D.: Starea de azi a satului românesc. Întâiele concluzii ale cercetărilor intreprinse în 1938 de echipele regale studenţeşti. Sociologie românească, Anul III, nr. 10–12, octombrie-decembrie, 431–436 (1938)

Gusti, D.: Starea de azi a satului românesc. In Golopenţia, A., Georgescu, D. C. 60 de sate româneşti – cercetate de echipele studenţeşti, vol. I- Populaţia, pp. V–XI. Institutul de Ştiinţe Sociale al României, Bucureşti (1941)

Limona, R.: Populaţia Dobrogei în perioada interbelică, 240 pp. Semănătorul- Editura-online, Bucureşti (2009)

Madgearu, V.N.: Evoluţia economiei româneşti după primul război mondial, 404 pp. Independenţa economică, Bucureşti (1940)

Manuilă, S.: România şi revizionismul - Consideraţii etnografice şi demografice. Arhiva pentru ştiinţă şi reformă socială, Anul XII, nr. 1–2, 55–82 (1934)

Manuilă, S.: Ştiinţa de carte a populaţiei României. Arhiva pentru ştiinţă şi reformă socială, XIV, 931–959 (1936)

Manuilă, S., Georgescu, M.: Populaţia României. In Gusti, D. (ed.) Enciclopedia României, vol. I, Bucureşti, pp. 133–160. Bucureşti (1938)

Measnicov, I.: Raportul dintre ştiinţa de carte şi numărul învăţătorilor în România. Sociologie românească, Anul II, nr. 2–3, februarie-martie, 112–119 (1937a)

Measnicov, I.: Mortalitatea populaţiei rurale ţărăneşti. Sociologie românească, Anul II, nr. 4, aprilie, 154–167 (1937b).

Measnicov, I.: Migraţiunile interioare în România. Sociologie românească, Anul IV, nr. 7–12, iulie-decembrie, 392-411 (1942)

Nistor, I.S.: Stema României: istoria unui simbol; studiu critic, 350 pp. Editura Studia, Cluj-Napoca (2003)

Pătraşcu, D.V., Manea, L.D.: Unirea Dobrogei cu România (14 noiembrie 1878). Buridava. Studii şi materiale, IX, 236–242 (2011)

Popa, M. D. (ed.):. Dicţionar Enciclopedic, vol. II. D-G. Editura Enciclopedică, Bucureşti (1996)

Preda-Mătăsaru, A.: Tratatul între România şi Bulgaria, semnat la Craiova la 7 septembrie 1940: trecut şi prezent, 398 pp. Editura Lumina Lex, Bucureşti (2004)

Radu, R.: Dimitrie Gusti şi Muzeul Naţional al Satului. Economia. Seria Management, Anul X, Nr. 2, 159–168 (2007)

Stinghe, H., Toma, C.: Despre germanii din Dobrogea, 280 pp. Editura Ex Ponto, Constanţa (2007)

Şandru, D.: Politica sanitară a României între cele două războaie mondiale. Vrancea-Studii şi comunicări, vol. III, 259–276 (1980)

Trebici, V.: Commentaria in demographiam (I). Argumentum. Sociologie Românească, Serie nouă, Anul VI, nr. 1–2, 109–120 (1995)

Vidican, A.: Alimentaţia copiilor de vârstă şcolară la Căianul-Mic-Someş. Sociologie românească, Anul III, nr. 7–9, iulie-septembrie, 381–382 (1938)

Vulcănescu, M., Diaconu, M.: Şcoala sociologică a lui Dimitrie Gusti, 189 pp. Editura Eminescu, Bucureşti (1998)

Printed in the United States
By Bookmasters